PERSPECTIVES IN PRIMATE BIOLOGY

ADVANCES IN BEHAVIORAL BIOLOGY

Volume 1 • BRAIN CHEMISTRY AND MENTAL DISEASE
Edited by Beng T. Ho and William M. McIsaac • 1971

Volume 2 • NEUROBIOLOGY OF THE AMYGDALA
Edited by Basil E. Eleftheriou • 1972

Volume 3 • AGING AND THE BRAIN
Edited by Charles M. Gaitz • 1972

Volume 4 • THE CHEMISTRY OF MOOD, MOTIVATION, AND MEMORY
Edited by James L. McGaugh • 1972

Volume 5 • INTERDISCIPLINARY INVESTIGATION OF THE BRAIN
Edited by J. P. Nicholson • 1972

Volume 6 • PSYCHOPHARMACOLOGY AND AGING
Edited by Carl Eisdorfer and William E. Fann • 1973

Volume 7 • CONTROL OF POSTURE AND LOCOMOTION
Edited by R. B. Stein, K. G. Pearson, R. S. Smith, and J. B. Redford • 1973

Volume 8 • DRUGS AND THE DEVELOPING BRAIN
Edited by Antonia Vernadakis and Norman Weiner • 1974

Volume 9 • PERSPECTIVES IN PRIMATE BIOLOGY
Edited by A. B. Chiarelli • 1974

Volume 10 • NEUROHUMORAL CODING OF BRAIN FUNCTION
Edited by R. D. Myers and René Raúl Drucker-Colín • 1974

Volume 11 • REPRODUCTIVE BEHAVIOR
Edited by William Montagna and William A. Sadler • 1974

PERSPECTIVES IN PRIMATE BIOLOGY

Edited by

A. B. Chiarelli

Institute of Anthropology
University of Turin
Turin, Italy

PLENUM PRESS • NEW YORK AND LONDON

Library of Congress Cataloging in Publication Data

NATO Advanced Study Institute on Comparative Biology of Primates, Turin, 1972.
Perspectives in primate biology.

(Advances in behavioral biology, v. 9)
"Lectures given at the NATO Advanced Study Institute on Comparative Biology of Primates held in Montaldo (Turin), June 7-19, 1972."
Includes bibliographies.
1. Primates—Congresses. I. Chiarelli, A B II. Title. [DNLM: 1. Primates. W3 AD215 v. 9 / QL737.P9 P467]
QL737.P9N35 1972 599'.8 74-10968
ISBN 0-306-37909-0

Lectures given at the NATO Advanced Study Institute on Comparative Biology of Primates held in Montaldo (Turin), June 7-19, 1972

A Division of Plenum Publishing Corporation
227 West 17th Street, New York, N.Y. 10011

United Kingdom edition published by Plenum Press, London
A Division of Plenum Publishing Company, Ltd.
4a Lower John Street, London W1R 3PD, England

Printed in the United States of America

Preface

The present volume is the result of a NATO Advanced Study Institute held in Montaldo, Turin (Italy), between the 7 and 19 June 1972.

The aim of the Study Institute has been the development of a general philosophy for the science of Primatology. Lecturers were selected from those scientists deeply involved and interested in this field. The course intended to serve students and researchers using primates in medical and biological research, but especially those interested in the natural history of the group and in human biology.

In the past the study of primates was largely limited to determine the origin of the human species. Today, however, interest in them extends far beyond this narrow focus. In terms of both practical human purposes and theoretical interests, the study of primate biology and behaviour is of ever increasing importance. Their close comparative relationships with man has proved of such great value to human biology and medicine that their numbers and kinds are quickly dwindling. For this reason, one of the main focuses of the A.S.I. was on their reproductive biology and conservation.

During the meeting days a broad series of lectures on specific topics of comparative anatomy, physiology, endocrinology, reproductive physiology, genetics and molecular biology, cytogenetics and behaviour were delivered by leading primatologists.

The papers more specific in topic have been published in appropriate scientific journals, mainly in the Journal of Human Evolution, see especially issue 3(6). The more generalized lectures have been here collected to provide a reading for students interested in Primatology and Anthropology.

The heterogeneity of the contributions here collected is apparent: the unifying concept is the value of the study of the living primates in biological researches and their vulnerable resource.

The appeal for conservation of non-human primates which was prepared and approved by all the participants of the Advanced Study Institute is therefore the most important contribution and conclusion of the volume.

The authors and the editor are obliged to Miss Chiara Bullo for her excellent work in coordinating the papers and in editing them.

A. B. Chiarelli
University of Turin
Institute of Anthropology

Turin, 20th Sept. 1973

Contents

EMBRYOGENESIS IN VITRO: AN EXPERIMENTAL MODEL FOR THE UNDERSTANDING OF REPRODUCTIVE PHYSIOLOGY AND DEVELOPMENT IN MAMMALS

Anil B. Mukherjee
State University of New York at Buffalo
School of Medicine
Department of Pediatrics
Buffalo, New York (U.S.A.)

During the last decade, problems of early embryonic development have undergone active investigation. The success of studies in this field are largely due to the availability of proper tissue culture techniques and media which allow for the growth and manipulation of mammalian embryos *in vitro*. Although a great amount of useful and important information has been amassed, a paucity of knowledge still exists, particularly with respect to pre-fertilization phenomena, eg. oocyte maturation and sperm capacitation. The intricacies of the mechanism of fertilization itself are poorly understood as are the earliest events of development and differentiation of the mammalian embryo during the initial cleavage stages. Since the pattern of development of embryos derived from laboratory animals is similar to that of domestic animals, most probably, information obtained by studying the mouse, rat and rabbit may possibly be extrapolated to primates and even humans (7). Therefore, experimentation with such species will continue to supply us with the lion's share of data concerning mammalian reproductive physiology and development.

This presentation will discuss attempts at devising a system in the mouse, which would yield a viable blastocyst following maturation of ova, capacitation of sperm and fertilization, completely *in vitro*. The subsequent transplantation of such developing blastocysts to proper recipients would yield normal, fertile offspring. In this manner, some of the earliest events in the processes of fertilization and embryogenesis could be studied under controlled conditions.

IN VITRO FERTILIZATION

Fertilization of mammalian ova *in vitro* has been successfully carried out in the rabbit (4, 12, 48) and the hamster (2, 54), but development of fertilized ova beyond the two cell stage in these species has not been achieved (53). In the mouse, however, Brinster et al. (5) have observed in vitro fertilization and embryonic development to the blastocyst stage using explanted fallopian tubes in organ culture as a support. More recently, Whittingham (51) has successfully fertilized mouse eggs in vitro, the transplantation of which into pseudopregnant recipient mothers, yielded 17 day old fetuses. Our initial efforts at in vitro fertilization concentrated on attempts with mature tubal oocytes and capacitated spermatozoa retrieved from the uterus following a succesful mating.

The mice used in all the studies to be described were of the ICR/Ha (albino) and C57BL/6J strains (black) obtained from West Seneca Laboratories of the Roswell Park Memorial Institute, Buffalo, New York. Large numbers of ova are readily obtainable following superovulation of mature mice (6-8 weeks of age) by an intraperitoneal injection of five international units (I.U.) of pregnant mares' serum gonadotrophin (PMS) followed, 48 hours later, by an injection of 5 I.U. of human chorionic gonadotrophin (HCG). Ten to twelve hours after the second injection, the mice are sacrificed by cervical dislocation and the oviducts excised and uncoiled. Unfertilized tubal ova are flushed from the oviducts into a petri dish (60 mm diameter) by injecting 0.5 ml of normal saline into the ampullary-isthmal junction with a 30-gauge needle (Fig. 1). The eggs are then isolated from cellular debris by a finely drawn capillary micropipette and transferred to a second petri dish containing 0.5 ml of tissue culture medium, diluted with normal saline to half the original concentration. The composition of the medium was almost identical to that of Whitten and Biggers (50) (oocyte medium) with the exception that sodium pyruvate concentration was 0.55 mM rather than 0.33 mM. The pH of the medium was maintained at approximately 7.4 by intermittent administration of 5% CO_2 in air.

Capacitated spermatozoa, as defined by Austin (1) and Chang (10), were obtained from the uteri of mice, three to four hours after mating with a proven fertile male. The uteri are dissected out and their contents released into a petri dish containing 1 ml of half strength growth medium at 37°C. The petri dish is immediately agitated to prevent coagulation of the sperm. Approximately 0.3 ml of sperm suspension was added to a cavity slide containing a number of unfertilized ova in medium. The mixture of the eggs and sperm is layered over with paraffin oil (Fisher Scientific, viscosity 125/135) so that the well is filled and the slide incubated for six to eight hours at 37°C in a 5% CO_2 atmosphere. Ova, without the addition of sperm, were cultured as controls.

After incubation, both control and treated ova are throughly washed with normal saline and placed in culture using the methods of Brinster (6) and Chang (13). The petri dishes are examined after 24 hours for cleaving zygotes and the two-cell embryos are placed

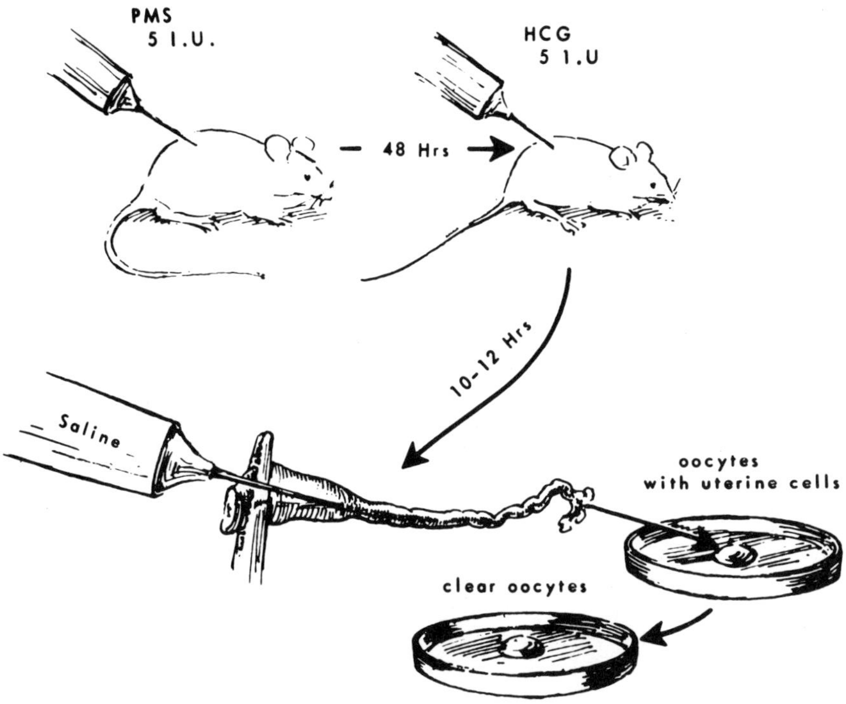

Figure 1. Induction of superovulation.

into new petri dishes with fresh medium (full strength). These zygotes develop to the blastocyst stage (64-130 cells) after approximately three days in culture (Fig. 2). Although some control eggs fragment during their culture period, no actual cleavage was observed in the unfertilized ova.

The blastocysts are then implanted into pregnant or pseudopregnant females (Fig. 3). The recipient mothers were made pseudopregnant through mating with proven sterile, vasectomized males. Both the vasectomized males and the recipient mother were always of opposite coat color to the implanted embryos. In some experiments, normally pregnant mice were used as recipient mothers instead of pseudopregnant animals. These matings were designed so that the progeny derived from the *in vitro* and *in vivo* fertilization were clearly distinguishable by coat color.

The appearance of the vaginal plug was taken as evidence of mating and counted as day zero of pseudopregnancy. After three days of growth *in vitro*, the blastocysts were collected into a glass capillary micropipette connected to a small rubber tube. A surgical

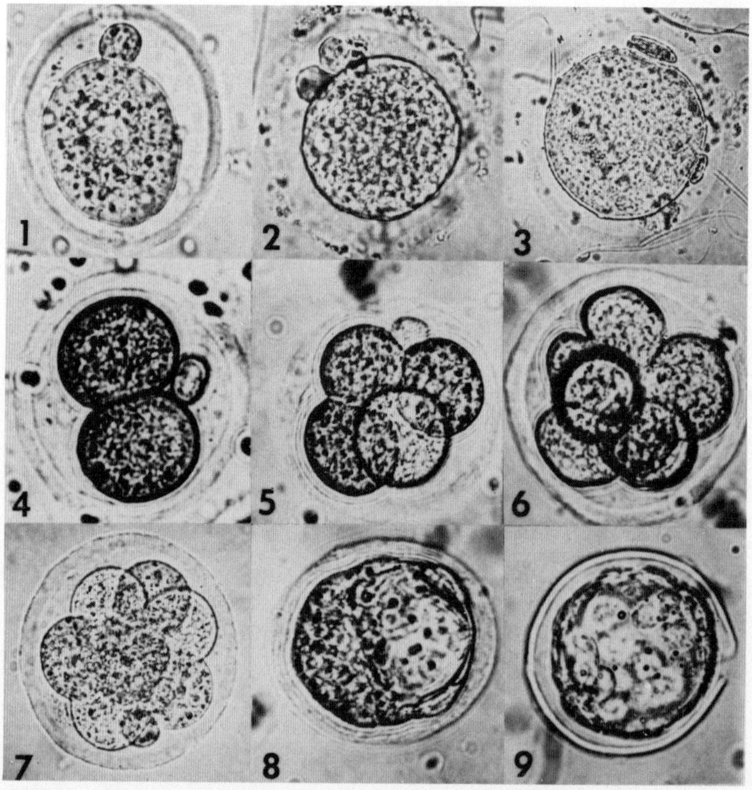

Figure 2. *In vitro* fertilization of mouse oocytes,
(1) Mature oocyte with first polar body;
(2-3) Sperm penetration and production of second polar body;
(4-7) Two-cell to eight-cell stages of embryos;
(8) Early blastocyst;
(9) Hatching blastocyst (note the broken zona).

incision was made on the dorsal surface of the recipient mouse, the uterotubal junction punctured by the micropipette so that the open tip reached the lumen of the uterus, and the blastocysts were expelled into the uterus (Fig. 4). Figure 5 shows implanted embryos in only one uterine horn of a pseudopregnant female; the second horn was not used. Birth usually occured 20-21 days after *in vitro* fertilization.

Table 1 shows the pooled results of six experiments. Of 253 eggs exposed to capacitated spermatozoa, 67 were fertilized, and appeared as two-cell embryos; 25 of these developed to blastocysts. Twenty-

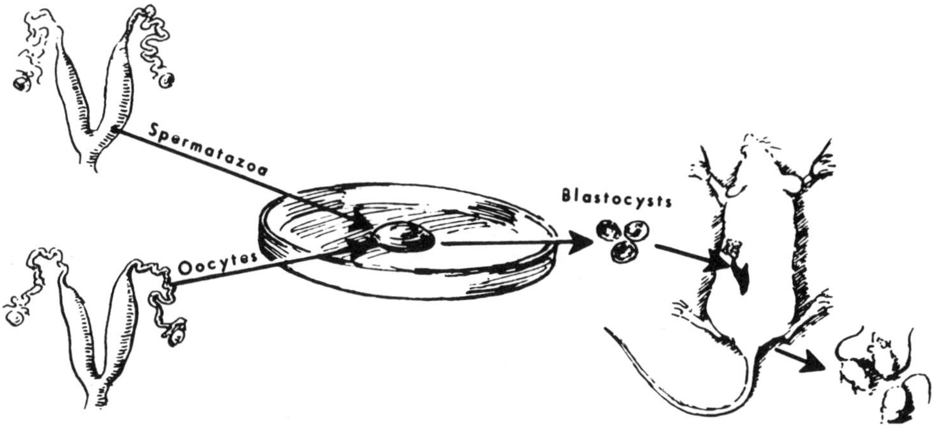

Figure 3. Diagramatic representation of the procedure for transplantation of blastocysts.

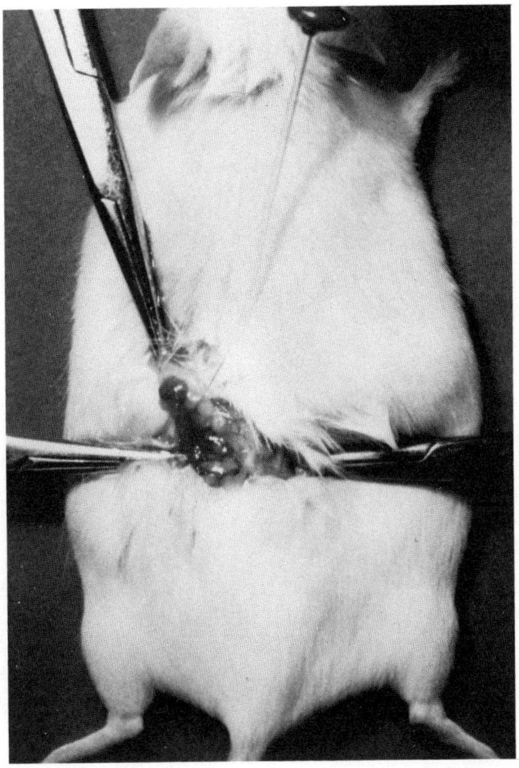

Figure 4. Technique of the introduction of blastocysts to the recipient mother.

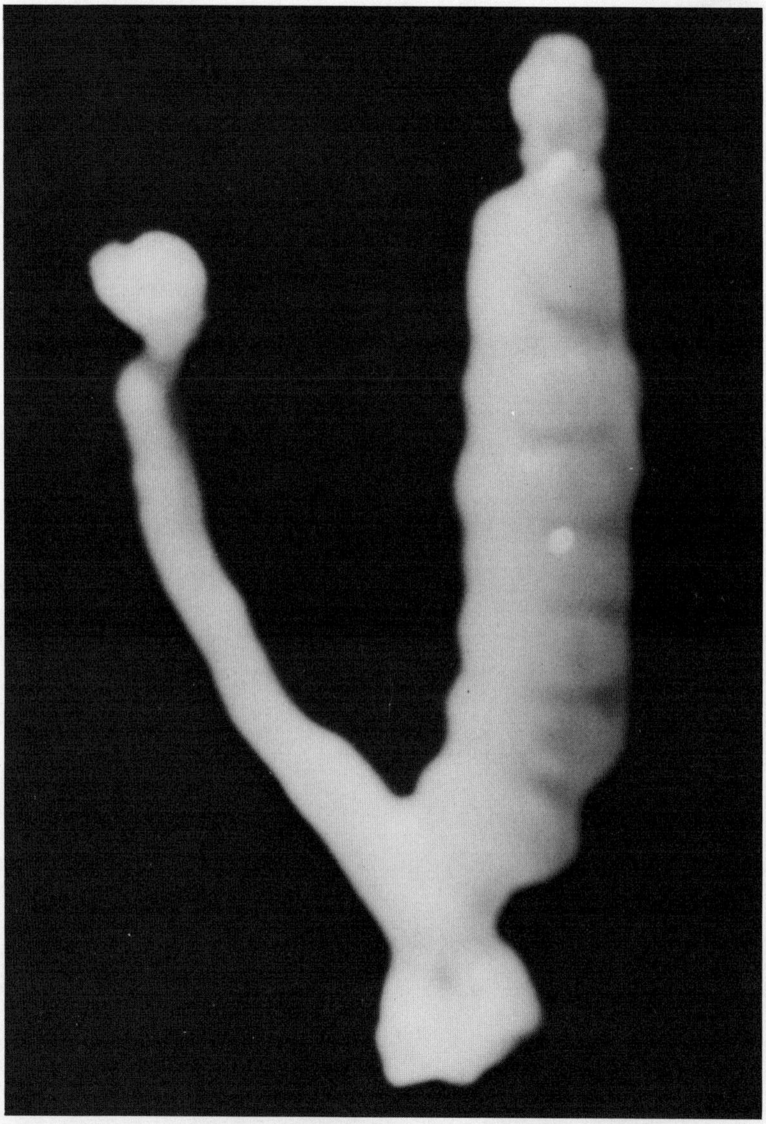

Figure 5. Recipient uterus eight days after transplantation. Note the presence of embryos in only one horn of the uterus.

TABLE 1: IN VITRO FERTILIZATION AND IMPLANTATION

Donor Strains*		Number of oocytes cultured	Number of two-cell embryos	Number of blastocysts	Number of blastocysts transplanted	Recipient	Number of progeny obtained
Ova	Sperm						
A	A	130	34	13	13	B	5 (A)
B	B	123	33	12	10	A	6 (B)
Total		253	67	25	23		11

*A = ICR/Ha (albino) B = C57BL/6J (black)

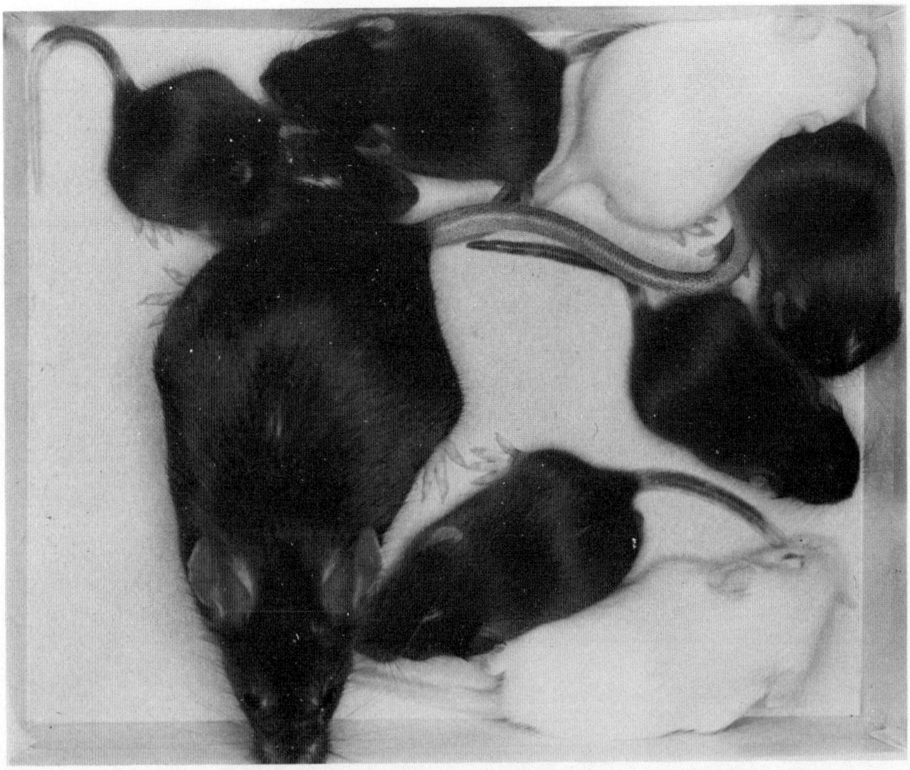

Figure 6a. Progeny from in vitro fertilization. Three blastocysts were introduced into the uterus of a normally pregnant recipient. The two white mice were born.

three blastocysts were transplanted into recipient mothers and eleven apparently normal offspring were obtained. Although only approximately 5% of the fertilized ova developed to the blastocyst stage, 51% (11) of the implanted blastocysts gave rise to progeny. Therefore, the most difficult part of this system compared to the controls appears to be the three day culture period necessary for blastocyst development. Nine of the eleven mice were born to pseudopregnant mothers and the remaining two to normally pregnant females. Figures 6a and 6b illustrate litters of in vitro developed blastocysts as live born offspring. In experiment three, two females derived from in vitro fertilization were mated with fertile males and delivered normal progeny.

These experiments conclusively demonstrate that in vitro fertilization of tubal mouse oocytes with capacitated spermatozoa results in viable embryos, which upon transplantation to proper recipients give rise to normal progeny.

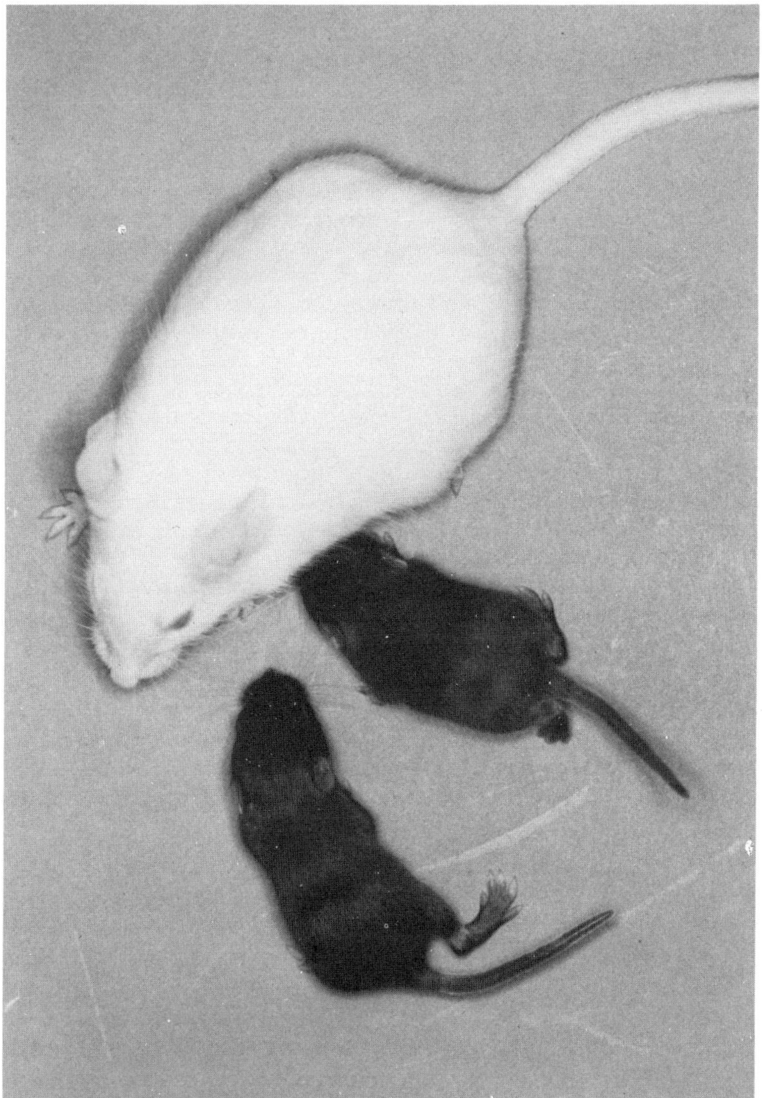

Figure 6b. Progeny from in vitro fertilization: Four blastocysts were introduced into the uterus of a pseudopregnant recipient. The two black mice are the result of this transplantation.

IN VITRO CAPACITATION OF EPIDIDYMAL SPERMATOZOA

Although the above experiments indicated that in vitro derived blastocysts could be obtained, implanted and subsequently developed to normal mice, the system was still dependent upon the animal for the production of mature ova and capacitated sperm. Therefore, our efforts next centered on attempts to capacitate epididymal sperm in vitro.

Existing evidence suggests that mammalian spermatozoa undergo functional changes during their passage through the female genital tract which enable them to penetrate the ovum. This phenomenon was described independently by Austin (1) and Chang (10) and generally is known as sperm "capacitation". The morphological criteria used by these authors was the lost staining properties of the acrosome from the sperm head. To date, it has been possible to determine the need for capacitation in only a few mammalian species, such as the cow, hamster, rat and mouse (3). However, the actual factors responsible for sperm capacitation are not well delineated. Previous experiments in the golden hamster have suggested that the active capacitating factors are not derived from the oviduct or uterus (2) but from the follicular fluid released into the genital tract during ovulation (55). Recently, in vitro capacitation of mouse (27) and hamster (56, 25) spermatozoa has been achieved by treatment with heat inactivated bovine follicular fluid. Our attempts centered on the use of human follicular and tubal fluids as a possible capacitating treatment on spermatozoa derived from the rat, mouse and man (41).

Samples of fresh spermatozoa were obtained from mature Wistar rats (albino) and the C57BL/6J and ICR/Ha mice strains. The animals were sacrificed by cervical dislocation and the epididymis was teased apart in 0.2 ml of normal saline. Fresh ejaculates from normal fertile human volunteers were also collected and diluted with 0.5 ml of normal saline. Thin films of spermatozoa from all three species were spread directly on microscope slides to serve as controls.

Tubal and follicular fluids were collected by the technique of Lippes et al. (30) from human patients undergoing voluntary tubal ligation and other surgical procedures. 0.2 ml aliquots of the sperm samples were placed in petri dishes (60 mm diameter) and incubated with 0.3 ml of each of the following fluids:

1. Follicular fluid - without heat inactivation
2. Follicular fluid - heat inactivated (56° for 30 minutes)
3. A mixture of heat inactivated follicular fluid: tubal fluid (1:2)
4. Tubal fluid
5. Normal saline

These mixtures were agitated for 15-20 seconds and covered with a thin layer of paraffin oil (Fisher Scientific - viscosity 125/135). The culture dishes were then incubated at 37°C in an atmosphere of

5% CO_2 in air for four to six hours. After incubation a few drops of the sperm genital tract fluid preparation were placed on a microscope slide in a thin film.

A single solution procedure was used to stain the spermatozoa for the presence or absence of the acrosome (8). The stain consisted of an aqueous solution of 5% Aniline Blue and 5% Eosin. The sperm-containing slides were stained for 15 minutes, then rinsed in tap water, followed by dehydration in a series of alcohols (70%, 95% and absolute ethanol), immersed in Xylene and mounted in permount. Under these conditions, the acrosome stains blue and the sperm nucleus red. Figure 7 illustrates the capacitation of mouse sperm following treatment with genital tract fluids. For each treatment 1,000 spermatozoa were counted at random from several slides and the presence or absence of acrosome staining was noted and scored.

Table 2 shows the effects of various fluid treatments on rat, human and mouse spermatozoa. In the direct preparations, with no pretreatment by the genital fluids, all sperm retained staining of the acrosome. In man and rat no significant difference in the frequency of acrosome staining loss was observed when sperm treated with uninactivated follicular fluid were compared to the saline treated controls (human - 7% vs. 0%; rat - 3.5% vs. 1%). However, in the mouse, 20.5% of the sperm lost the acrosome staining when treated with uninactivated follicular fluid, as opposed to 1.5% in the untreated controls. On the other hand, treatment with tubal fluid alone led to a loss of acrosome staining in human (58.5%), and rat sperm (40.2%), while in the mouse, the effect was comparable to non-heat inactivated follicular fluid (26% vs. 20.5%).

Since non-heat inactivated follicular fluid led to the immobilization of the sperm, heat inactivation of this fluid (56°C for 30 minutes) was used to counteract this effect. This phenomenon had already been demonstrated for bovine follicular fluid (56). Treatment of sperm with heat inactivated follicular fluid led to considerable loss of acrosomal staining in all species, although not to the same degree (Tab. 2). Increased loss of acrosome staining of approximately 10 fold (67.6 vs. 7.0%) and 15 fold (43.6 vs. 3.5%) was observed in human and rat sperm, respectively. The increase in the mouse, however, was only 11.5% (32 vs. 20.5%). Apparently, heat inactivation alters human follicular fluid in such a way as to enhance the loss of acrosome staining reaction in the sperm of rats, mice and humans. Although heat inactivated follicular fluid, as well as non-treated tubal fluid alone, caused loss of acrosome staining, the most effective treatment in all three species was a combination of the two in a ratio of 1:2 (heat inactivated follicular fluid: tubal fluid). Such a treatment yielded the loss of acrosome staining in 80% of human; 74% rat and 50.3% of mouse spermatozoa.

In addition to the loss of acrosome staining, the functionality of <u>in vitro</u> capacitation was tested in the mouse by using sperm treated with various fluids and their combinations for <u>in vitro</u> fer-

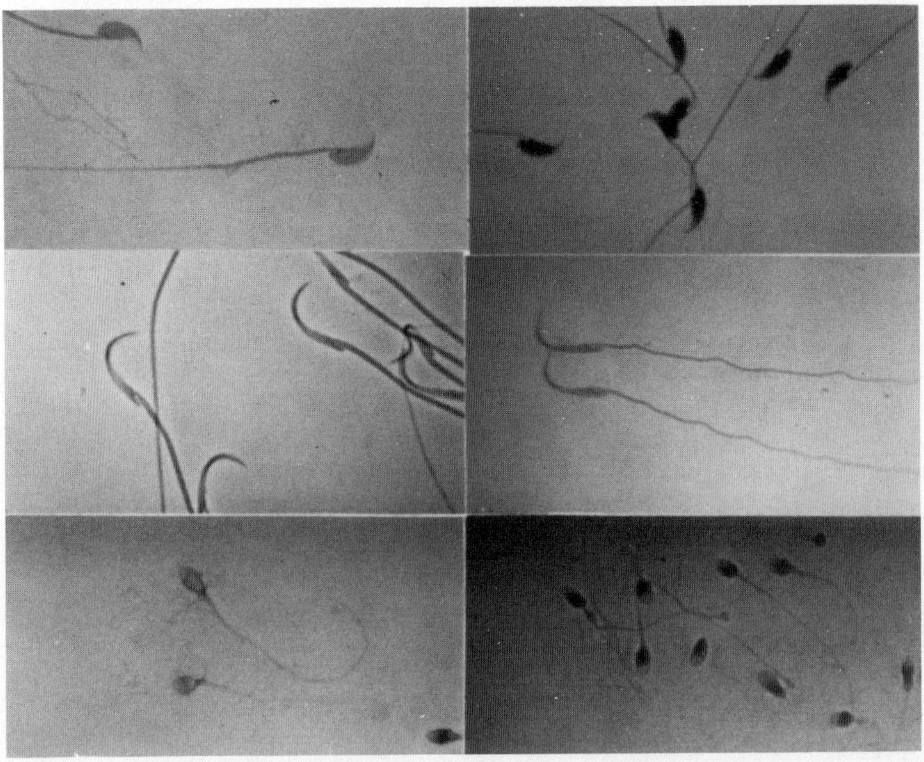

Figure 7. Staining reaction of mouse, rat and human spermatozoa before and after exposure to human follicular and tubal fluids. Top row-- mouse; Middle row--rat; Bottom row--human. On the left are untreated sperm and on the right, sperm treated with inactivated human follicular fluid: tubal fluid (1:2). Note the loss of acrosome staining reaction after treatment.

tilization. Table 3 shows the results of pooled data on in vitro fertilization from approximately equal numbers of matured tubal oocytes from two mouse strains (C57BL/6J and ICR/Ha). Control sperm (treated with normal saline or exposed to tissue culture medium) did not effect fertilization. Although five apparent two-cell embryos were observed after exposure to culture medium, four of these were abnormal, representing fragmentation of oocytes. Similarly, two such fragmented oocytes were observed after utilization of saline treated sperm.

TABLE 2: INDUCTION OF ACROSOME STAINING LOSS BY VARIOUS TREATMENTS WITH HUMAN FOLLICULAR AND TUBAL FLUIDS

Treatment	% Sperm Without Acrosome Staining*		
	Human	Rat	Mouse
Control (Direct)	0.0	0.0	0.0
Control (Saline)	0.0	1.0	1.5
Follicular fluid (without inactivation)	7.0	3.5	20.0
Follicular fluid (inactivated at 56°C for 30 minutes)	67.6	43.6	32.1
Inactivated follicular fluid: tubal fluid (1:2)	80.0	74.0	50.0
Tubal fluid	58.6	40.2	26.0

*At least 1,000 sperm counted in each case.

TABLE 3: BIOLOGICAL EVIDENCE OF CAPACITATION THROUGH *IN VITRO* FERTILIZATION*

Treatment of Sperm	No. of eggs	Two-cell embryos	Blastocysts
None	86	0 (5 fragmented)	0
Saline	95	0 (2 fragmented)	0
Media	93	5 (1 normal; 4 abnormal)	0
Follicular fluid (without inactivation)	89	0	0
Follicular fluid (inactivated at 56°C for 30 minutes)	108	23	15
Inactivated follicular fluid: tubal fluid (1:2)	121	72	56
Tubal fluid	88	14	7

*The data represent pooled results of experiments using both ICR/Ha and C57BL/6J strains of mice.

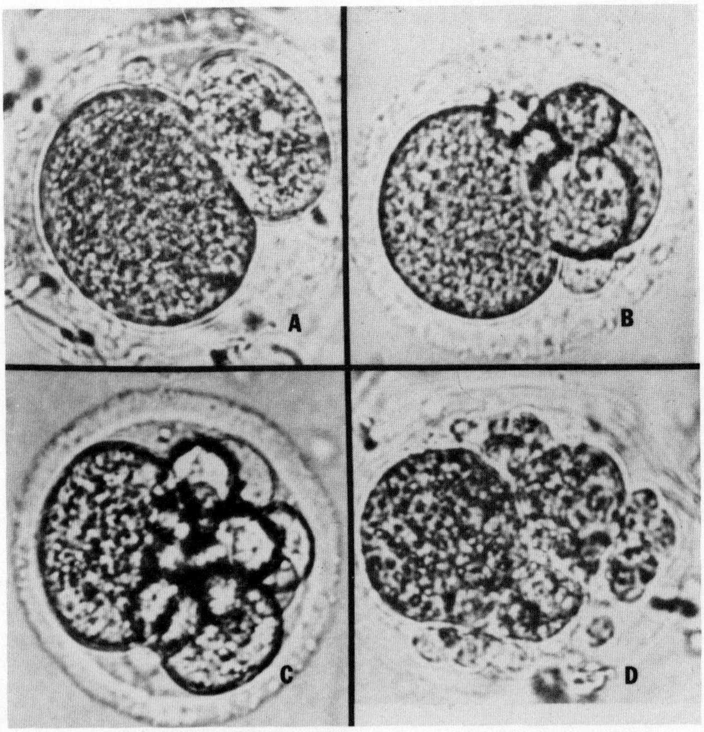

Figure 8. Abnormal development of an embryo resulting from *in vitro* fertilization of partially capacitated spermatozoa. All the pictures were of the same embryo taken sequentially at different times in culture.

No blastocysts were obtained from these abnormal embryos.

On the other hand, sperm treated with tubal fluid alone were capacitated, as evidenced by the appearance of normal two-cell mouse embryos, in 15% of the oocytes tested by *in vitro* fertilization (14/88). Of these, seven (50%) developed to normal blastocysts in three days of *in vitro* growth. The use of heat inactivated follicular fluid alone resulted in two-cell embryos in 21.3% (23/108) of oocytes with 65.2% (15/23) developing to blastocysts. Again, the most effective treatment was a 1:2 mixture of inactivated follicular fluid: tubal fluid, yielding 59.5% (72/121) two-cell embryos of which 77.7% (56/72) developed to blastocysts.

Use of partially capacitated spermatozoa (treated for two hours or less) in *in vitro* fertilization resulted in abnormal development of embryos (Fig. 8). Attempts at *in vitro* fertilization of human ova

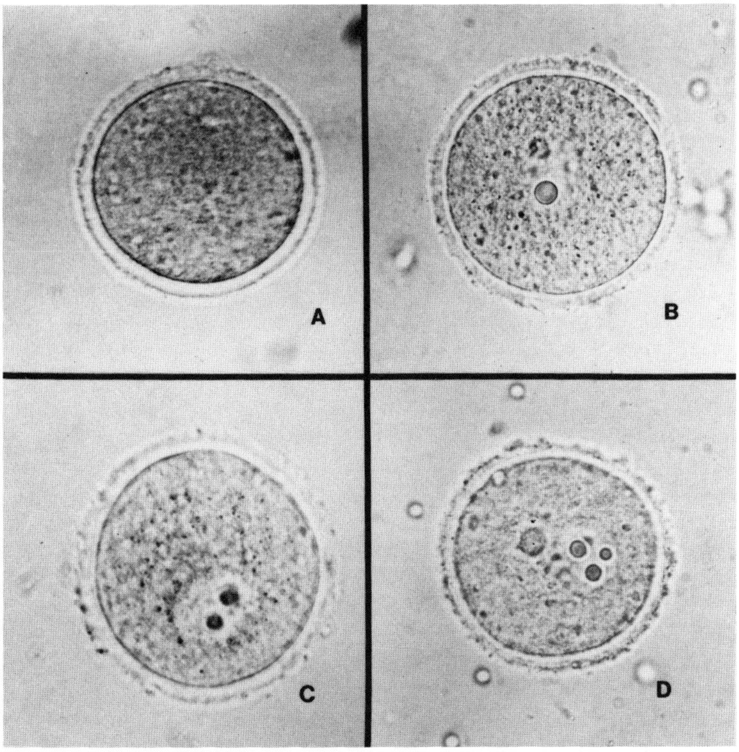

Figure 9. Immature oocytes collected directly from the ovary of mature mice. Note the four types of oocytes. (A) No germinal vesicle visible. (B) Germinal vesicle with one nucleolar structure. (C) Germinal vesicle with two nucleolar structures. (D) Germinal vesicle with three nucleolar structures.

with sperm treated by a 1:2 mixture of heat-inactivated follicular fluid: tubal fluid proved unsuccessful. From a total of 12 ova mixed with treated sperm, the only positive result was the formation of pronuclei in one egg.

IN VITRO MATURATION OF FOLLICULAR OOCYTES

The forgoing experiments demonstrated the feasibility of sperm capacitation _in vitro_. The next problem, to complete the _in vitro_ system, was the maturation of follicular oocytes.

Follicular oocytes were obtained from the ovaries of six to eight week old ICR/Ha and/or C57BL/6J inbred mice. Approximately 30 to 50 immature oocytes, with a well-defined germinal vesicle (Fig. 9), were obtained from each mouse by teasing the ovarian follicles in 1:1 mixture of normal saline and Ham's F-10 tissue culture medium at pH 7.2. The oocytes were separated from cellular debris and cultured from eight to ten hours in a drop of Ham's F-10 or modified Whitten and Bigger's medium with sodium pyruvate concentration of 0.60 mM instead of 0.33 mM in a 60 mm plastic petri dish under paraffin oil (viscosity 125/135) at 37°C in an atmosphere of 5% CO_2 in air.

One of the causes of abnormal fetal development may be preceding errors during the meiotic divisions. Only a few studies have concerned the direct examination of gametes (15, 16). In our laboratory, various types of abnormalities were observed during the cytological study of some 3,000 ovarian oocytes during in vitro maturation. Some oocytes, directly upon release from the follicle, were seen to possess varying numbers of nucleolus-like structures in the germinal vesicles (0-4) (Fig. (), while those with only a single nucleolar structure in the vesicle appeared to mature normally. Most of the abnormalities, however, involved the distribution of chromatin material between the polar body and the functional oocyte at the various meiotic divisions. Four types of errors were seen (Fig. 10) and their distribution in 3,000 oocytes examined were as follows:

1. Equal amount of chromatin in each segment - four oocytes (0.16%)
2. Large polar body possessing all the chromatin - fifteen oocytes (0.50%)
3. All the chromatin extruded into the polar bodies - four oocytes (0.16%)
4. All the chromatin retained within the oocyte - two oocytes (0.06%).

Such abnormalities were present in approximately 0.88% of the immature oocytes examined. Naturally, fertilization of these eggs would lead to aneuploid offspring.

The functionality of in vitro oocyte maturation was tested by the use of such eggs in fertilization tests. The oocytes were examined at the end of the culture period to check for the presence of metaphase II configuration indicative of maturity. It was found, however, that at 13-15 hours of culture the majority of the oocytes had extruded the first polar body and the chromatin was in the metaphases II stage. The sequential development of the oocytes are represented in Figure 11a, 11b.

Approximately $1x10^5$ mouse spermatozoa (0.3 ml) capacitated in vitro were added to a petri dish containing 50 oocytes matured in vitro as described above. The mixture of eggs and sperm was completely covered with a layer of paraffin oil. The petri dish was incubated at 37°C in 5% CO_2 in air. After 24 hours, the cultures were examined for two-cell embryos, which were transferred to fresh medium. In three days, the two-cell embryos developed to the blastocyst stage

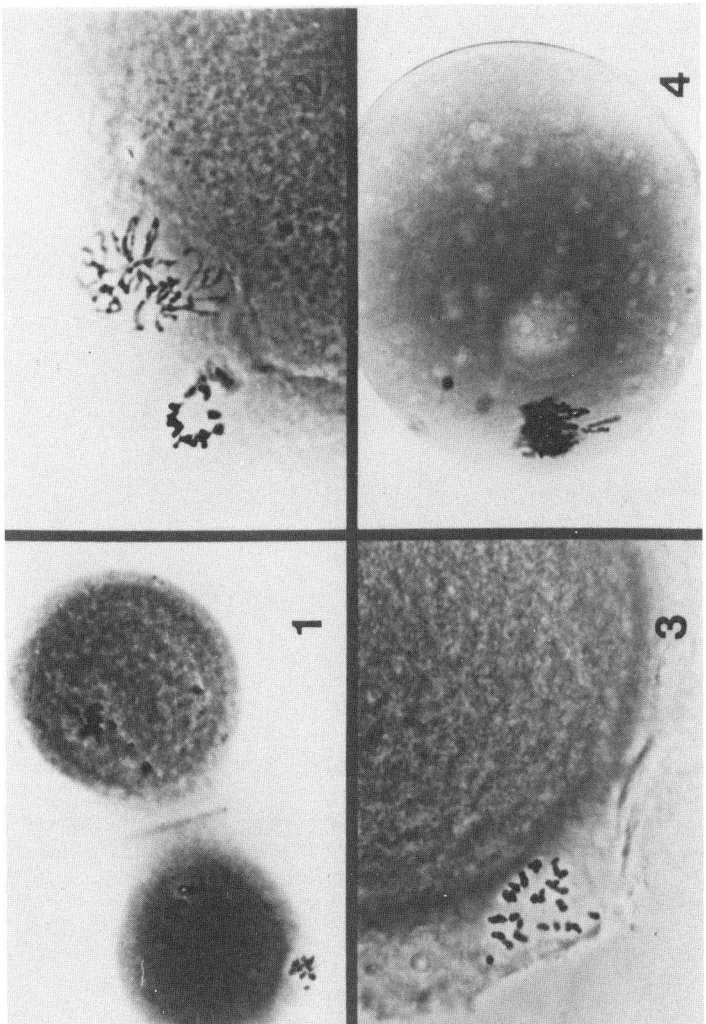

Figure 10. Anomalies in oocyte maturation in the mouse. (1) Equal amount of chromatin in each segment. (2) Large polar body with all the chromatin. (3) All the chromatin extruded into two polar bodies. (4) All chromatin retained within the oocyte.

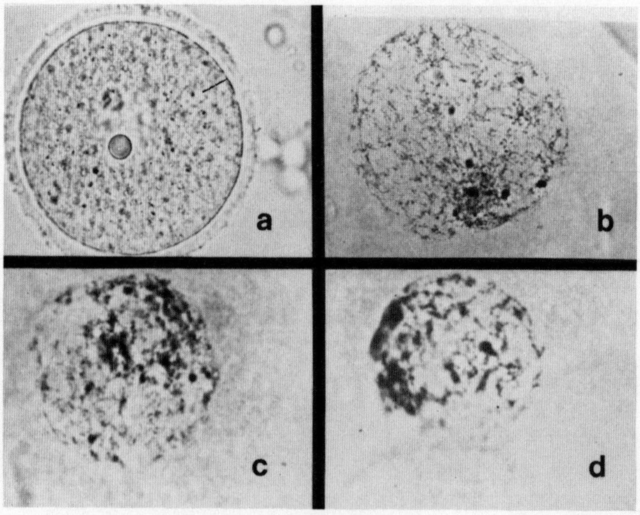

Figure 11a. Normal maturation of mouse ovarian oocytes *in vitro*: a. Live oocyte upon liberation from an ovarian follicle; b. Characteristic nuclear stage (dictyate) of oocytes upon liberation from the follicle. Two large chromocenters are adjacent to the high nucleolus; c. Condensing chromatin I. At the start of culture; d. Condensing chromatin II. Typical after two hours of culture.

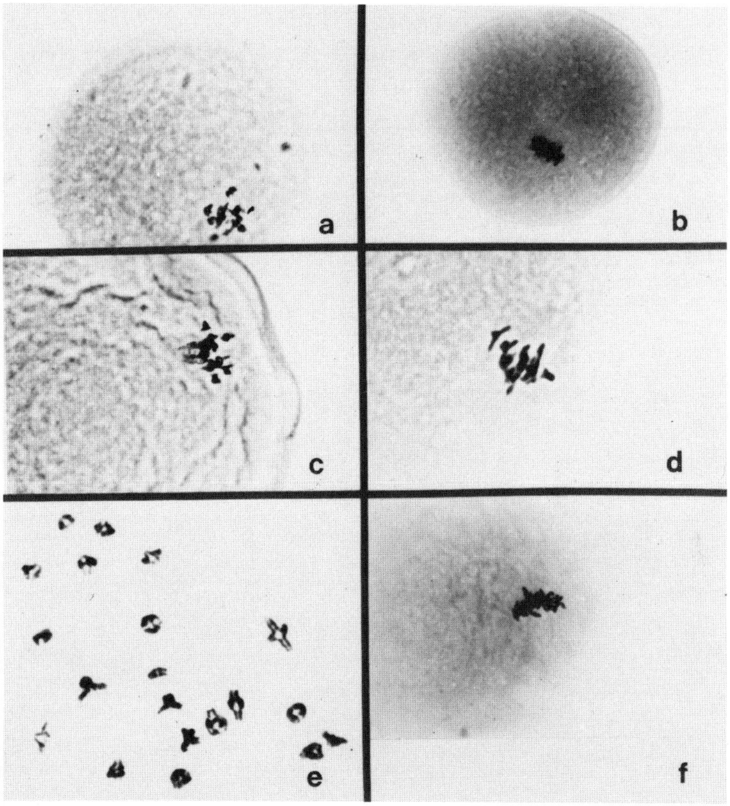

Figure 11b. Normal maturation of mouse ovarian oocytes in vitro: a. Condensing chromatin III. Three hours of culture; b. Metaphase I. Nine to ten hours of culture; c., d., e. Metaphase I, typical of nine to ten hours of culture; f. Metaphase II. Typical of 12-16 hours of culture. The first polar body is degenerated.

and transplantation of such blastocysts to proper recipients, as described above, was then performed. Table 4 shows the pooled results of twelve experiments. Of 1,200 oocytes cultured, 720 shed first polar body and were considered matured. In vitro fertilization of 600 of these oocytes was attempted and 138 (23%) apparently normal two-cell embryos were obtained. Forty-eight embryos developed to blastocysts (Fig. 12) and their transplantation to surrogate mothers resulted in twenty-two normal progeny (13:9) which were subsequently found to be fertile.

TABLE 4: COMPLETE IN VITRO FERTILIZATION & DEVELOPMENT OF MOUSE EMBRYOS:

Number of ovarian oocytes (immature)	Number of matured oocytes (with 1st polar body)	Number of mature oocytes used for in vitro fertilization	Number of two-cell embryos	Number of blastocysts	Number of transplanted	Number of progeny
1200	720	600	138 (23%)	48 (35%)	48	22 (13♂:9♀)

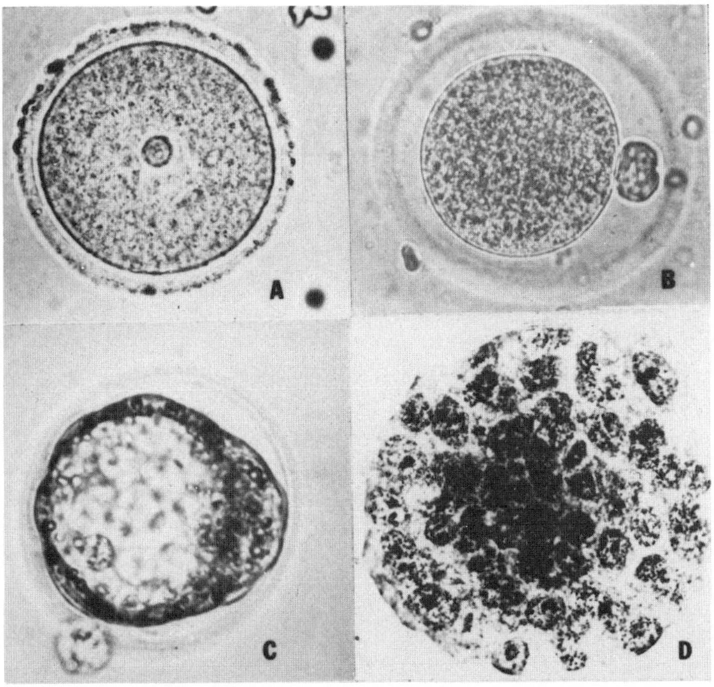

Figure 12. In vitro fertilization of oocytes matured in culture and sperm capacitated in vitro: (A) Immature oocyte; (B) Mature oocyte; (C) Live blastocyst; (D) Fixed blastocyst to show the cell number.

DISCUSSION

(a) Current problems of in vitro embryogenesis. In vitro meiosis of the mammalian oocyte has been studied for quite some time. The advances in tissue culture procedures and formulation of support media have led to a great deal of sophistication in this area so that, at present, the maturation of oocytes from various species including mouse, rat, hamster, sheep, monkey, rabbit, cow and man is now feasible (11, 18-21, 28, 29, 36, 38, 43-47, 49, 57). Recently, the successful fertilization of in vitro matured oocytes by normally capacitated spermatozoa in the mouse, followed by subsequent transplantation of the developing embryo to proper recipients, resulted in 14 or 15 day old fetuses. Obviously, an in vitro system, as described above, would provide an experimental model which could be applied to the detailed and critical analysis of various areas of reproductive physiology and development biology. However, one important goal in this area of research is the development of a system which would allow observation of the entire process, beginning with male and female gametogenesis to the development of a normal animal in vitro and be totally independent of parental participation and maternal environment.

Such a system would necessitate great progress in new techniques and approaches. The greatest obstacle which is presently obvious is the stage from in vitro fertilization to the development of the blastocyst. The experiments described above indicate that the greatest attrition in our system occurs during this period. The efficiency of this stage of embryonic development will undoubtedly improve as innovations in support medium are made to enhance the survival of early cleavage embryos.

The sustenance of post-blastocyst embryos in vitro must also be accomplished and some modicum of progress has already been achieved in this area of developmental biology. Hsu (26) recently reported the initiation of cardiac function (heart beat) in post-blastocyst mouse embryos maintained in vitro. Zapol et al. (58) have described an artifical placenta capable of maintaining a premature lamb fetus. The stable metabolic state of this fetus for several days indicates that extra-uterine survival is within the realm of feasibility. These are impressive studies but progress towards a complete in vitro system demands the prior solution of a multitude of more basic issue.

Perhaps one of the more pressing difficulties to overcome is the perfection of various tissue culture media necessary for the different stages of embryonic development. Media for the initial periods eg. maturation of oocytes, capacitation of sperm, fertilization and growth to blastocysts, have already been achieved to some degree (50). For later periods, the approach will necessitate an extremely detailed analysis of the uterine environment, and its changes during pregnancy. This must begin with the fractionation of genital tract fluids, follicular and tubal. Such analyses have already been attempted by Lippes et al. (30) who have provided excellent data on the identification and possible function of constituents of human tubal fluid. Some protein components of oviductial fluid in primates, including man, have also been determined (33, 34, 37). Such data on normal uterine metabolism is prerequisite to the adaptation of tissue culture media which may serve a similar purpose. The elucidation of those physiologic changes of the endometrium and uterus during gestation is also critical before in vitro techniques can be established to mimic these functions. Although these problems, at present, may seem overwhelming, the technology necessary for their solution is available and it may be only a matter of time until workable solutions will be forthcoming.

(b) Future possibilities and potential applications of in vitro embryogenesis. Some pratical applications of an artifical system of fetal development have already been made in the field of livestock improvement (32). Embryo transfer techniques have already been successfully employed in the transfer of cattle and sheep from different parts of the world using unrelated animals as carriers (17). Prenatal mortality due to advanced maternal age has been investigated through embryo transfer studies. Sex ratios of rabbit litters can now be controlled by sex chromatin determination of excised trophoblastic cells prior to transplantation of the pre-sexed fetuses into a recipient mother. Examination of the morphological sex at a later stage of embryonic development was in agreement with the early prediction (22).

Although sperm penetration of the ovum in vitro has been described, this is but one event in the complex process of fertilization. Very little is known concerning the molecular aspects of fertilization, eg. the onset of de novo synthesis of nucleic acids, membrane changes, and shifts in cellular metabolism from the gametic stage to that of the pronucleus and zygote. An in vitro system would obviously allow detailed studies of these phenomena (42).

Gene action in the X-chromosome of mammals is believed to be regulated by the inactivation of one of the two X-chromosomes in each somatic cell of females. This X-inactivation hypothesis (31) has been proved for the most part; however, the matter of timing of X-chromosome inactivation is unclear. The application of an in vitro system to early stages of embryonic cleavage, coupled with autoradiography or biochemical techniques, would facilitate the identification of the precise cell stage, in various species at which the second female X-chromosome becomes inactivated and initiates its late DNA replication pattern.

The approach of nucleus transplantation, so elegantly developed in amphibians (24), could be applied to investigate the ontogenetic commitment of the oocyte cytoplasm. Introduction of diploid somatic nuclei into single cell embryos would provide such a system. Graham (23) has demonstrated that the introduction of an adult spleen nucleus into e fertilized mouse egg, prior to the first cleavage division, resulted in synchronous division of the injected and zygotic nuclei. Such experimentation, may allow investigation into the very complex relationships between nucleus and cytoplasm, the interaction of multiple genomes in a single cell, and intragenomic relationships by using cells with proper genetic markers.

(c) The role of drugs in the fetal environment. The use of drugs during pregnancy has, in the past, been directed towards the treatment of the disease of the mother with little concern for the fetus as a drug recipient. However, these therapeutic endeavors have been associated with unexpected and often tragic results in the developing fetus for whom the drug was not intended.

The technique of in vitro fertilization and development of embryos offers the unique opportunity to develop an assay system for screening possible mutagenic and teratogenic effects of drugs and other exogenous agents. The great advantage of this system is the ability to mature and obtain functional gametes *in vitro* under carefully controlled conditions. Following this technique, treatment of sperm (prior) to or following capacitation) or the ovum (during its maturation *in vitro*) prior to fertilization may be attempted. Additionally, any particular stage in zygotic development could then be treated. Treated embryos may be examined at various stages, pre- or post-implantation, for parameters such as chromosomal aberrations, congenital malformations, increased resorption frequency, and if allowed to progress to term and mature, the animals can be screened for carcinogenesis. A wide variety of chemical and viral agents, as well as physical treatments (eg. radiation or nutritional effects), may be assayed in this manner.

(d) Genetic engineering. Experiments could be designed, using the *in vitro* technique, to attempt genetic transformation by injection of purified DNA. It has been observed that introduction of amphibian oocyte cytoplasm induced transplanted nuclei to swell and replicate their DNA (23). When injected into an oocyte, the cytoplasm replicates and acts as a template for mRNA production. As a simple test for the mammalian system, DNA can be isolated and purified from tissue culture cells of C57BL/6J (black) and ICR/Ha (albino) following the method of Marmur (35). DNA from black mice could be injected to single-cell zygotes, obtained by *in vitro* fertilization of sperm and egg from albino mice. These embryos could be grown to the blastocyst stage and transplanted to white recipients and vice versa. Coat color will then be the indication of transformation.

(e) Enzyme induction. Homozygous Gunn rats are genetically deficient in the enzyme UDP-bilirubin glucuronyl transferase (UDP-glu-transferase), an inducible enzyme in both humans (52) and in other animals (9). DNA could be isolated from the normal rat fibroblast cultures and injected into single cell zygotes of homozygous Gunn rats obtained for *in vitro* fertilization. These embryos, after development to the blastocyst stage, could be transplanted to the recipient homozygous Gunn rat. If analysis of the progeny shows the absence of jaundice and presence of the enzyme, the true concept of genetic engineering would be supported. Preliminary experiments in our laboratory, using unfertilized mouse oocytes as well as early embryos, show that such micromanipulation is possible and such oocytes and embryos can develop normally following injection of foreign substances (39).

The above outlined experiments represent but a few examples of the myriad of problems and questions in genetics and reproductive physiology that could be approached by exploiting the technique of *in vitro* embryogenesis. Although this system has yet to be vastly improved and refined, from the standpoint of increased yields, the present technique is capable of serving as a productive research tool.

SUMMARY

During the past decade considerable progress in developing suitable embryo culture media and improving techniques for manipulating embryos *in vitro* have allowed problems of mammalian reproductive physiology and developmental biology to be investigated. A complete *in vitro* technique has been developed in the mouse for capacitation of epididymal spermatozoa, maturation of ovarian oocytes and fertilization of such gametes to give rise to viable embryos, which, when transplanted to proper recipients result in normal fertile progeny. This technique, although requiring perfection, has many future possibilities and potential applications as a research tool in solving problems of mammalian reproductive physiology and developmental biology. This technique may be useful in the investigation of (a) gene action in the X-chromosomes and initiation of late DNA replication pattern in one of them during embryogenesis; (b) nuclear transplantation and the ontogenetic commitment of oocyte cytoplasm; (c) the role of drugs, mutagens and carcinogens in the fetal environment; (d) genetic engineering eg. enzyme induction and (e) many problems involving fertility and sterility.

ACKNOWLEDGMENTS

The author wishes to thank Drs. Maimon M. Cohen and Ronald G. Davidson for valuable suggestions and critical review of the manuscript. This work was supported in part by Project No. 417 of the Maternal and Child Health Service DHEW, The Lalor Foundation and the GRS grant RR05493 of the NIH.

References

1. Austin, C.R. 1951. Aust. J. Sci. Res., 4, 581-589
2. Barros, C. and Austin, C.R. 1967. J. Exptl. Zool., 166, 317-323
3. Bedford, J.M. 1970. Biol. Reprod., 2, 128-158
4. Brackett, B.G. and Williams, W.L. 1965. J. Exptl. Zool., 160, 271
5. Brinster, R.L. and Biggers, J.D. 1965. J. Reprod. Fertil., 10, 277
6. Brinster, R.L. 1968. Mammalian embryo in culture. In: The Mammalian Oviduct, Hafez and Blandan (eds.). The Univ. of Chicago Press, Pp. 419-444
7. Brinster, R.L. 1971. Mammalian embryo metabolism. In: The Biology of the Blastocyst, R.J. Blandan (Ed.) Univ. of Chicago Press, Pp. 303-318
8. Casarett, G.W. 1953. Stain Tech., 28, 125-127
9. Catz, C. and Yaffe, S.J. 1968. Ped. Res., 2, 361
10. Chang, M.C. 1951. Nature, 168, 697
11. Chang, M.C. 1955. J. Exptl. Zool., 128, 379-405
12. Chang, M.C. 1959. Nature, 184, 466
13. Chang, M.C. and Pickworth, S. 1968. Egg transfer in the laboratory animal. In: The Mammalian Oviduct, Hafez and Blandan (Eds.) The Univ. of Chicago Press, Pp. 389-405
14. Cross, P.C. and Brinster, R.L. 1970. Biol. Reprod., 3, 298-307
15. Donahue, R.P. 1968. J. Exptl. Zool., 169, 237-249
16. Donahue, R.P. 1970. Cytogenetics, 9, 106-115
17. Dziuk, P.J. 1968. Egg transfer in cattle, sheep and pigs. In: Mammalian Oviduct, Hafez and Blandan (Eds.) The Univ. of Chicago Press, Pp. 407-417
18. Edwards, R.G. 1962. Nature, 196, 466-450
19. Edwards, R.G. 1965. Nature, 208, 349-351
20. Edwards, R.G., Bavister, B.D. and Steptoe, P.C. 1969. Nature, 221, 632-635
21. Foote, W.D. and Thibault, C. 1969. Ann. Biol. Anim. Bioch. Biophys., 9, 329-349
22. Gardner, R.L. and Edwards, R.G. 1968. Nature, 218, 346
23. Graham, C.F., Arms, K. and Gurdon, J.B. 1966. Develop. Biol., 14, 349
24. Gurdon, J.B. 1968. Sci. Amer., 219, 24

25. Gwatkin, R.B.L. and Anderson, O.F. 1969. Nature, 224, 1111-1112

26. Hsu, Y.C. 1971. Nature, 231, 100-102

27. Iwamatsu, T. and Chang, M.C. 1969. Nature, 224, 919-920

28. Jagiello, G.M. 1969. J. Cell Biol., 42, 571-574

29. Kennedy, J.F. and Donahue, R.P. 1969. Science, 164, 1292-1293

30. Lippes, J., Gonzales-Enders, R., Pragay, D. and Bartholomeu, W. 1972. Contraception, 5, 85-103

31. Lyon, M.F. 1961. Nature, 190, 372-373

32. Mann, T. 1969. Nature, 224, 649-654

33. Marcus, S.L. 1964. Surgical Forum, 15, 381

34. Marcus, S.L. and Saravis, C.A. 1965. Fertil. Steril., 16, 785

35. Marmur, J. 1961. J. Mol. Biol., 3, 208

36. McGaughey, R.W. and Polge, C. 1971. J. Exptl. Zool., 176, 383-396

37. Moghisi, K.S. 1970. Fertil. Steril., 21, 821

38. Moricard, R.D. and DeFonbrune, P. 1937. Arch. Anat. Microscop., 33, 113-138

39. Mukherjee, A.B. 1972. Micromanipulation of pre-implantation stages of mouse embryos (unpublished)

40. Mukherjee, A.B. and Cohen, M.M. 1970. Nature, 228, 472-473

41. Mukherjee, A.B. and Lippes, J. 1972. Canad. J. Genet. Cytol. 14, 167-174

42. Mukherjee, A.B. 1972. Nature 237, 397-398

43. Ohno, S., Klinger, H.P. and Atkin, N.B. 1962. Cytogenetics, 1, 42-51

44. Ohno, S. 1964. Life history of female germ cells in mammal. In: The Second International Conference on Congenital Malformations. Natl. Found. New York, Pp. 36-40

45. Pincus, G. and Enzmann, E.V. 1935. J. Exptl. Med., 62, 665-675

46. Pincus, G. and Saunders, B. 1939. Anat. Res., 75, 537-545

47. Sreenan, J. 1968. VI Congr. Intern. Reprod. Anim. Insem. Artif. Paris, Vol. 1, 577-580

48. Thibault, C. and Dauzier, L. 1961. Ann. Biol. Anim. Bioch. Biophys., 1, 277

49. Whitten, W.K. 1956. Nature, 177, 96

50. Whitten, W.K. and Biggers, J.D. 1968. J. Reprod. Fertil., 17, 399-401

51. Whittingham, D.G. 1968. Nature, 220, 592

52. Yaffe, S.J., Levy, G., Pharm, D., Matsuzawa, T. and Baliah, T. 1966. New Eng. J. Med., 275, 1461

53. Yanagimachi, R. and Chang, M.C. 1964. J. Exptl. Zool., 156, 361

54. Yanagimachi, R. and Chang, M.C. 1963. Nature, 200, 281

55. Yanagimachi, R. 1969a. J. Reprod. Fertil., 18, 275-286

56. Yanagimachi, R. 1969b. J. Exptl. Zool., 170, 269-280

57. Yuncken, C. 1968. Cytogenetics, 7, 234-238

58. Zapol, W.M., Kolobow, T., Price, J.E., Vurek, G.G. and Brown, R. L. 1969. Science, 166, 618.

COMPARATIVE NEUROANATOMY OF PROSIMIAN PRIMATES: SOME BASIC CONCEPTS BEARING ON THE EVOLUTION OF UPRIGHT LOCOMOTION

D. E. Haines, H. M. Murray*, B. C. Albright*, and G. E. Goode*
Department of Anatomy, West Virginia University School of Medicine, Morgantown, West Virginia, 26506;
*Department of Anatomy, Medical College of Virginia HSD-VCU, Richmond, Virginia, 23298, U.S.A.

"From a consideration of the details of comparative anatomy of living forms and from the evidence now available from fossil hominoids, it appears reasonably certain (and, indeed, is agreed by authorities who hold widely differing views on a number of phylogenetic details) that the most important single factor in the evolutionary emergence of Hominidae as a separate and independent line of development was related to the specialized functions of erect bipedal locomotion."

Le Gros Clark, 1964

I. BASIC STRUCTURE OF NERVE TISSUE[1]

The mammalian nervous system is composed of (1) nervous elements, which are highly specialized for irritability and conductivity, and (2) supportive non-nervous elements. The nervous elements are called neurons and the supportive elements are the neuroglial cells and a limited amount of connective tissue.

The neurons are specialized for the conduction of impulses and the neuroglial cells serve supportive, nutritional, and phagocytic

[1]This brief introduction is indicated for those who have little or no knowledge of the anatomy of the nervous system. The general morphological points emphasized in this section will assist the uninitiated in understanding the more detailed aspects of subsequent discussions.

roles. The structural and functional unit of the nervous system is the neuron which consists of a cell body and its processes (Fig. 1). Dendrites are usually short and branching and convey impulses toward the cell body while the axon is usually unbranched, can be extremely long, and conveys impulses away from the cell body toward the synapse. The synapse is the apposition of the axon of one neuron to the dendrites, cell body or axon of a second neuron, the intervening space being the synaptic cleft (Fig. 1).

The cell body of a neuron is characterized by its size (6-130 micra in diameter), its shape, and the staining characteristics of the granular endoplasmic reticulum (E-R). The granular E-R usually stains violet with most commonly used neuro-stains, and in stained sections it is usually referred to as the Nissl substance. The cell body is the trophic center of the neuron and provides the metabolic products necessary to maintain the integrity of the cell body and its processes. If the axon of a cell body is cut, or if the cell body is destroyed, all portions of the axon distal to the lesion will degenerate (Fig. 1). This distal degeneration of the axon can be stained by special techniques thus providing a unique research tool. For example, a tract can be severed or a group of cell bodies destroyed and the extent of the axons can be followed after special staining for the degeneration products of the lesioned axons. A lesion of the axon may cause a characteristic pattern of dispersion and dissolution of the Nissl substance referred to as chromatolysis (Fig. 1).

A group of cell bodies in the central nervous system (CNS) is called a nucleus, and the axons of the cell bodies are grouped into bundles called fasciculi or tracts (Fig. 1). Some tracts, such as the corticospinal, are extremely long whereas others are short. In the peripheral nervous system (PNS) a group of cell bodies is called a ganglion and is usually surrounded by a heavy layer of connective tissue.

The dorsal roots of the spinal cord (Fig. 6) convey <u>afferent</u> information not only because it is sensory, but also because the impulses on the fibers are traveling toward the spinal cord. <u>Efferent</u> (or motor) impulses are traveling on fibers exiting the spinal cord via the ventral roots. The terms afferent and efferent, when referring to pathways in the CNS, can be interpreted to mean "fibers arriving" or "fibers exiting" from any given area (Fig. 1).

Embryology: Neural Tube and Brain Vesicles

The nervous system appears on the dorsal side of the developing embryo as a specialized area of the ectoderm (neural ectoderm)

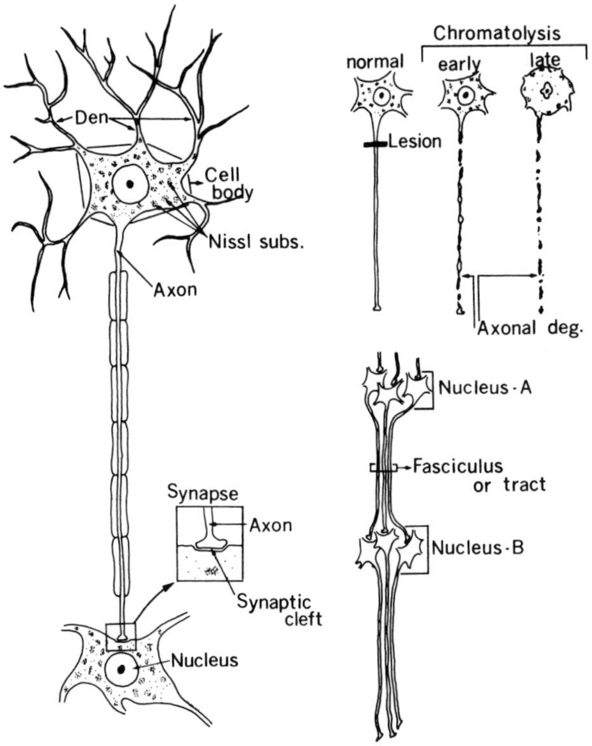

Figure 1. Semidiagrammatic illustration of a normal multipolar neuron, neuronal degeneration, and the relationships of cell bodies to tracts. Note the characteristic shape of the cell body and the orientation of the dendrites (Den) and axon. When a lesion is placed in the axon or the cell body, the peripheral portions of the axon degenerate. If the cell body is destroyed the axonal degeneration would be essentially the same. In the lower right drawing the axons of Nucleus A form a tract. This tract represents the efferents of Nucleus A and the afferents of Nucleus B.

designated as the neural plate (Fig. 2). As the embryo differentiates the lateral edges of the neural plate become elevated to form the neural folds and the resulting groove on the dorsal midline of the embryo is the neural groove (Fig. 2). The neural folds join in the dorsal midline to form the neural tube which has a prominent cavity, the neurocoele (Fig. 2). A small cluster of cells become detached from the dorsolateral aspect of the neural tube to form the future

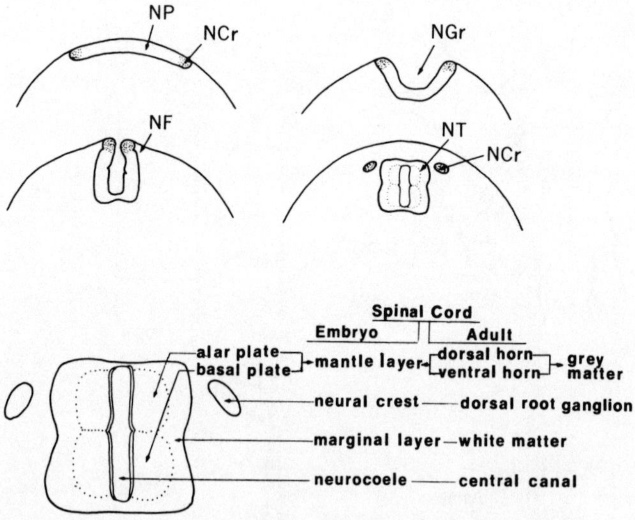

Figure 2. Diagrammatic illustration of the early stages of development of the neural tube. The embryonic and adult homologous structures of the spinal cord are indicated. Compare with Figure 6.

dorsal root ganglia (PNS). The caudal portions of the primitive neural tube will differentiate into the spinal cord of the adult, and the rostral portions will undergo rapid and elaborate differentiation to form the brain of the adult.

The rostral or cephalic end of the differentiating neural tube shows three primary brain vesicles by the second to third week. From caudal to rostral these are the rhombencephalon (hindbrain), the mesencephalon (midbrain), and the prosencephalon (forebrain) (Fig. 3). By the fifth week these brain vesicles have undergone further development into the five divisions seen in the adult brain (Figs. 3,4, Table 1).

Through this time of differentiation the primitive neurocoele is forming the central canal of the adult spinal cord and the ventricles of the adult brain (Fig. 4, Table 1). All primates have the four ventricles noted in Table 1 and all, with the exception of some higher primates and man, have a patent central canal of the cord throughout life.

Table I. A flow diagram to show the derivitives of the neural plate, primary and final brain vesicles. The principal structures for the five adult divisions are listed, and that portion of the ventricular system associated with each division is indicated.

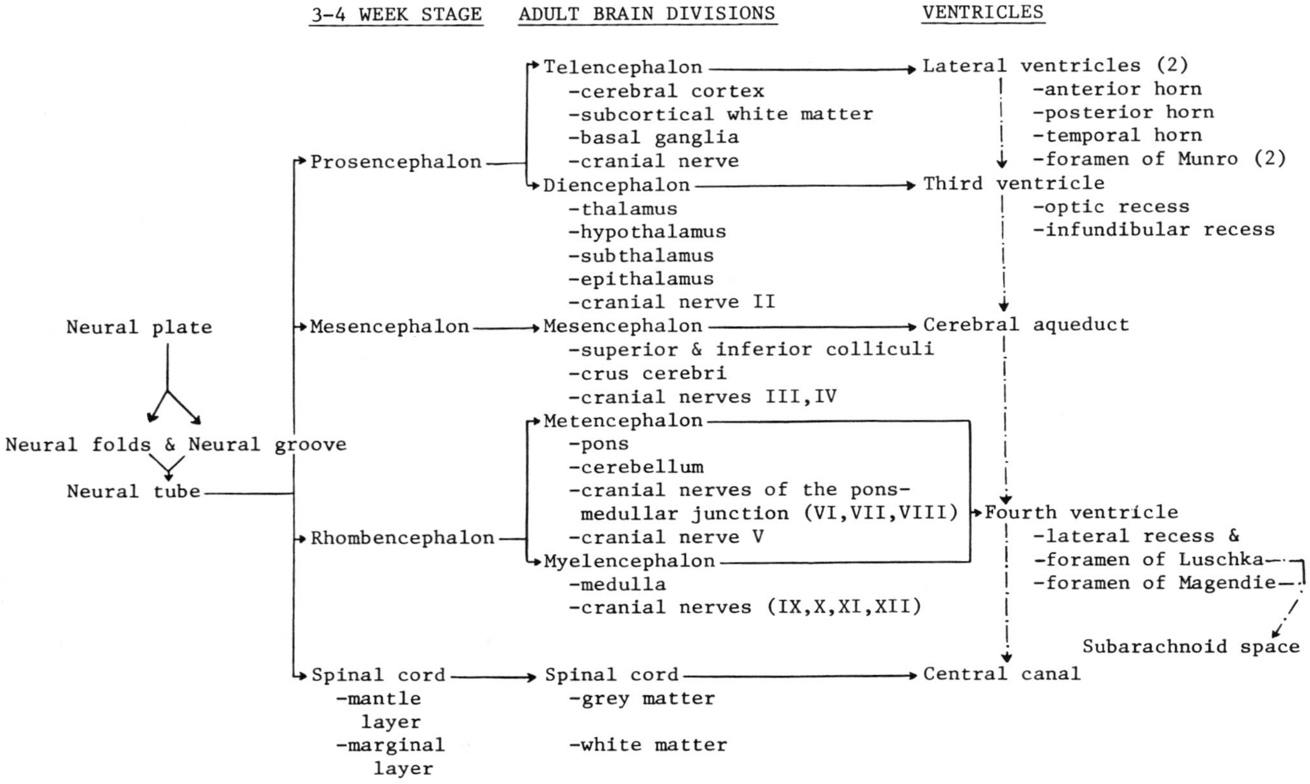

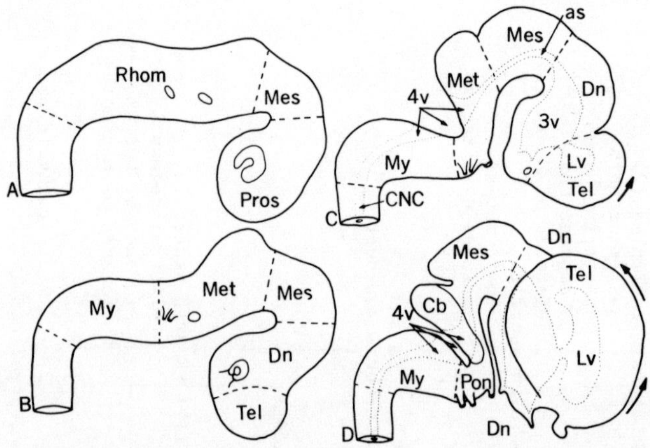

Figure 3. Semidiagrammatic illustration to show the development of the rostral portions of the neural tube (A, B). The three primary brain vesicles differentiate into the five divisions of the adult (C, D). The brain continues to develop and the telencephalon enlarges caudally over the diencephalon (arrows). The approximate position of the ventricular system in relation to gross anatomical landmarks is indicated (C, D). Only one of the two lateral ventricles is illustrated.

CNS, PNS, and ANS

The mammalian nervous system can be divided into a variety of structural and functional subdivisions. At the most general level it is divisible into the central nervous system (CNS), the peripheral nervous system (PNS), and the autonomic nervous system (ANS). The CNS and PNS are structural divisions. The brain and spinal cord comprise the CNS. The PNS includes all nerves to and from the appenendages and viscera, all peripheral motor (autonomic) and sensory ganglia, and all spinal and cranial nerves. The ANS is a functional subdivision having centers in the CNS and PNS which are related to its function. The ANS has nuclei in the brain stem and spinal cord which send their axons into the peripheral nervous system where they synapse on cell bodies in peripheral motor ganglia (Fig. 5). Efferent fibers from the cell bodies of these ganglia will terminate on either smooth muscle, cardiac muscle, or glandular epithelium. The terminal fibers of the ANS will only innervate one or more of the three effector tissues mentioned above regardless of where these efferent autonomic fibers travel in the body. The CNS and PNS carry a wide variety of sensory and motor modalities. However, the ANS carries only motor impulses to visceral effectors.

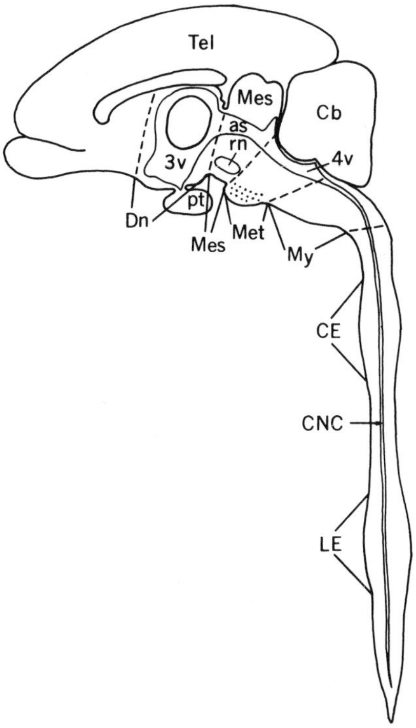

Figure 4. Midsaggital view of the brain and spinal cord of G. senegalensis. Note the five divisions of the adult brain and the comparative positions one to the other. The red nucleus is shown and the stippled area is the basilar portion of the pons. In this drawing the spinal cord is foreshortened.

Spinal Cord

The characteristic appearance of the neural tube has become altered in the adult spinal cord. The mantle and marginal layers of the neural tube have given rise to the grey and white matter of the cord, the central canal is small, and the relations of the dorsal and ventral roots are clarified. The grey matter contains neuron cell bodies, some of which contribute to principal ascending and descending pathways. The grey matter is also divisible into a dorsal horn primarily associated with incoming sensory information and a ventral horn that contains large cell bodies which send their axons out of the spinal cord to innervate somatic voluntary musculature. The white matter of the cord is divided into three regions:

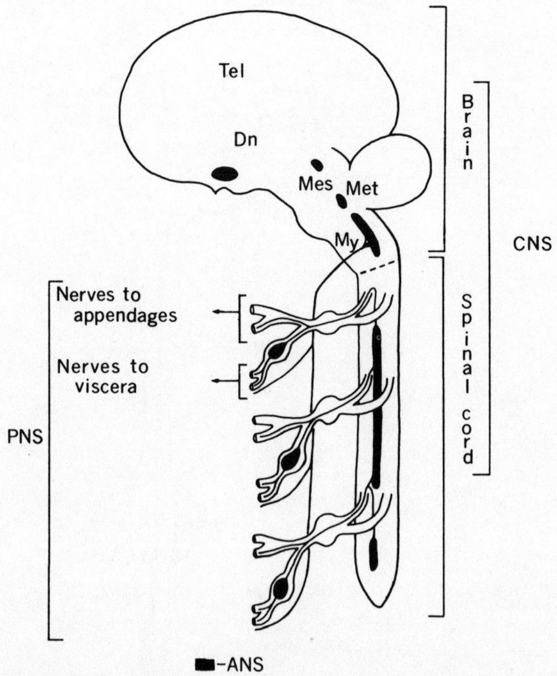

Figure 5. Diagrammatic illustration to show the central nervous system (CNS), the peripheral nervous system (PNS), and the autonomic nervous system (ANS). The CNS is brain and spinal cord, and the PNS is composed of dorsal and ventral roots and all nerves to appendages and viscera. The darkened areas in the CNS represent autonomic nuclei and in the PNS autonomic ganglia. The solid dark lines show the route of autonomic fibers. The cranial nerves are not shown.

a dorsal funiculus, a lateral funiculus and a ventral funiculus (Fig. 6). The first contains primarily ascending (sensory) tracts and the latter two contain both ascending (sensory) and descending (motor) tracts.

The dorsal horn receives afferent impulses originating from peripheral receptors. Following a synapse the cell bodies of the dorsal horn send this information into ascending spinal pathways or via collateral branches to motor neurons of the ventral horn for reflexes. However, the principal influence on cell bodies of the ventral horn is through large descending pathways coming from motor

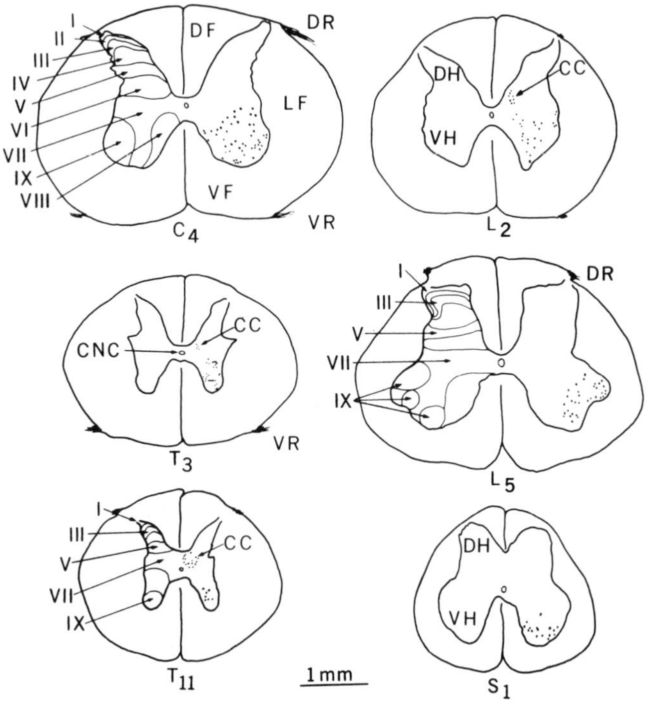

Figure 6. Tracings of cross sections of the spinal cord of G. senegalensis. Note the characteristic appearance of the respective levels. Roman numerals represent the lamination pattern of the grey matter (see text for discussion).

areas of the cerebral cortex and from nuclei of the brainstem associated with motor function.

The size and cell population of the ventral horn changes from one spinal level to another according to the demand from the somatic musculature. For example, the ventral horn is large in lower thoracic and lumbar levels due to the demand of the large amount of somatic musculature of the hindlimb. However, in mid-thoracic levels the ventral horn is small since there is comparatively little somatic musculature requiring innervation (Fig. 6).

The main areas of the cord which will be considered in more detail in subsequent discussion are Clarke's column (nucleus dorsalis), the spinocerebellar tracts, the corticospinal tracts, and the rubrospinal tracts. These are nuclei and pathways in the cord that are intimately related to postural orientation and locomotion.

Divisions of the Adult Brain

It is noted above that the five divisions of the adult brain are determined by their embryological sequence. The prosencephalon differentiates into the telecephalon and diencephalon, the mesencephalon remains unchanged, and the rhombencephalon differentiates into the metencephalon and myelencephalon. In the adult many important centers can be identified within one particular division.

Telencephalon

The telencephalon is composed of the cerebral cortex, the basal ganglia, and large bundles of subcortical white matter. The cerebral cortex receives all sensory information from the body (e.g., touch, pain, vision, taste, etc.) and gives rise to motor tracts (e.g., corticospinal) which are responsible for voluntary motor activity (Fig. 7). The corticospinal tracts originate from the motorsensory cortex and descend through all divisions of the brain as an uncrossed tract. At the junction of the medulla and spinal cord most of the motor fibers cross (pyramidal decussation) and will eventually influence the voluntary musculature on the side of the body contralateral to their origin. The basal ganglia are subcortical groups of cell bodies associated with pathways which phylogenetically are old and express an indirect influence on motor function. The subcortical white matter is composed of fibers passing between the cortex and lower centers of the neuraxis (e.g., corticopontine, corticorubral, corticospinal, etc.) or of fibers passing from one area of the cortex to another (e.g., corticocortical).

Diencephalon

The diencephalon serves as a center for the relay of information coming from a wide variety of peripheral sensory receptors onto the sensory cortices of the parietal, temporal, and occipital lobes (Fig. 7). Many of the nuclei of the diencephalon serve as relay centers for the feedback of information from lower centers onto the cortex. For example, the ventrolateral nucleus of the thalamus receives fibers from the ipsilateral basal ganglia and the contralateral cerebellar nuclei and sends fibers back to the basal ganglia (corticostriate) and thalamus (corticothalamic). Consequently a variety of centers (thalamus, basal ganglia, etc.) influence the cortex and vice versa.

Mesencephalon

The mesencephalon has relay centers for reflexes related to auditory and visual input, motor centers for the control of extra-

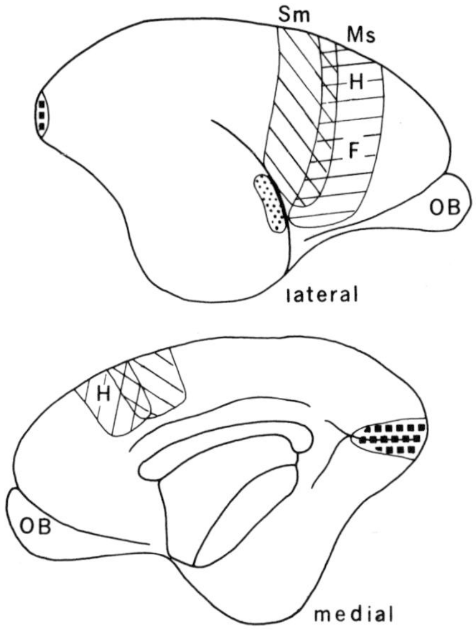

Figure 7. Lateral and medial views of the functional areas of cerebral cortex of an idealized prosimian. Note the overlap of the motor and sensory cortices. Stippling represents the auditory area and the black squares represent the primary visual cortex.

ocular muscles (cranial nerves III and IV) and important bundles of descending fibers from the cortex located in the crus cerebri. A characteristic structure of the midbrain, and one intimately related to motor function, is the red nucleus. This is a fairly prominent center located in the ventral portion of the midbrain (Fig. 4). This nucleus receives fibers from the cerebellum (crossed), basal ganglia and cortex (both uncrossed) and sends its efferent fibers to the spinal cord as the crossed rubrospinal tract.

Metencephalon

The metencephalon is the pons and cerebellum. The dorsal area of the pons (tegmentum) has the nuclei of cranial nerves VI and VII, portions of the nuclei of cranial nerves V and VIII, and other nuclei.

The ventral portion of the pons (basilar) is partially composed of nuclei which send their fibers across the midline to enter the contralateral cortex of the cerebellum through the middle cerebellar peduncle. Neurons of the cerebellar cortex project onto the deep cerebellar nuclei and the cell bodies of the deep nuclei in turn send their fibers out of the cerebellum via the superior cerebellar peduncle (SCP). As these fibers exit in the SCP they decussate and some ascending crossed fibers pass to the red nucleus and thalamus, while some descending crossed fibers pass to the reticular formation and inferior olivary nucleus. The cerebellum as a whole is functionally concerned with coordination of motor activity, the maintenance of posture and equilibrium, and proper regulation of muscle tone (synergy). The cerebellum receives information from peripheral receptors, integrates this information with feedback from higher centers, and sends out fibers to higher centers (eventually reaching the cortex) and lower centers (eventually reaching spinal levels) which express an important influence on locomotion.

Myelencephalon

The myelencephalon is the most caudal of the brain segments of the adult and is commonly referred to as the medulla oblongata or medulla. The nuclei of cranial nerves IX, X, and XII are located in the medulla and portions of cranial nerves VIII and XI are also important medullary centers. Ascending posterior spinocerebellar fibers pass along the dorsolateral surface of the medulla to enter the inferior cerebellar peduncle (ICP). The ICP also carries fibers into the cerebellum from other important medullary centers (e.g., inferior olivary nucleus, reticular nuclei). The corticospinal tract which has descended through the subcortical white matter of the telecephalon, the crus cerebri of the midbrain, and the basilar portion of the pons, is now found as the pyramids of the medulla.

II. INTRODUCTION TO THE PROBLEM

During the course of phylogeny which ultimately led to the higher primates and man many important morphological and physiological changes took place. These changes presumably left the animals of each succeeding phase more adequately adapted to their new ecological niche. These factors provided a new pool of characteristics for the continuing evolution of subsequent forms.

One of the more significant series of developments were those ecological changes and morphological adaptations which resulted in the advent of upright posture and bipedal locomotion (64). The fossil record has provided considerable information concerning osteological changes which may be related to evolving locomotor patterns. With the exception of excellent studies of neuroevolution using endocasts (89,90) little of no information is available on possible neuroanatomical changes (specializations?) that took place during the course of evolution specifically related to upright locomotion.

There is no question but that muscle function played an important role in the evolutionary development of upright locomotion. There are many spinal cord and brain centers specifically related to postural orientation and locomotion which either receive afferent information directly from skeletal muscle or directly influence its effector response. When attempting to deduce a sequence of events that may have taken place in the course of phylogeny the problems are: What extant forms may possibly provide the best clues? and, What structures in these forms should be considered for comparative study? The 'sequence of events' under consideration are: What neuroanatomical specializations took place in the central nervous system that may have contributed to the eventual evolution of upright posture and bipedal locomotion? Not only were osteology and myology undergoing modification but it is hypothesized that central nervous system centers related to postural orientation and locomotion were also undergoing a specific series of changes. These neuroanatomical specializations were undoubtedly concurrent with (and probably highly dependent on) morphological adaptations taking place in the pelvis and lower limbs. It has been shown in a wide variety of vertebrate forms that the nervous system can undergo distinct specializations related to specific functions.

The animals considered relevant to this problem are the tree shrews[2] and members of the family Lorisidae, namely Galago, Nycticebus and Perodicticus. These animals represent examples of morphological baselines from which many subsequent changes may have taken place.

Since the time of Carlsson (14) the taxonomic position of Tupaia has been pondered by many authorities. They are considered either to have affinities with the primates (11,12,63,99), no direct affinities with the primates (13,107), or affinities with neither the primates nor insectivora (40,70,103). In their total morphological pattern (63) the tree shrews are considered, by many, to be a living representative of the primitive stock from which primates evolved. Romer (93) has suggested that the lack of specialization in the tree shrews may imply that these animals are close to the

[2]The authors are well aware of the controversies on the taxonomy of the tree shrews and their inclusion in this report does not imply that they should or should not be considered as a member of the Prosimii.

ancestry of "all the placental orders" not just the primates. Even though the future will undoubtedly bring more definitive evidence to bear on the question of Tupaia taxonomy, it can be said that the total morphology of these animals is transitional between the sub-primate mammals and the lower primates. In this respect they provide an extant form (consequently a complete spectrum of "soft anatomy") in which functional morphological states, which may represent some pre-primate levels of development, can be studied. Also of considerable importance is the fact that the tree shrews, whether terrestrial or arboreal, move most effectively and most commonly in a quadrupedal manner (52,99). In the present context the tree shrews are considered transitional quadrupedal mammals occupying the morphological gap between pre-primate mammals and the lower primates.

Some members of the Prosimii constitute a locomotor group which has been characterized as the "vertical clingers and leapers" (VC&L) (75,76). These extant animals have morphological and behavioral characteristics in common which given them an advantage for this particular mode of locomotion.

Napier and Walker (76) postulated that the VC&L share certain osteological similarities with primates of the Eocene and they concluded that VC&L may represent the earliest locomotor specialization in primates. They also suggested that VC&L was preadaptive to some or all subsequent patterns of primate locomotion. This hypothesis has been soundly criticized (18,105). It is only fair to point out that these criticisms are based on essentially the same evidence used by Napier and Walker to develop their concept. Before any hypothesis, which has been developed to explain a chain of events that may have taken place in the course of phylogeny, is accepted or discredited it must be subjected to a varity of experimental tests. Le Gros Clark (63,64) has stressed the importance of the "total morphological pattern" when attempting to ascertain the relative significance of phylogenetic events or trends. This concept implies that detailed comparative studies on extant forms may provide information on morphological modifications that took place in the soft anatomy of extinct forms. It is profitable to study osteology (and paleo-osteology), myology and behavior, and then to extrapolate possible trends (100). However, locomotion (past and present) is not only a function of osteological relationships and muscle mass, but also of the afferent and efferent innervation of specific muscle groups (e.g., extensors, flexors, etc.).

Within the family Lorisidae there are the so-called VC&L and several forms which show a distinct type of quadrupedalism. Included in the latter locomotor type are Nycticebus coucang and Perodicticus potto. These Lorisidae move in a slow deliberate manner either on a horizontal or vertical support. They are quadrupeds in that they utilize all four appendages on the substrate when progressing from one point to another. The slow loris and potto progress in this manner when either casually exploring immediate environment or fleeing a stressful encounter.

The lesser bushbaby, Galago senegalensis, is a prime example of a so-called vertical clinger and leaper. The hindlimb of the lesser Galago is relatively long, muscular and so modified to give this animal a mechanical advantage for its typical saltatory locomotor pattern (43,44,76,101,102). Galago will progress quadrupedally when examining its immediate surroundings but will resort to vertical leaping when investigating peripheral territory or if fleeing a stressful situation.

In the animals discussed above there is a transitional quadruped (Tupaia) that may be an extant example of the stock from which primates and possibly all placentals evolved. (The locomotor pattern of Tupaia is presumably an example of the quadrupedalism of the rodent-like primates of the Paleocene). Also represented is the family Lorisidae which contains a prime example of the so-called VC&L category and a distinct form of quadrupedalism. The former is Galago and the latter is Nycticebus and Perodicticus. Even though the slow loris and potto may represent a type of specialized climber (88) their quadrupedal mode of locomotion is presumably near that point from which subsequent specializations may have taken place.

Comparative studies on the higher primates have provided much valuable data on the various anatomical adaptations related to upright posture. However, these animals have evolved, to varying degrees, away from the theoretical transitional stage or baseline point. The tree shrew and some members of the family Lorisidae in their phylogeny, morphology and locomotor style appear to be close to the baseline point from which subsequent change took place. As the locomotor patterns of the Paleocene and Eocene were evoling from quadrupedalism to some upright phase there were obvious osteological and myological adaptations. It is postulated that there were distinct and characteristic neuroanatomical modifications of sensory and motor centers, and pathways which occurred either in response to or concurrently with these gross anatomical changes. Consequently, the modifications of the nervous system were partially a reflection of evolving locomotor patterns. Using extant forms which represent significant (yet basic) levels of phylogenetic development: What were some of the early neuroanatomical adaptations? and, What pool of characteristics was thus formed from which subsequent change could have taken place?

III. SENSORY CENTERS AND PATHWAYS RELATED TO POSTURE

Spinal Cord

The spinal cord is that portion of the central nervous system that (1) receives sensory information from the major portion of the body and conveys this information to relay nuclei, and (2) conducts descending motor impulses to ventral horn (motor) neurons. The ventral horn neuron serves as the final peripheral motor pathway.

As mentioned earlier (Basic Structure of Nervous Tissue) the spinal cord is composed of grey and white matter. The white matter consists primarily of long ascending and descending fiber tracts. Ascending fibers may originate from nuclei of the grey matter. Descending fibers originate from various motor centers within the brain and brainstem and influence the activity of ventral horn cells.

The arrangement of nuclei in the grey matter has been extensively studied and divided into ten laminae (79,92). The dorsal horn consists of laminae I through VI; the intermediate region is designated as lamina VII and the ventral horn is divided into laminae VIII and IX. Lamina IX is composed of large multipolar (motor) nuerons which lie in ventral and ventrolateral regions of the ventral horn. Lamina X consists of grey matter surrounding the lateral aspects of the central canal (Fig. 6).

The area of the spinal cord that receives from and contributes to a single spinal nerve is called a spinal segment. Except for the cervical nerves, the numbers assigned to spinal segments and nerves correspond to the number of vertebrae. All prosimians have eight cervical spinal nerves. The Tupaiidae and Galaginae have thirteen thoracic levels and the Lorisinae are highly variable having 14 - 16 thoracic levels (97).

Peripheral sensory impulses originating in response to deep pressure, changes in joint position and muscle length are classified as proprioception. Proprioception refers to sensory input brought about by movement of the body and not by externally applied stimuli. Consequently, integration of information entering the CNS in response to limb or trunk movement is necessary to insure proper limb placement and coordination of fine truncal adjustments. It is reasonable to assume that a change from a generalized locomotor pattern (quadruped) to a more specialized, more complex pattern (semi-erect and erect bipedalism) would require an increase in the specificity of the neural structure and functional capacity of the cord.

Variations in the size and location of spinal cord enlargements have been correlated with locomotor behavior (6,59). The pigeon with fourteen cervical segments has a well-formed lower cervical enlargement corresponding to the innervation of the wings (59). The ostrich, a runner with rudimentary wings, has an immense lumbar swelling (104). In prosimian and higher primates differences in cervical and lumbar enlargements are more difficult to detect. Lassek (59) reported differences in grey-white ratios in monkey and man. The monkey showed a proportionately greater increase in grey matter in cervical and lumbar enlargements. Despite this, he reported that the grey-white matter volumes in both were "...quite similar .....indicating an equality of function in the anterior and posterior limbs."

At the present there is little available literature on the prosimian spinal cord. Le Gros Clark (62) reported that the spinal cord of the pen-tailed tree shrew (Ptilocercus lowii) was 88 mm in length and extended "...down to the back of the sacrum." He described the cord as having a distinct cervical but a faint lumbar enlargement. The spinal cord in the lesser Galago was reported to be 11 cm in length and to extend to the caudal region of L6 (50). In contrast to the pen-tailed tree shrew, the lesser Galago spinal cord has a distinct lumbar swelling (3.5 mm in diameter) which is approximately 1 mm greater than the diameter of the thoracic cord (50). The cervical enlargement is slightly greater than the lumbar enlargement. Despite this, the lumbar swelling in the lesser bushbaby has a greater increase in grey matter (6). These authors related the lumbar enlargement in the lesser Galago to their jumping abilities. In the lower lumbar and sacral segments the posterior funiculus (dorsal columns) comprises approximately 25% of the total white matter area (6). Corresponding percentages in Cebus and man are 26% and 39% respectively (4,6).

The spinal cord of the greater bushbaby is proportionately similar in size to that of the lesser Galago. In both the cord extends to the same segment (L6) and characteristically has a very short cauda equina. In the slow loris and potto the cord extends to the lower portion of L1 and shows a long cauda equina.

At the present, the similarities in spinal cord gross structure, differentiating loris and potto from the saltatory Lorisidae (Galago) can only be considered in terms of genetic inheritance. It is most probable, however, that there are definitive cytoarchitectural differences related more specifically to the divergent patterns of locomotion.

Nucleus Dorsalis (Clarke's Column)

The nucleus dorsalis is a distinct cell column in the spinal cord located dorsal and lateral to the central canal. This nucleus is responsible for relaying sensory information from neuromuscular spindles and Golgi tendon organs to the cerebellum.

One of the characteristic features of the column is a greater population of cells in lumbar levels than in middle or high thoracic segments. In man the column extends from C8 to L3 (24). The extent of the column is a rather consistent feature in mammals even though there is slight variation among genera (1,19,21,92). Degeneration and electroneurophysiological studies have demonstrated that axons from Clarke's column neurons enter the ipsilateral lateral funiculus and form the dorsal spinocerebellar tract (DSCT). In the _Macaca mulatta_ DSCT fibers terminate in the vermal and paravermal cortex of anterior lobe and portions of the vermis of the posterior lobe of the cerebellum (115). Based on its characteristic distribution, Clarke's column may be correlated with voluntary and reflex function of the hindlimb and lower trunk.

In the lesser _Galago_ (1) Clarke's column (nucleus dorsalis) extends from C8 to L3. Based on distinct differences in size, cells within the nucleus can be placed in one of three classifications (Classes A, B, or C). Class C neurons are elongated or fusiform and measure 50-87 μ in length and 20-37 in width. Class B neurons measure 25-49 μ along their major axis and may be either stellate, elongate or fusiform. Class A cells measure less than 25 μ and may be oval, triangular, or elongate. The increase in cell density in Clarke's column in lower thoracic or upper lbmbar levels is due to increases in Class B and C cells. Class A cells do not show any appreciable change in population throughout the column. Class B and C cells are believed to be the principle cells of origin for the dorsal spinocerebellar tract (10,35,36,68). Clarke's column neurons in the tree shrew are morphologically similar to those in the lesser _Galago_ and do not warrant additional description.

Even though Clarke's column in _Tupaia_ and _Galago_ have similar caudal terminations (L3) some important differences in distribution are observed (Fig. 8). In lesser _Galago_ the column has its greatest cell population in segments T12 or L2 and abruptly terminates at the mid-L3 level. Between T12 and T11 there is a significant decrease in cell population (Fig. 8). A population of approximately 700 cells remains constant T6 at which point there is a second sharp decrease in cell number. From T5 to the rostral end of the column (C8) there is a gradual decrease in cell density.

The cell population of Clarke's column in _Tupaia_ (Fig. 8) peaks at L2 and as in the bushbaby, terminates abruptly at the mid-L3 level. From L2 to T12 there is a distinct decrease in cell population and from T12 to the rostral termination of the column (T2) there is a gradual decrease in cells (Fig. 8).

Variations in the distribution of any sensory relay nucleus (e.g. nucleus dorsalis) can be attributed to differences in muscle mass and, consequently, in peripheral sensory requirements. When comparing the lesser Galago with the tree shrew both of the above variables exist. The hindlimb of the bushbaby is not only larger but is also mechanically suited for vertical leaping and saltatory locomotion. When the supporting base is reduced from four to two (such as in the case of the quadruped versus the biped) and distal and proximal musculature of the limb and the lower trunk must continuously make the fine reciprocal intermuscular adjustments. This activity enables the animal to maintain upright posture. As mentioned earlier, most of these postural adjustments are reflex in nature and depend on proprioception (muscle length) as one of the main sources of peripheral information. When a bipedal locomotor pattern is superimposed on this more complex posture the need for a finer, more appropriate peripheral sensory system is greatly increased. It seems probable based on locomotor differences, that Galago senegalensis would show a more specialized proprioceptive system when compared with Tupaia. The differences in the distribution of cells in Clarke's column in these two genera is interpreted as an indication of relative locomotor development. The saltatory mode of Galago requires increased input from hindlimb and reciprocal stabilization of the lower trunk. Both of these somatic requirements are reflected in the cell composition of the nucleus (Fig. 8). The quadrupedal locomotor pattern of Tupaia does not require excessive proprioceptive input from hindlimb and little or no stabilization of the trunk. This lack of somatic specialization is reflected in the cell population of Clarke's nucleus in Tupaia (Fig. 8) especially when compared to the bushbaby.

Dorsal Column Nuclei

The dorsal funiculus of the spinal cord is composed of two paired fasciculi, gracilis and cuneatus (Figs. 9-11). Both fasciculi are referred to as the dorsal column pathways and convey impulses which originate from various cutaneous mechanoreceptors. This sensory information is interpreted as various qualities of tactile sensation particularly fine touch (two point) discrimination. Axons within the fasciculi terminate within the gracile and cuneate nuclei of the medulla. Axons originating from these nuclei cross the midline, form the medial lemnisci and terminate within a specific relay nucleus of the thalamus. Subsequently, dorsal column sensory information is conveyed to the somatosensory cortex via thalamocortical projections. In addition to the nucleus gracilis and cuneatus, the dorsal column nuclei include a more laterally located nucleus, the lateral cuneate

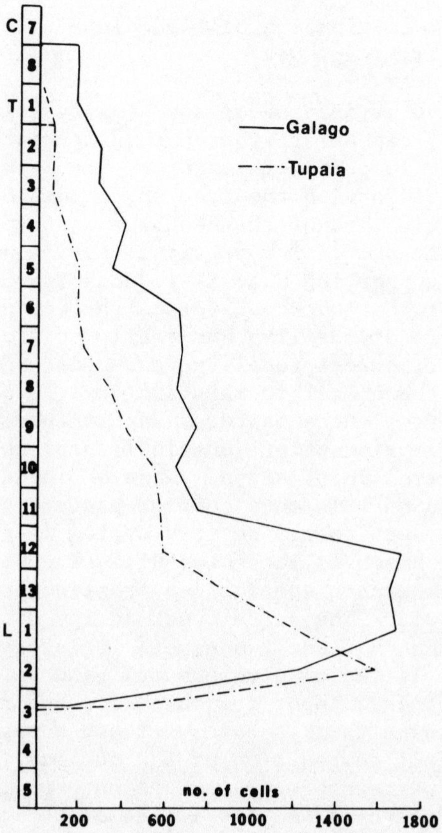

Figure 8. Graph of cell distribution of Clarke's column in Galago senegalensis and Tupaia glis. Note the elevated cell population in Galago between T6 and T11 and the broad peak from T12-L1. Contrast with the tree shrew (see text for discussion).

nucleus (LCN). Even though the LCN does not contribute fibers to the medial lemniscus it is considered to be a lateral extension of the cuneate nucleus.

The lateral (accessory or external) cuneate nucleus receives fibers from upper thoracic and cervical cord levels that ascend within the fasciculus cuneatus. In contrast to the other dorsal column nuclei, the LCN receives impulses originating primarily from muscle spindles and tendon organs and project to the cerebellar cortex via cuneocerebellar fibers (10,81,82). Cells within the LCN have been described as large multipolar neurons resembling the cells of Clarke's column of the spinal cord (19,25,82).

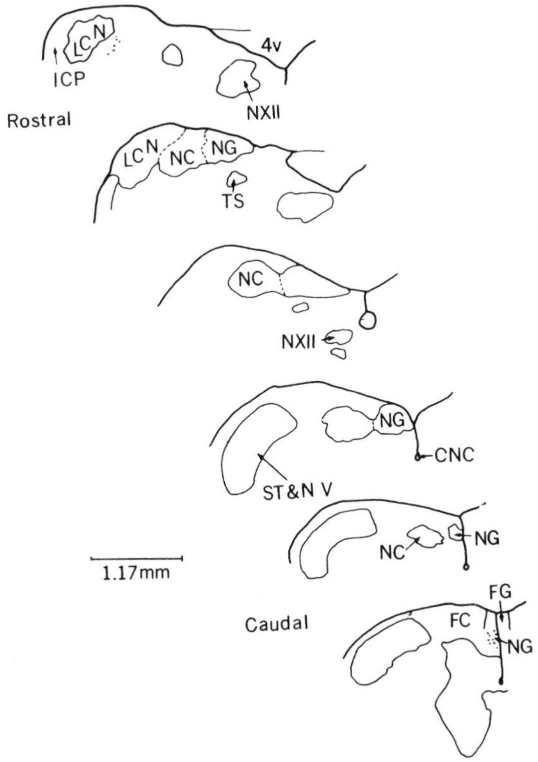

Figure 9. Projected tracings of dorsal column nuclei in Tupaia from caudal to rostral. Each tracing was made at intervals of approximately 600 micra.

The fasciculus and nucleus gracilis receive fibers from the caudal, sacral, lumbar and low thoracic segments and can be functionally correlated with the hindlimb and lower trunk. The fasciculus cuneatus and medial cuneate nucleus receive fibers from upper thoracic or cervical cord levels and are therefore concerned with upper trunk and forelimb activities. In primates, e.g., Saimiri, Callithrix, and Cebus (32), and nonprimates, e.g., shrew mice and hedgehog (3), there is an additional midline nucleus in the medulla (nucleus of Bischoff). When first identified, Bischoff (5) associated this nucleus with the functions of the tail. In many primates this correlation is not always possible. In particular, the spider monkey (Ateles) does not have such a nucleus even though he possesses a highly developed prehensile tail (3,20).

From a comparative neuroanatomical standpoint, the dorsal column-medial leminscal system developed in response to an increased sensory requirement apparent in locomotor phylogeny. From comparative studies on hedgehog and tree shrew it was suggested that the dorsal column pathway had a major influence on the general to specific differentiation of the ventroposterior and posterior nuclear complex of the thalamus (49,96).

Tilney (106) reported volumetric measurements of the dorsal column nuclei in primates from *Tarsius* to man. The nucleus cuneatus is in most cases the larger of the two nuclei. The difference in size may be a reflection of the various forelimb or hindlimb locomotor specializations. For example, the nucleus cuneatus in the gibbon is considerably the larger, while the reverse is true in the marmoset and gorilla.

In *Macaca* the nucleus gracilis begins at the level of the pyramidal decussation as a small group of cells in the ventral aspect of fasciculus gracilis. Rostrally, the nucleus assumes a triangular shape and reaches the dorsal surface of the medulla. The nucleus cuneatus contains two cell types and begins slightly rostral to the nucleus gracilis. The lateral cuneate nucleus begins at more rostral levels and can be seen as a separate nuclear group. The cells of the LCN are larger than those in nucleus gracilis or cuneatus and reportedly resemble those of the Clarke's column (30).

In contrast to *Macaca*, the gracile nucleus in the spider monkey can be divisible into two definite subgroups, the pars dorsomedialis and ventrolateralis (20). Both regions contain similar cell types and extend caudally to the rostral region of the first cervical segment. With reference to the specific organization of the nucleus and fasciculus gracilis, Chang and Ruch (20) concluded that in the spider monkey there is a specialized dorsal column system related to the presence of a "...fifth hand." In addition, the fact that the nucleus cuneatus in *Ateles* is the smaller of the two nuclei can be attributable to the rather "... imperfect development of the hand...".

In *Tupaia* the gracile nucleus begins in the rostral level of the first cervical segment as a small midline nucleus within the fasciculus gracilis (Fig. 9). Approximately 500 μ from this point the cuneate nucleus is formed as a small lateral aggregation of cells. Rostrally both nuclei increase in size, become joined, and approach the dorsal surface of the medulla. The lateral cuneate nucleus beings as an isolated cell group located on the dorsolateral surface of the medulla approximately 2 mm rostral to the first appearance of the cuneate nucleus. In *Tupaia* the LCN enlarges and becomes contiguous with the cuneate nucleus (Fig. 9). The cuneate nucleus extends approximately 400 μ rostral to nucleus gracilis and terminates in the caudal level

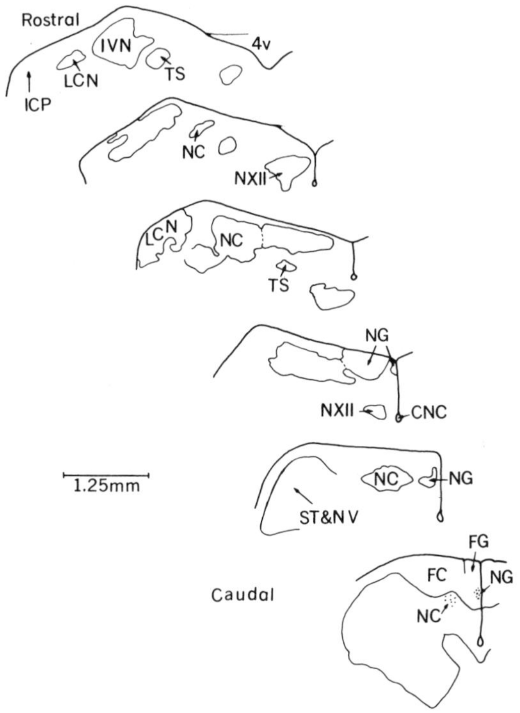

Figure 10. Projected tracings of dorsal column nuclei in Galago senegalensis from caudal to rostral. Each tracing was made at intervals of approximately 600 micra.

of the inferior vestibular nucleus. The LCN continues rostrally approximately 400 μ where it terminates lateral to the inferior nucleus of VIII and medial to the inferior cerebellar peduncle (Fig. 9). Cells of the lateral cuneate nucleus are larger than those of the other nuclei, contain moderately clumped Nissl substance and appear fusiform, elongate, oval, or triangular. Similar cells may be observed scattered throughout the cuneate nucleus and in transitional regions. Cell types within the gracile and cuneate nuclei are predominantly oval with dispersed Nissl substance. Both nuclei show a predominance of large cells in their caudal portions and small cells rostrally. On occasion a small group of stellate cells may be observed in the ventrolateral region of the nucleus cuneatus. Even though a similar arrangement of cells has been reported in Potto (32)their presence in Tupaia as a definitive cellular division is questionable.

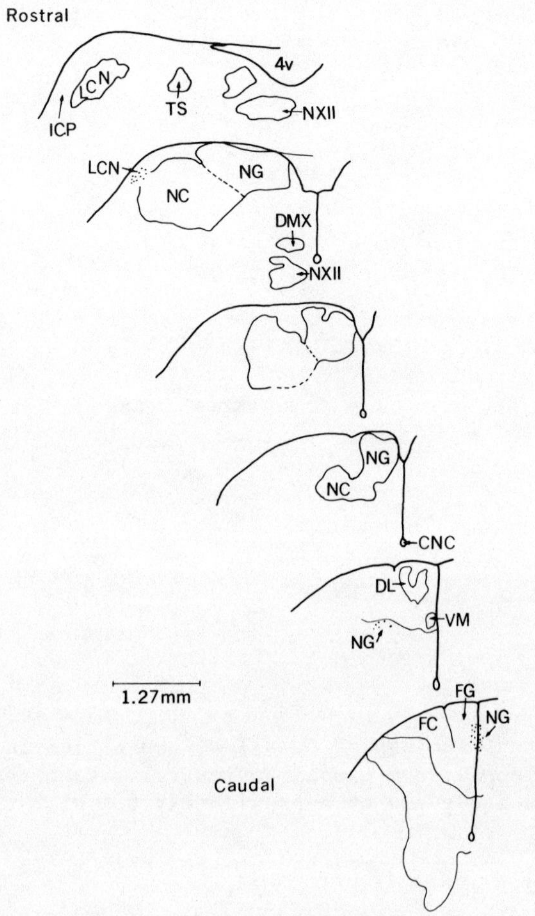

Figure 11. Projected tracings of dorsal column nuclei in *Nycticebus coucang* from caudal to rostral. Each tracing is spaced at an interval to show the major changes in the nuclei.

In the lesser bushbaby the dorsal column nuclei are well differentiated and appear larger in proportional cross sectional area than those in *Tupaia* (Fig. 10). Cells within the nucleus gracilis and nucleus cuneatus are similar to those observed in *Tupaia*. As in *Tupaia* the caudal areas of the gracile-cuneate complex has a predominant population of large cells and small cells are most numerous rostrally. However, at the rostral pole of the cuneate nucleus a few large neurons are characteristically present in its dorsolateral aspect. The appearance of the larger cells coincides with the level where the LCN is well differentiated. As in *Tupaia* cells in the LCN appear slightly larger than the predominant cell types of the other nuclei. Some LCN neurons appear similar in shape (though considerably smaller) to those in Clarke's column in the lesser bushbaby (1). At a level through the rostral part of the pyramidal decussation the caudal and medial portions of the gracile nuclei are apposed to the midline and appear to form a midline nucleus. This group of cells extends caudally into the upper levels of C1. Whether this small midline group of cells is homologous to the nucleus of Bischoff (3,5,32) or has a particular functional significance await the results of further study.

In the slow loris the nucleus gracilis begins adjacent to the midline in the rostral levels of C1 (Fig. 11). This group of cells is presumably in the same position as the midline nucleus described in Potto (32). At the caudal aspect of the pyramidal decussation the midline cells shift to a more ventral position and a dorsolateral cell group can be observed. Rostral to this level the two divisions join and form a well differentiated nucleus gracilis. Just prior to their union the nucleus cuneatus appears lateral to the ventral portions of the gracile nucleus (Fig. 11). At further rostral levels the nuclei gracilis and cuneatus fuse and remain contiguous throughout most of their course (Fig. 11). Cells within both nuclei appear similar in size, shape and staining characteristics. In the most rostral levels of the cuneate nucleus there appears to be an increase in larger cells. At this level, however, it is difficult to clearly differentiate the LCN from the cuneate nucleus. Because of this the exact rostral termination of the cuneate nucleus can not be easily determined. The LCN in *N. coucang* terminates in a manner similar to that described above for *Tupaia* and *Galago* (Fig. 11).

IV. THE CEREBELLUM

Gross Morphology

It is not profitable in the present report to discuss in detail the gross anatomy of the cerebellum of the tree shrew and the Lorisidae. This information appears elsewhere (37,38,41), therefore, the following discussion is purposefully brief.

The mammalian cerebellum develops from bilateral thickenings of the alar portion of the rhombencephalic lip (57,58). These thickenings join in the dorsal midline and enlarge caudally over the developing pons and medulla, the intervening space forming the future fourth ventricle (Fig. 3). The first embryological fissure which separates the caudally located flocculonodular lobe from the rostrally located corpus cerebelli (Figs. 3,12,18). The second fissure to appear in the embryo is the primary fissure which divides the corpus cerebelli into anterior and posterior lobes. The anterior lobe of the corpus cerebelli is located between the anterior medullary velum and the primary fissure, and in the adult primate it is differentiated into five lobules, designated I to V from rostral to caudal. The posterior lobe of the corpus cerebelli is located between the primary and posterolateral fissures and is differentiated into lobules VI to IX of the adult (Fig. 12,13). The portion of the cerebellum located caudally to the posterolateral fissure is the flocculonodular lobe and represents lobule X. The cerebellum consists of ten lobules in Tupaia and all Prosimii studied to date and has a midline vermis portion and lateral hemispheric portions. The vermian lobules are indicated I to X from rostral to caudal, and the hemispheric portion of each respective vermian lobule is indicated by the prefix H. The hemispheres of the primate cerebellum are composed of lobules HII to HX. The first vermian lobule (lingula) is unique for a variety of reasons, one of which is the fact that it has little or no hemisphere in primates (57).

Anterior Lobe

The cortex of the anterior lobe of the corpus cerebelli receives afferent input from the spinal cord and brainstem centers. Some of these centers, namely the lateral cuneate nucleus and Clarke's column have been discussed above. The fact that several centers (e.g. nucleus dorsalis project directly to the anterior lobe cortex may be reflected in the relative development of this lobe.

In the so-called "basal" and "progressive" insectivores the anterior lobe is poorly developed (Fig. 12). It is small and practically unfissured in Sorex. The anterior lobe in Elephantulus is indented by the caudal portion of the mesencephalon and it is only slightly more fissured than in the "basal" insectivores. In both of these animals the anterior lobe is not only small but also shows a poorly developed or absent hemispheric region (Fig. 12). In Ptilocercus and Tupaia the anterior lobe appears somewhat larger from the rostral aspect (Fig. 12). The mesencephalon indents the anterior lobe, similar to that seen in the elephant shrew, and in both of these tree shrews the hemispheres are poorly developed. (Fig. 12).

Galago and Tarsius are vertical clingers and leapers. Nycticebus and Perodicticus may be considered as specialized quadrupedal climbers. In these genera the anterior lobe of the cerebellum is distinctly larger than in the tree shrews (Fig. 12). Of particular note are the facts that the mesencephalon does not impinge on the anterior lobe so severely, and there are distinct and comparatively well developed hemispheric portions of the anterior lobe (Fig. 12). The lateral cerebellar margin of the anterior lobe is apposed to the

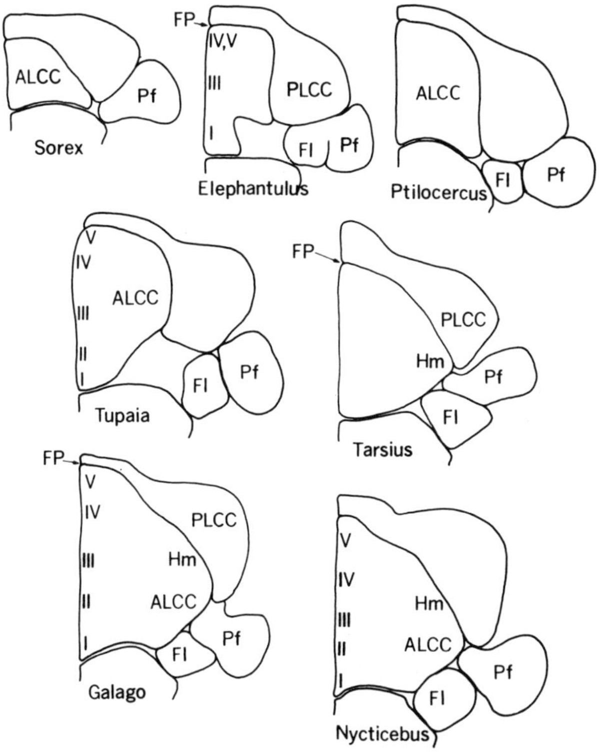

Figure 12. Anterior view of the cerebellum of basal and progressive insectivores, tree shrews, and some Prosimii. All specimens are drawn to about the same size. The Roman numerals represent the approximate positions of the lobules of the anterior lobe. The illustrations of <u>Sorex</u> and <u>Elephantulus</u> are modified from Brauer (9), and that of Tarsius from Woollard (114). The drawing of <u>Ptilocercus</u> is modified from Le Gros Clark (62).

flocculus or paraflocculus (usually both) in all prosimians examined to date. In nonprimate mammals (<u>Sorex</u>, <u>Elephantulus</u>) and in the tree shrews the lateral margin of the anterior lobe is, at best, only slightly apposed to the flocculus. In most forms it does not extend even this far lateral (Fig. 12).

Posterior Lobe

The posterior lobe of the corpus cerebelli in all primates is proportionally larger than the anterior lobe (Fig. 13). On the posterior surface of the cerebellum the border between the vermis and hemisphere is clear due to the deep and distinct paramedian fissure (Fig. 14). The clarity of division between the vermis and hemisphere is lost on the anterior lobe since the paramedian fissure does not extend onto the anterior lobe. In the basal insectivore, Sorex, the entire posterior lobe is small, the hemisphere is distinctly abbreviated and the paramedian fissure is vague (9). The small size of the cerebellar hemisphere accentuates the appearance of the large paraflocculus. The progressive insectivore Elephantulus (9) shows a larger hemisphere which tends to decrease the apparent size of the paraflocculus (Fig. 13). The vermis of the elephant shrew is larger and the area of the paramedian fissure more prominent.

The paraflocculus and flocculus of Ptilocercus and Tupaia are distinct separate entities. The vermis is large, the paramedian fissure is deep and the hemispheres enlarged (Fig. 13). As the posterior lobe vermis and hemispheres enlarge and the paramedian fissure deepens the clarity of continuation of the lobules of the vermis with their hemispheric counterparts is obscured in Tupaia and the Lorisidae (37,41). The Lorisidae show posterior lobe hemispheres that are comparatively large, well differentiated and overshadow the flocculus and paraflocculus. The vermis is pronounced and the paramedian fissure deep (Fig. 13,14). The absolute size of the paraflocculus in the Lorisidae appears to be smaller than that of the Tupaiidae. In Tarsius the hemisphere is of moderate size when compared to the lesser Galago and Woollard (114) shows the paraflocculus to be quite small (Fig. 13). The hemisphere of the posterior lobe receives a majority of its afferent input from pontine nuclei in the basilar portion of the pons. In the tree shrews reported to date the pons in small. In the Lorisidae the basilar area of the pons is slightly larger in those genera with an enlarged cerebellar hemisphere (42).

Flocculonodular Lobe

The nodulus is the vermal portion (X) and the flocculus is the hemispheric portion (HX) of the flocculonodular lobe. The flocculonodular lobe is by far the smallest lobe of the adult cerebellum. In all the animals mentioned above, with the exception of Sorex, the flocculus is a distinct structure separate from the paraflocculus. The nodulus is of moderate to small size in most genera. The connection between the nodulus and the flocculus is a small bundle of fibers called either the floccular or flocculonodular peduncle. This peduncle is distinct in Tupaia, Galago, Nycticebus, and Perodicticus (37,41).

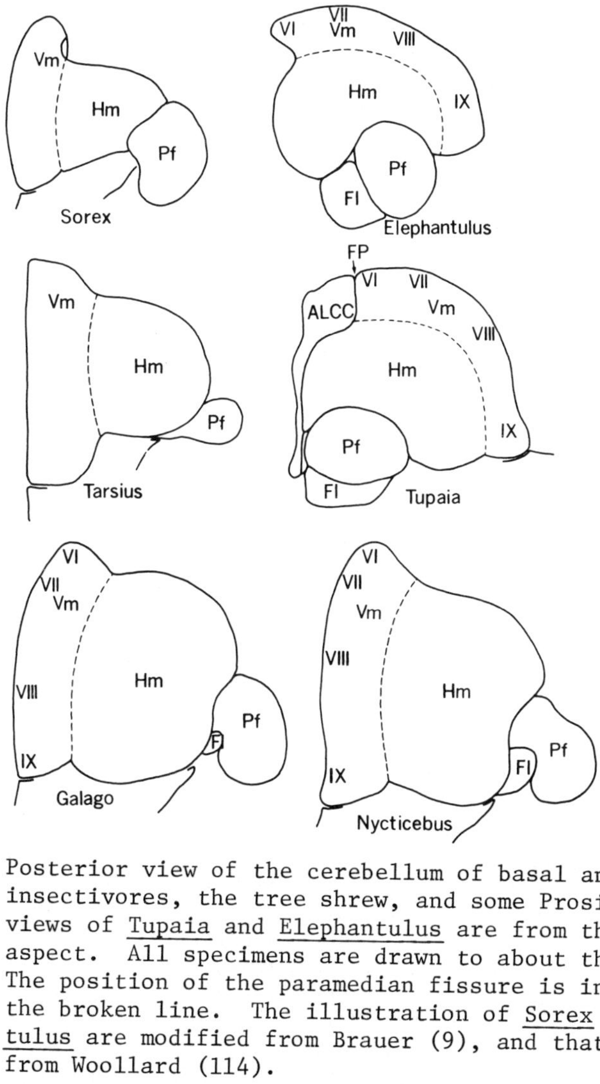

Figure 13. Posterior view of the cerebellum of basal and progressive insectivores, the tree shrew, and some Prosimii. The views of Tupaia and Elephantulus are from the lateral aspect. All specimens are drawn to about the same size. The position of the paramedian fissure is indicated by the broken line. The illustration of Sorex and Elephantulus are modified from Brauer (9), and that of Tarsius from Woollard (114).

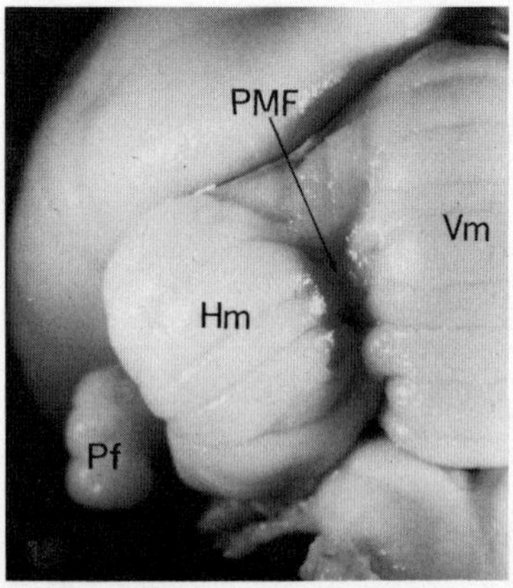

Figure 14. Caudal view of the cerebellum of *Perodicticus potto*. Note the position of the paramedian fissure, and compare with Figure 13. This fissure is comparatively the same in location and depth in the tree shrews and all the Lorisidae. 3X

Cerebellar Cortex: Afferent and Efferent Connections

The histologic structure of the cerebellar cortex of the Tupaiids and the Lorisidae is essentially similar to that of other higher primates. The cortex is clearly organized into an outer molecular layer and an inner granular layer separated by a middle layer of large Purkinje cells. Even though there are some finer differences (unpublished observations) the important concept is that between genera the primate cerebellar cortex is remarkedly consistent in its organization.

Based on a variety of studies, many of which have been done on nonprimate forms, the cerebellum can be divided into three functional areas. The archicerebellum (or vestibulocerebellum) is that portion of the cerebellar cortex that receives afferents from the vestibular complex. It includes the nodulus, uvula (IX), and dorsal and ventral portions of the paraflocculus (HVIII and HIX).

The paleocerebellum (or spinocerebellum) are those areas of the cerebellar cortex that receive their primary afferent input from the spinal cord. These impulses come from the cell bodies of origin in the spinal cord grey that contribute to the dorsal and ventral spinocerebellar tracts. The areas of the cerebellum which receive spinocerebellar fibers are the vermis of the anterior lobe (lobules I to V), portions of the anterior lobe hemisphere (lobules HII to HV), a small contribution to lobule VI, vermis of posterior lobe (VIII, IX), and a small amount to the paraflocculus (HVIII). The external cuneate nucleus projects mainly to hemispheres of anterior and posterior lobes (HV, HVI, HVII and HVIII). This nucleus also projects some fibers to vermian lobules VII and VIII. The inferior olivary nucleus also projects extensively to all areas of the cortex of the vermis and hemispheres (for review see ref. 10).

The neocerebellum (or pontocerebellum) are those areas of the cortex which fibers from the pontine nuclei. In general the rostral pontine nuclei project to the anterior lobe and rostral portions of the posterior lobe. The middle and caudal pontine nuclei project to the posterior lobe cortex. The pontine nuclei project to all areas of the cortex in a specifically localized pattern (10).

The cerebellar cortex, after receiving its afferent input, integrates the information and projects it onto the deep cerebellar nuclei via the axons of the Purkinje cells. These projections are collectively termed cerebellar corticonuclear connections. The Purkinje cells are the largest neurons in the cortex, multipolar in nature and have axons which arise from their basilar aspects, descend through the subcortical white matter to enter into a synaptic relationship with the cell bodies of the deep cerebellar nuclei. A carefully placed lesion in the cortex will result in destruction of the Purkinje cells and result in a characteristic pattern of degeneration of their descending axons. In this manner the functional relationship between specific areas of the cerebellar cortex and deep nuclei can be studied.

The early concept of the cerebellar corticonuclear connection stated that the medial (or vermal) functional zone projected only to the medial nucleus, the paravermal functional zone projected onto the interposed nuclei, and the lateral functional zone projected only onto the lateral nucleus (Fig. 15). The cortex of the anterior lobe projected onto the more anterior portions of the respective nuclei, and the cortex of the posterior lobe onto the more caudal portions (for review see refs. 27,28).

Using more modern silver methods for degenerating fibers it has been shown that the original zonal concept is no longer completely valid. Studies in Macaca (28) have shown that the lateral functional zone projects mainly to the lateral nucleus with lesser projections onto the interposed nuclei (Fig. 15). In Macaca the medial functional zone projects only onto the medial cerebellar nucleus and sends some projections into the vestibular nuclei (25). In Galago and Tupaia the cerebellar corticonuclear fibers of the so-called lateral functional zone project onto the lateral nucleus and send some fibers into the lateral areas of the interposed nuclei (Fig. 15). The

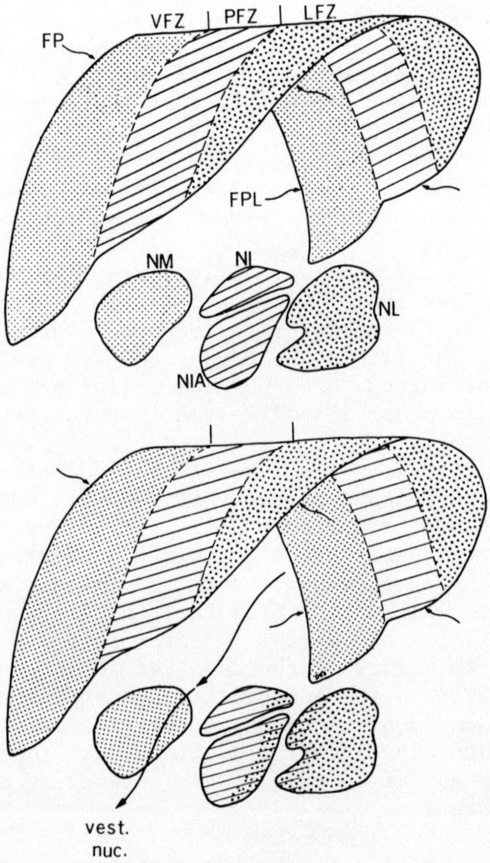

Figure 15. Diagrammatic representation of the zonal concept of cerebellar corticonuclear projections. See text for discussion.

medial areas of the interposed nuclei are free from degeneration regardless of the location of the lesion so long as it is in the lateral zone. The cortex of the paravermal functional zone in bush-baby and tree shrew projects mainly onto the interposed nuclei with lesser contributions to the medial most areas of the lateral nucleus. Although this overlap is not clear cut it nevertheless is consistently seen (Fig. 15). The medial functional zone of *Galago* and *Tupaia* projects onto the medial cerebellar nucleus similar to that seen in *Macaca* (28). However, there are additional fibers from the vermal cortex that project rostrally through the medial nucleus to the ipsilateral vestibular nuclei. In *Galago* and *Tupaia* these cerebello-vestibular fibers project mainly to the spinal, lateral, and superior vestibular nuclei from vermal lesions. Lesions of the paravermal and lateral zones produce extremely sparse degeneration in the superior and lateral vestibular nuclei (unpublished observations). This tendency for the so-called functional zones to project onto more than one deep cerebellar nucleus has also been reported in other nonprimate mammals (27,34).

Cerebellar Nuclei and Cerebellar Efferents

The cerebellar nuclei of *Tupaia*, *Galago*, *Nycticebus* and *Perodicticus* are divisible into four masses on either side of the midline. From medial to lateral they are the nucleus medialis (NM), nucleus interpositus posterior (NIP), nucleus interpositus anterior (NIA), and nucleus lateralis (NL) (39,41).

The NM of *Tupaia* is an irregular oval-shaped mass of cells located dorsal to the fourth ventricle. The cell population of the NM is an intermixed population of small and medium sized cell bodies. The interposed nuclei (NIP and NIA) are only partially separated from each other and in their caudal and ventral area are conjoined to the NL (Fig. 16). The NL in addition to having an area in common with the NIP and NIA is comparatively small and irregular (Fig. 16).

The NM of the Lorisidae is large in all genera and located dorsolateral to the fourth ventricle. The NM is smallest in the lesser bushbaby and largest in potto (Fig. 17). The area between the NM and interposed nuclei is occupied by large bundles of small, medium and large neurons. Both portions of the interposed nuclei are closely associated with the lateral nucleus throughout (Fig. 17). The NIA and NIP collectively form a relatively large group of cells (Fig. 17).

The lateral cerebellar nucleus of *Nycticebus* is similar to that in *Perodicticus*, and the NL of greater and lesser *Galago* are similar. In the former the NL is slightly medially concave, appears to show a primitive hilus, and is joined to portions of the NIA and NIP. In the latter the NL is more distinctly separate from the interposed nuclei, distinctly undulatory in appearance and the hilus region is well formed (Fig. 17).

The cell bodies of the deep cerebellar nuclei give rise to axons which exit the cerebellum via the superior cerebellar peduncle (SCP), cross the midline and terminate in several contralateral centers. Principal centers which receive crossed fibers via the SCP are the red nucleus of the mesencaphalon, the ventrolateral thalamic nucleus, and the inferior olivary nucleus. The red nucleus in turn gives rise to a descending motor tract. The ventrolateral nucleus of the thalamus, after receiving from the contralateral cerebellar

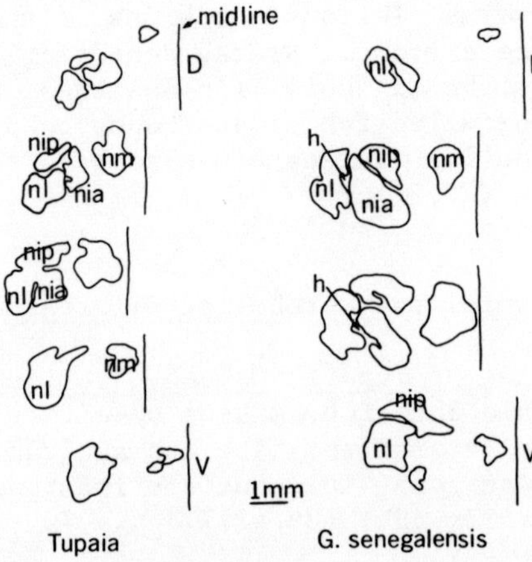

Figure 16. Tracings of horizontal sections of the cerebellar nuclei from Tupaia glis and Galago senegalensis. The midline is indicated and the illustrations are from dorsal (D) to ventral (V).

nuclei, projects to the motor cortex. Contributions to the red nucleus and VL nucleus of the thalamus are through the ascending crossed limb of the SCP. The descending crossed limb of the SCP contains fibers from the interposed nuclei which descend to the inferior olivary nucleus and other areas.

Cerebellum and Locomotion

It can be gleaned from the brief discussion above that the cerebellum is an important center in a series of pathways intimately related to locomotor function and postural orientation. The cerebellum receives afferent information from Golgi tendon organs and neuromuscular spindles through the nucleus of Clarke in the spinal cord. It also receives from the lateral cuneate nucleus, pontine nuclei, and other areas. The cerebellar cortex projects onto the deep nuclei which in turn project to the contralateral red nucleus, ventrolateral thalamic nucleus, inferior olivary nucleus and other regions.

The vermis and hemisphere of the anterior lobe receive the majority of spinal cord projections consequently considerable interest has been expressed in the physiological effects of anterior lobe lesions. Lesions of the anterior lobe in squirrel monkey caused difficulty in jumping. The animal would no longer participate

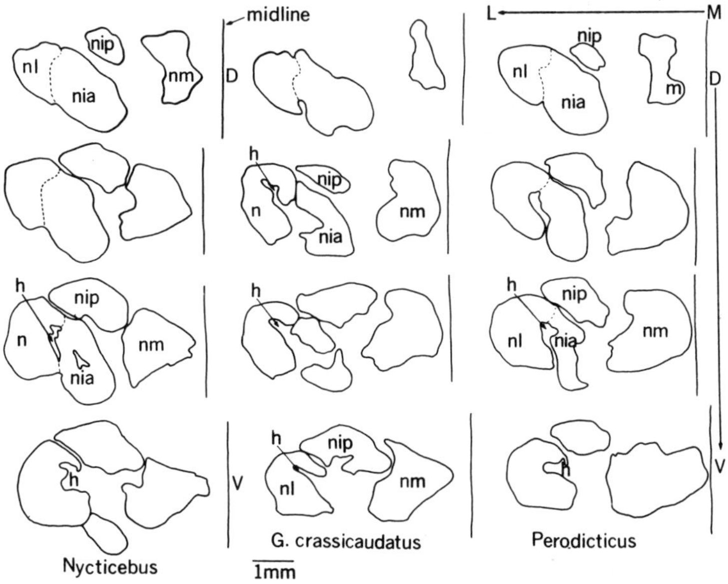

Figure 17. Tracings of horizontal sections of the cerebellar nuclei of Nycticebus coucang, Galago crassicaudatus, and Perodicticus potto. The midline is indicated and the illustrations are oriented from dorsal (D) to ventral (V).

in a "series of rapid jumps around the cage" which was its normal behavior (84). Others have suggested that there is little or no effect of anterior lobe lesions. In Macaca and rat lesions of the anterior lobe result is transient momentary stiffening (25,69). Crosby et al., (25) also report that in monkey with subtotal destruction of the anterior lobe there were no "overt cerebellar signs."

Due to the fact that Galago progresses in a saltatory manner it is relatively easy to detect locomotor dysfunction. In two animals (Galago senegalensis) extensive lesions were placed in the vermis of the anterior lobe (Fig. 18). In one animal the lesion was superficial and affected only portions of lobules II - V. This animal showed transient tremor at rest and only a slight amount of asynergy.

In a second animal the entire vermis was destroyed, and this animal showed marked locomotor difficulties. A persistent ataxia and resting tremor were present. This animal had great difficulty in executing its characteristic vertical locomotor pattern. When the animal attempted to leap it showed noticeable hyperextension of the hindlimbs and dysmetria. If this animal was placed on the floor

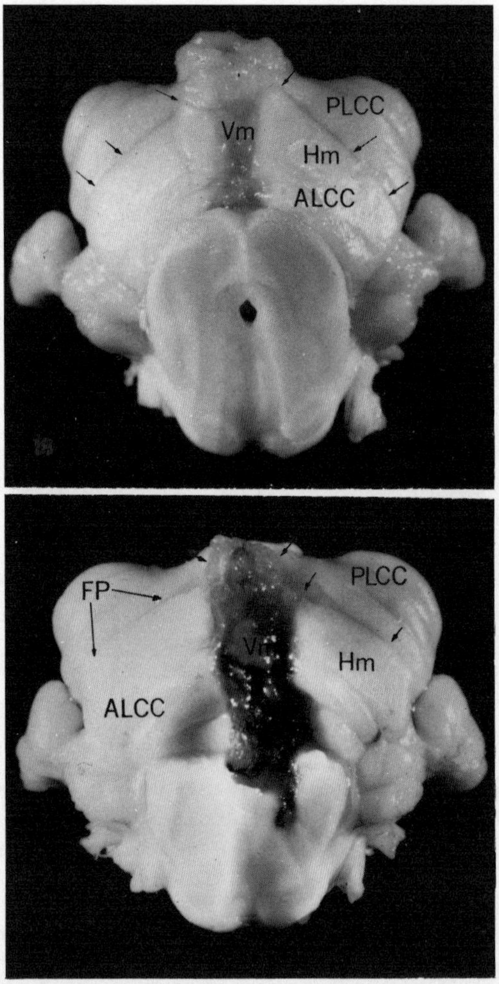

Figure 18. Anterior view of the cerebellum of G. senegalensis showing subtotal (upper) and total (lower) destruction of the vermis. The arrows indicate the course of the primary fissure. 3X.

2 meters from a nesting box with a small opening (approximately 6 x 8 cm) he would jump toward the box. Instead of easily negotiating the opening in a single motion (which a normal animal can do with ease) this animal would jump toward the hole, crash into the front of the box, fall back a few cm, and then crawl in. Even after repeated

attempts the pattern was the same. This animal showed its characteristic locomotor dysfunction until it was sacrificed at 15 days.

Destruction of the deep cerebellar nuclei and the superior cerebellar peduncle produce notable locomotor dysfunction (16, 17, 116). Lesions of the lateral cerebellar nucleus result in ataxia, tremor, moderate asynergia and dysmetria. Minor dysmetria, some intention tremor, past pointing, flexion of the head to the ipsilateral side are seen in lesions of the interposed nuclei. When the posterior portions of the interposed nucleus are destroyed the rhesus monkey shows "poor usage of the ipsilateral upper limb" (116). Lesions of the fastigial nucleus produce ataxia, leaning and falling toward the side of the lesion, and unsteady gait.

Stereotaxic lesions of the superior cerebellar peduncle in rhesus produce hypokinesia and ataxia gait. Little or no tremor is seen. Surgical destruction of the peduncle produces more severe and lasting deficits. However, this is probably due to involvement of other immediately adjacent structures.

The sum total function of the cerebellum can be summarized in a word, synergy. The cerebellum governs the smooth coordinated execution of voluntary muscle movements by receiving afferents from the periphery, integrating this with feedback information from the brainstem, and ultimately influencing the ventral horn cells through cortical and brainstem centers primarily responsible for motor function.

V. MOTOR CENTERS AND PATHWAYS

Red Nucleus and Rubrospinal Tract

The limited caudal extent of the corticospinal tract and the lack of ventral motor horn contacts from this tract in *Tupaia*, and the sparse termination of this tract on ventral motor horn cells in the prosimians which have been studied (slow loris, greater bushbaby, lesser bushbaby) pose questions relating to motor function. A prime question is: What is the origin of supraspinal (suprasegmental) control over the activity of the ventral motor horn neurons in the spinal cord? One pathway which has been credited with such control in a variety of animals is the rubrospinal tract. As mentioned above in Basic Structure of Nerve Tissue, the rubrospinal tract arises from neurons in the red nucleus.

The red nucleus is a large ovoid column of cells located in the ventral part or tegmentum of the mesencephalon. It extends rostrally from approximately the caudal margin of the superior colliculus to the caudal level of the diencephalon (Fig. 4) (54) and appears oval to round in shape in transverse section. In the rhesus monkey it is traversed throughout its rostro-caudal length by fibers of the superior cerebellar peduncle and in its middle portion be emerging root fibers of the third cranial nerve (15). In several prosimians (slow loris, greater and lesser bushbaby, and potto) the third nerve exits the brainstem from the interpeduncular fossa caudal to the mammillary body (Fig. 19).

Three types of neurons have been identified in the red nucleus of mammals. The large cells are characterized as being multipolar neurons (50-90 μ) with abundant Nissl bodies in their cytoplasm. The medium cells are described as stellate or fusiform (30-40 μ), and the small-sized cells are about 20 μ (for review see Massion, 72).

Hatschek (45) studied the red nucleus of various mammals and subdivided it into a large-celled or magnocellular part and a small-celled or parvocellular part. He found that the large cells predominated in lower mammals, while the small cells predominated in man. Subdivision of the red nucleus into a caudal magnocellular part and a rostral parvocellular part is generally accepted (54,72,77,85). The magnocellular part of the red nucleus is thought to be of relatively large and equal importance from the marsupials to the monkeys, but regresses in the anthropoids and especially in man, with a concomitant increase in size and importance of the parvocellular part (45,46,77,108,111).

The red nucleus is an important relay center in relation to its afferent and efferent tracts. Afferents to the nucleus come mainly from the contralateral deep cerebellar nuclei via the superior cerebellar peduncle and from the ipsilateral cerebral cortex. There are no direct sensory afferents from the spinal cord to the red nucleus in the macaque and man (7,73).

Of the cerebellar nuclei which project to the red nucleus, the lateral nucleus is thought to send fibers to the parvocellular part, the interposed nucleus to the magnocellular part, while the fastigial or medial nucleus does not send any fibers to the red nucleus. Fibers from the interposed and lateral nuclei of the macaque were found by Rand (91) to be distributed to both parts of the contralateral red nucleus. In a recent study on the NIA of the cat, a somatotopical organization of the efferent projections to the red nucleus was found (23). Two patterns of organization were noted: a caudorostral organization in the red nucleus which corresponded to a mediolateral arrangement in NIA and a mediolateral pattern in the red nucleus which corresponded to a caudorostral arrangement within the NIA. The projection from NIA was limited to the posterior two-thirds of the red nucleus and terminations were found on cells of all sizes within the nucleus. In the earlier study of the red nucleus of the cat, Pompeiano and Brodal (87) suggested that the somatotopical organization corresponded to a functional organization. A forelimb area was located in a dorsomedial portion of the caudal red nucleus and a hindlimb area was situated ventrolaterally. Further rostral, the forelimb area was situated dorsally while the hindlimb area was ventrally located.

Efferents of the red nucleus project as either descending or ascending pathways. Of the main descending tracts the rubrospinal is crossed while the rubro-olivary is uncrossed. The rubrospinal tract is classed as belonging to an indirect subcortical motor system (55) which is phylogenetically old. The major ascending tract, the rubrothalamic, projects ipsilaterally to the ventrolateral nucleus of the thalamus. It is generally accepted that the rubrospinal tract arises in the magnocellular part of the red nucleus whereas the rubro-olivary and rubrothalamic tracts originate in the parvocellular part.

Noback and Moskowitz (77) state that the "magnocellular nucleus contains more neurons in the lower than in the higher primates" and that "this is reflected in the reduction of the rubrospinal tract to a minor tract in the higher primates." The very small rubrospinal tract in man corresponds to the paucity of large cells in the red nucleus (108). In the chimpanzee and gorilla the rubrospinal tract contains more fibers than in man, but a significant increase in volume is seen only in the macaque. The rubrospinal tract of the cat resembles that of Macaca being only slightly larger; however, it is situated in the same position in the spinal cord, just ventrolateral to the lateral pyramidal tract (95). The position of the tract in the opossum spinal cord is similar to that in the cat (71). Verhaart (109) describes the fiber content and course of the rubrospinal tract in a series of primates and tree shrew. In Saimiri sciureus it is located ventral to the lateral pyramidal tract at high cervical levels of the spinal cord (his Fig. 2a). In the Prosimii (Lemur, Nycticebus, Galago) the author typifies the tract as smaller than the pyramidal tract at cervical levels (his Fig. 5a). The rubrospinal tract in Tupaia is also located in the lateral funiculus in association with the dorsal spinocerebellar tract (his Fig. 8a). Fibers of the tract in Saimiri, Lemur, Nycticebus and Galago are much larger and much more numerous than those of the pyramidal tracts in tree shrew (109). Rubrospinal fibers extend throughout the length of the spinal cord in the cat, opossum and monkey (55,71,87).

Lesion of the red nucleus or rubrospinal tract in the macaque results in transient locomotor deficits characterized by decreased muscle tone (hypotonicity), tremor, ataxia and asynergy (15,53,74,80). Stimulation of different areas of the nucleus induces flexion in either the contralateral fore- or hindlimb (86). Furthermore, stimulation of the nucleus induces inhibition of the ventral horn motor neurons of the contralateral extensor muscles. Thus the magnocellular part of the red nucleus can be thought of as an excitatory center for flexor muscles and an inhibitory center for extensor muscles. The

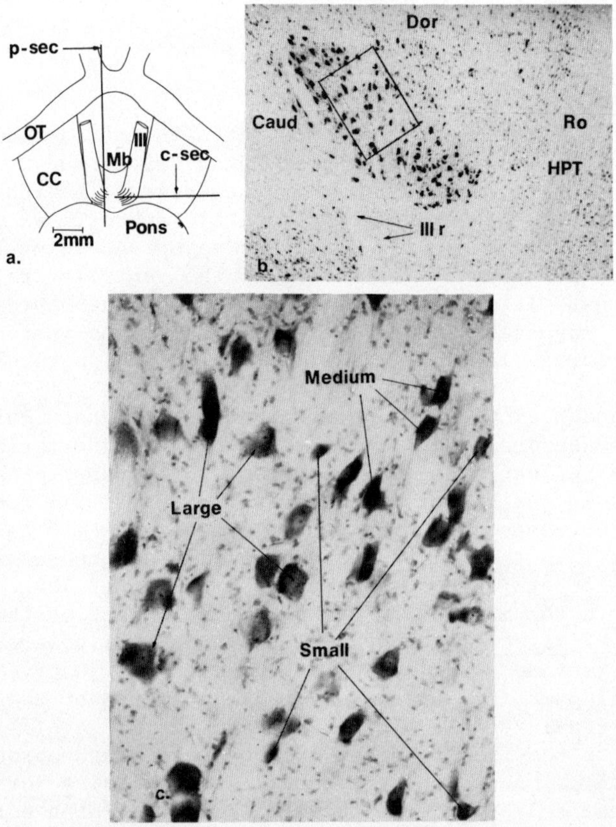

Figure 19. a. Diagram of the ventral view of a prosimian brain. p-sec: Approximate level of parasagittal section of the brain through the red nucleus as seen in Figs. 19b and 20a. c-sec: Approximate level of coronal section of the brain through the red nucleus as seen in Fig. 20b, c & d.

b. Photomicrograph of parasagittal section of _G. crassicaudatus_ showing the relationships of the caudal large celled area of the red nucleus with the rootlets of the oculomotor nerve (IIIr). Dor: dorsal, Caud: caudal, Ro: rostral. Abbreviations are the same in Fig. 20. Nissl stain 27X.

c. Higher power photomicrograph of area contained within the rectangle of Fig. 19b. Nissl stained section demonstrates the three types of neurons in the red nucleus. This is the characteristic appearance of neurons of the red nucleus in the tree shrew and the Lorisidae. 160X.

most important activity of the magnocellular part of the nucleus is the control of muscle tone of contralateral flexor muscles (72). It is generally accepted that it is the flexor muscles of proximal joints through which voluntary locomotor activity is initiated. Total locomotor behavior is expressed as synergistic flexor-extensor activity. In the quadrupedal position, the flexor muscles also act in a synergistic manner with the extensor muscles to support the body weight, the muscle synergism effecting rigidity of the limbs.

To date, there has been little or no literature concerning the prosimian red nucleus, its afferents or efferents. Preliminary data collected on the cytoarchitecture of the red nucleus of *Tupaia*, *Galago*, *Nycticebus* and *Perodicticus* indicate that the cell population consists of large (45-70 μ), medium (20-40 μ) and small (10-20 μ) neurons (Figs. 19,20). The large cells predominate in the caudal portion of the nucleus and decrease in numbers rostrally (Figs. 21,22). Medium and small cells are found throughout the rostral-caudal length of the nucleus, the small cells predominating in the most rostral portion. The nucleus cannot be easily divided into magno- and parvocellular parts nor can the rostral pole of the nucleus be clearly delimited. For this reason, the level of the habenulopeduncular tract has been tentatively determined to mark the rostral extent of the nucleus (Fig. 19).

Preliminary observations have been made on the length of the red nucleus in *Tupaia*, *Galago*, *Nycticebus* and *Perodicticus*. Its rostrocaudal extent is least in *Tupaia* (1400 μ), while in *G. senegalensis* and *N. coucang* it is somewhat larger (1600 μ). The nucleus in *G. crassicaudatus* and *P. potto* is almost twice as long (2400 μ) as that of *T. glis*.

The population of large cells in the caudal portion of the red nucleus of the lesser bushbaby is greater than in the tree shrew (Figs. 21,22). The total (large and medium) population of cells of these animals does not differ greatly, however. Although there are differences in the total population of cells in the red nucleus of the lesser and greater bushbaby, the distribution of large cells is similar (Fig. 21). There is no great difference in the distribution and number of large cells within the red nucleus of *G. crassicaudatus* and *P. potto*, while there is some difference in the total population of cells (Figs. 21,22). The distribution of large cells within the red nucleus of *N. coucang*, by comparison, shows a rather sharp increase in numbers caudally. The distribution and numbers of large cells within the caudal portion of the red nucleus show differences between *Tupaia* (a transitional quadruped) and *Galago* (VC&L), and also between *Galago* and *Nycticebus*, a quadruped that may represent one specialization in locomotion from the VC&L category.

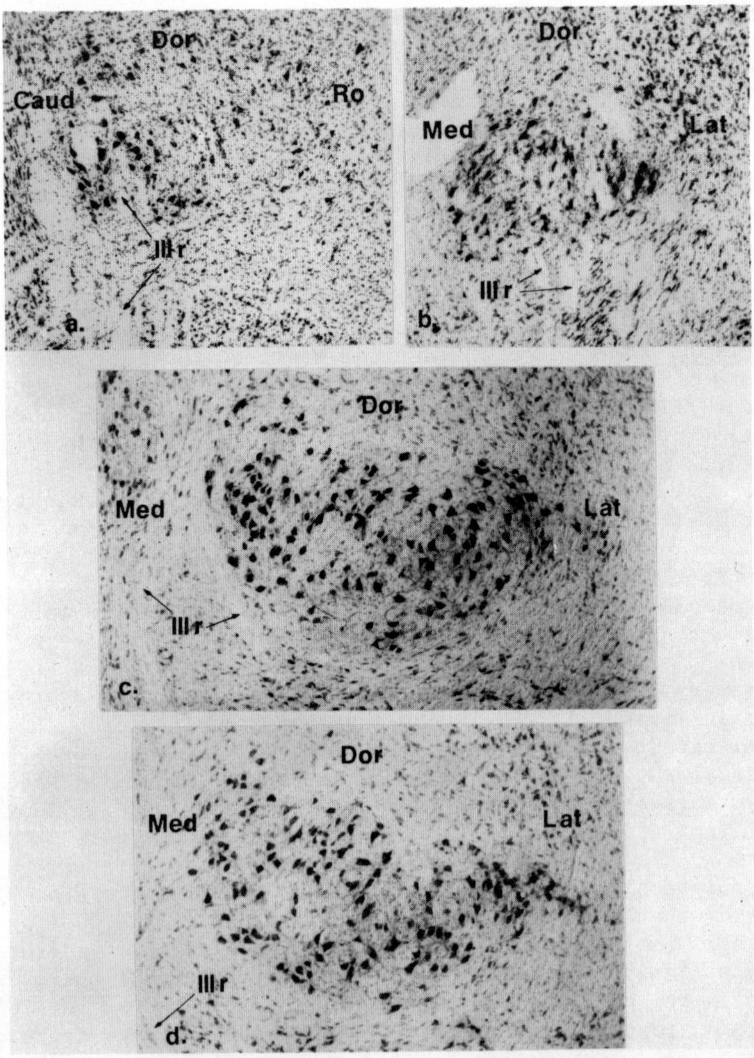

Figure 20. a. Photomicrograph of parasagittal section through the red nucleus of Tupaia. Note that the nucleus is not as well delineated as that of Galago. Nissl stain. 35X.

b. Coronal section through the caudal red nucleus of G. senegalensis. Nissl stain. 35X.

c. & d. Nissl stained coronal section through the caudal red nucleus of Nycticebus. Note the shape of the nucleus (d. slightly rostral to c.) and its relationship to the rootlets of the oculomotor nerve. 35X.

Cerebral Cortex

The differentiation of the forebrain from earlier stages of histogenesis (mantle and marginal layers of the neural tube) begins at the end of the second month of gestation. The cells of the mantle layer migrate peripherally and become organized into horizontal layers. The nerve cell bodies (mantle) are now superficial to the underlying white matter (marginal layer).

Although there are generally six layers making up the mature cortical grey matter in primates, this structure is not uniform throughout. In the somatomotor area (Fig. 23), there is a predominance of large pyramidal cells and giant cells of Betz. In the somatosensory area there is a characteristic layer of small granule-like cells in addition to the pyramidal cells. The pyramidal and giant cells of the motorsensory cortex project axons into the underlying white matter thus giving rise to some of the long descending tracts of the brainstem and spinal cord. The motorsensory cortex is therefore characterized by its cell type and by its projections.

In most nonprimate forms, the motorsensory cortex is an amalgam, with only slight separation, if any, of primary motor and sensory cortex (65). As one ascends the primate series, there is cortical separation of these functions which can be seen from the cytoarchitecture and projections of these areas.

In a prosimian primate, *Galago senegalensis*, there is a motorsensory cortical overlap. Rostrally there is a separate motor area and caudally, a separate sensory area (51) (Figs. 7,23). Other larger members of the family Lorisidae have a sulcus separating primary motor and sensory areas. *Perodicticus potto* has a definite "central sulcus" while the vertical component of sulcus "e" is the homologue in the slow loris (42,94). This cortical separation of function can also be seen in the termination of the long descending fiber systems of the spinal cord, specifically the corticospinal system.

Corticospinal System
General Concepts

An important concept in the evolution of the central nervous system (CNS) is encephalization. Encephalization is a phylogenetic expression related to the hierarchical organization of the nervous system. It is a process by which the cerebral cortex takes over the functions of lower centers. A more recent interpretation states that higher (newer) centers do not "replace" lower (older) centers, but act in parallel with them (78). It is with this understanding

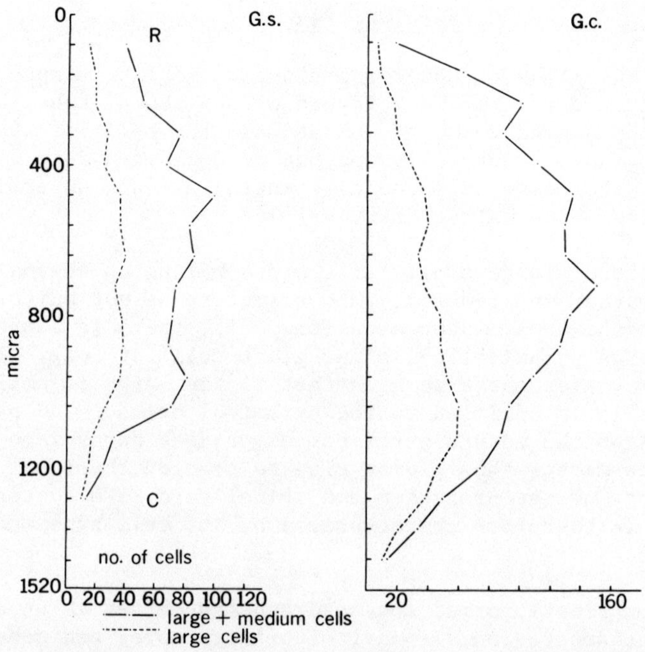

Figure 21. Rostral to caudal (R-C) distribution of large and medium sized cells within the red nucleus of Galago (see text for discussion).

that we can study the components of long descending tracts which affect lower motor neurons in the ventral horn of the spinal cord.

Descending central nervous system pathways affecting lower motor neurons may be classified into corticospinal and subcorticospinal projections (55). The direct system (corticospinal) is composed of axons originating from both motor and sensory cortical areas. Neurons of the somatosensory areas project predominantly

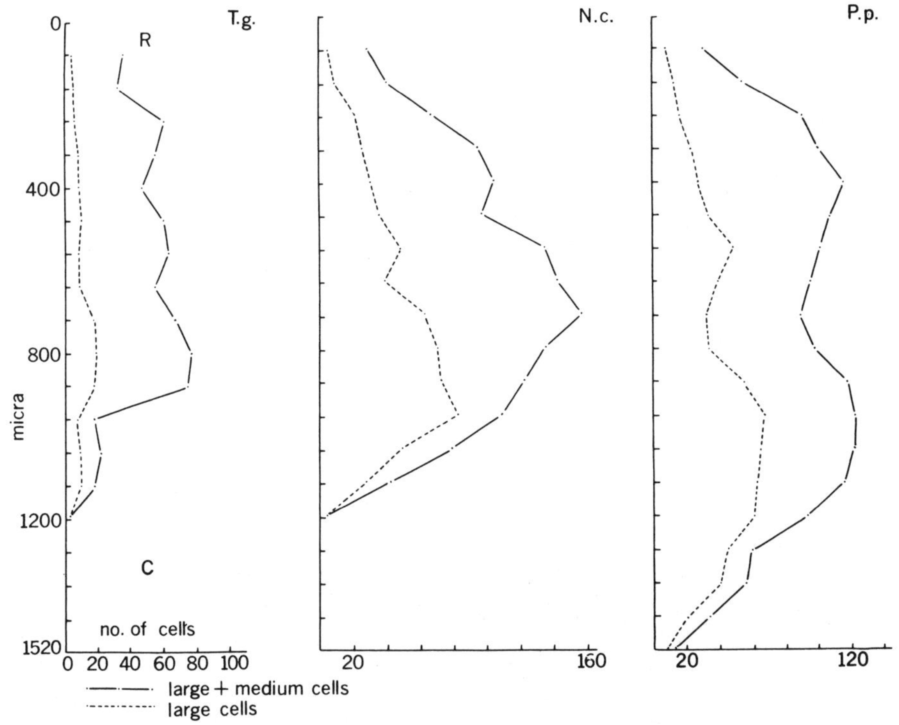

Figure 22. Rostral to caudal (R-C) distribution of large and medium sized cells in the red nucleus of Tupaia, Nycticebus, and Perodicticus. (see text for discussion).

to sensory neurons (i.e., spinal trigeminal nucleus, dorsal column and solitary nuclei, and to cells in the dorsal horn, lamina III-VI) in both non-primate and primate forms. Since these projections are present in non-primate forms, animals without a well developed corticospinal system, it would appear that these direct cortical projections to sensory relay nuclei are phylogenetically older than projections to motor nuclei. Neurons of the somatomotor areas

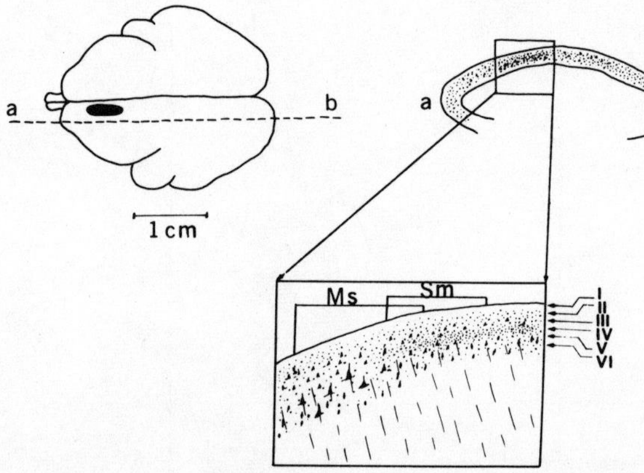

Figure 23. Sagittal section through the motorsensory cortex is represented in Galago senegalensis. The layers of the cortex are numbered: I-molecular layer, II-external granular, III-external pyramidal, IV-internal granular, V-internal pyramidal, VI-fusiform layer; note differences in cytoarchitecture in Motorsensory (Ms) and Sensorymotor (Sm) cortices. Long corticofugal projections originate from large pyramidal cells in deeper layers of these areas.

project predominantly to internuncial neurons which indirectly affect the motor neurons of the ventral horn. There are also direct projections from the motor cortex to dendrites and cell bodies of alpha motor neurons. (lamina VII-IX). It is this latter pathway which may provide a means for further refinement and improvement in direction, speed and appropriateness of movements in pectoral and pelvic limbs (60).

Corticospinal System of Lower Primates

The corticospinal system in the Malayan tree shrew has been reported by a number of investigators (47,49,98,109). Cortical localization in Tupaia follows the general schema found in other small placental mammals. There is a motorsensory amalgam with slight separation of motor function toward the rostral pole. Cortical topography (indentations and eminences) appears related to patterns of localization (66). Projections from the motorsensory cortex into the spinal cord are mainly contralateral in the dorsal funiculus (Fig. 24). A few fiber bundles are reported in the contralateral and ventral funiculi and in the ipsilateral dorsal funiculus as well. The major tract descends through thoracic segments, the aberrant bundles terminating in upper cervical segments. Preterminal and terminal degeneration is mainly in the medial portions of the dorsal horn (medial half of lamina V & VI). Relatively few fibers project dorsally to lamina III or ventrally to lamina VII. No degenerating axons have been reported crossing the dorsal or ventral grey commissures.

In a prosimian primate, Galago senegalensis, projections from the cortex can be traced through the brainstem, ipsilateral to the lesion, to the caudal medulla. Here most of the tract crosses to the contralateral lateral funiculus, a smaller portion descends ipsilaterally in the ventral funiculus. Fibers from the lateral tract can be traced coursing the spinal grey throughout the cord, but degeneration is most conspicuous in the cervical and lumbar enlargements. Degeneration is heavy at the base of the dorsal horn while some fibers course ventrally into lamina VII. Other fibers cross in the anterior grey commissure to terminate in the ipsilateral ventral horn. From the ventral tract fibers can be seen passing into the ipsilateral ventral horn, mainly into lamina VIII (Fig. 25).

A similar projection pattern is found in the slow loris, Nycticebus coucang (13). Although no ventral tract was reported from lesions in seventeen specimens, fibers of passage were reported in the dorsal and ventral grey commissures, axonal debris being most abundant in the medial part of the ventral horn. Some fibers in the ventral horn appeared to synapse directly with motor neurons (Fig. 24).

One may characterize the primate corticospinal system as follows. The motor sensory cortex projects its fibers into four spinal pathways forming crossed and uncrossed lateral and ventral tracts. The bulk of the system descends in the contralateral lateral funiculus, and is most consistent, coursing the length of the cord. The ventral tract is more variable terminating in high cervical levels in Tupaia (48), low thoracic in Galago (33), lower lumbar in Macaca (67) and Pan (31). Fiber termination is chiefly on cells of the contralateral dorsal horn and intermediate zone.

The projections of the motorsensory cortex to the spinal cord can then be organized into a dorsal sensory modulating component (into lamina IV-VI) and a ventral motor component (into lamina VII-IX). The motor component can be further organized into medial and lateral zones. Recent evidence in Macaca for this medial-lateral organization has been reported by Kuypers and Brinkman (56).

In addition to this generalized primate pattern, which can also be seen in some non-primate mammals, the primate trend is as follows. In prosimian primates there is a definite projection to

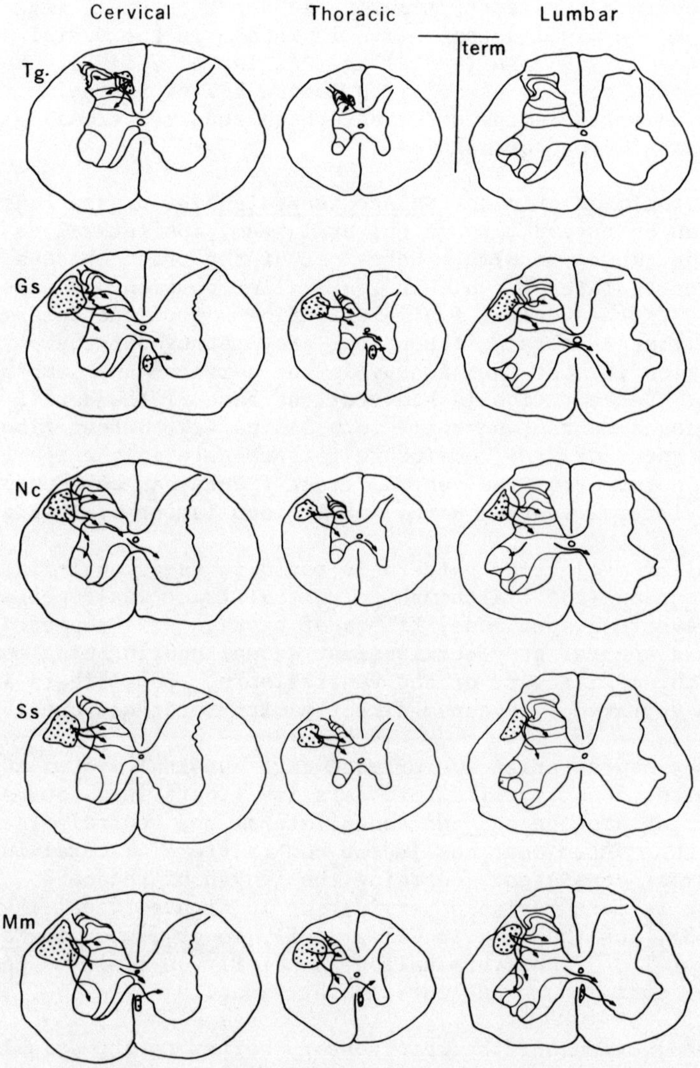

Figure 24. Diagram of representative cross sections of the cervical thoracic, and lumbar cord. Note the location of the corticospinal tracts and the laminae of termination. T.g.- Tupaia glis: G.s.- Galago senegalensis: N.c.-Nycticebus coucang: S.s.- Saimiri sciureus: M.m.- Macaca mulatta. The tract terminates (term) in lower thoracic segments in Tupaia.

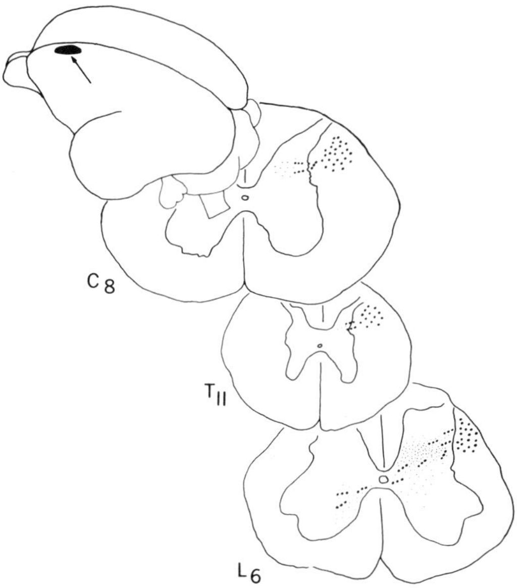

Figure 25. A lesion (arrow) was placed in the somatomotor cortex associated with the hindlimb. Little degeneration can be traced into cervical or thoracic segments. In the lumbar enlargement dense degeneration is seen at the base of the dorsal horn. Moderate degeneration can be traced into the intermediate zone (lamina VII) and crossing the anterior grey commissure into the ipsilateral ventral horn (lamina VIII).

the medial half of the ventral horn via the ventral tract or from commissural fibers from the lateral tract. This addition may be interpreted as the first cortical projection paralleling the phylogenetically older ventromedial subcorticospinal system (vestibulospinal and medial reticulospinal systems). Anthropoidea add to this trend toward cortical projections to the dorsolateral aspect of the ventral horn, paralleling the phylogenetically older dorsolateral subcorticospinal system - the rubrospinals.

Cortical control over the medial aspect of the ventral horn in some prosimians may be related to their adaptations of upright posture and locomotion (vertical clingers and leapers). The cortical control over the dorsolateral nuclei of the ventral horn in higher primates has been closely coupled with the history of prehension of the extremities and the development of tactile and proprioceptive sensations (83).

VI. DISCUSSION AND SUMMARY

Any attempt to ellucidate the early neuroanatomical specializations that may be related to evolving locomotor patterns is fraught with difficulties. Principal among these is the paucity of available information on the prosimian nervous system caudal to the level of the diencephalon. Many of the centers relevant to the question at hand are in the caudal brainstem and spinal cord, consequently much baseline information is not available.

It is immediately obvious that many centers related to postural and spatial orientation and locomotion have not been discussed. Some of the more significant areas are the vestibular nuclei and vestibulospinal tracts, the reticulospinal system, and organization of the basal ganglia and their afferent and efferent connections. All of these centers, to varying degrees, are components of the subcortical or indirect motor system. Their influence on the lower motor neuron of the spinal cord is indirect and secondary in nature. In the present study only some primary afferent and efferent centers related to locomotion are considered.

Although there have been a wide variety of studies reported on several different genera of tree shrews (2,26,29,113) the nervous system has been the object of limited effort. Therefore, for the reasons outlined above (Section II), it is necessary to rely heavily on the available information on and availability of Tupaia. In an overall sense the tree shrews represent phylogenetically transitional forms which show a quadrupedal mode of locomotion. Occasionally the tree shrews are referred to as "true quadrupeds" although the meaning of this is unclear. It is highly probable that the quadrupedalism of the tree shrews is similar to that of the rodent-like primates of the Paleocene.

It has been suggested that the vertical clinging and leaping category of locomotion was the first significant specialization above quadrupedalism and that VC&L was preadaptive to most or all subsequent types of primate locomotion (75,76). The inference of this concept is that the specialized quadrupedal-climber pattern of Nycticebus and Perodicticus may represent one example of a further specialization of VC&L baseline. The above approach was to consider some sensory and motor centers of the pre-primate mode (Tupaia), the first level of specialization (Galago), and one possible subsequent level (Nycticebus).

Clarke's nucleus (21,22) of the spinal cord receives its afferent input from neuromuscular spindles and Golgi tendon organs, and projects onto the ipsilateral cerebellar cortex (Section III). In Galago the distribution of cells from T_{12} to L_2 indicates that a significant amount of proprioceptive information comes from the lower limb. The increased number of cells between levels T_6 - T_{11} appears to imply that the lower trunk of Galago also has significant proprioceptive input. This suggests that the lower trunk of Galago is stabilized for his characteristic locomotor progression. The pattern of projections into Clarke's column following dorsal rhizotomy at selected hindlimb cord levels also supports this concept. Removal of dorsal root ganglia from the lumbosacral cord results in degeneration throughout the T_6 - T_{11} and T_{12} - L_2 levels of Clarke's nucleus in Galago (unpublished observations). This shows a more extensive overlap of hindlimb proprioception with the cells of the nucleus dorsalis than is seen in other forms. Consequently it would appear that in Galago there is complex integration of proprioception from hindlimb with that of the lower trunk.

The cells of Clarke's column project mainly to the vermal and paravermal functional zones of the anterior lobe of the cerebellum. The increased cell population in Clarke's column of the bushbaby is correlated with the well-developed cerebellar vermis and hemisphere of the anterior lobe. In Tupaia Clarke's column is only moderately developed, showing a sharp peak at L_2 and no elevated cell population in lower thoracic levels. This indicates a lesser proprioceptive input from the hindlimb and little, if any, neuroanatomical basis for stabilization of the lower trunk in the tree shrew. This lesser cell population in Clarke's column also correlates with the poorly developed anterior lobe of the cerebellum. The anterior lobe of the cerebellum of non-primate mammalian quadrupeds (specifically insectivores and tree shrews) is poorly developed (8,9,37,57,61,62).

The deep cerebellar nuclei receive information from the cerebellar cortex and project to contralateral brainstem centers (Sections IV,V). The cerebellar cortex of the tree shrews project onto primitive and poorly differentiated deep nuclei (39,41). In the Lorisidae the cerebellar hemispheres of both anterior and posterior lobes are well developed and project onto enlarged and clearly differentiated deep nuclei (39,41). The cerebellar nuclei are largest in Nycticebus and Perodicticus and smallest in Tupaia. In all Lorisidae the lateral nucleus shows a hilus but in the tree shrew the lateral nucleus shows no hint of a hilus area.

The dorsal column nuclei relay impulses of light tactile sense and conscious proprioception (Section III). There is proportionate increase in the size and differentiation of the dorsal column nuclei from Tupaia to Nycticebus. Without further evidence at the present time it can only be suggested that as locomotor patterns changed there was increased need for fine tactile sense to be relayed to an expanding sensory cortex.

There is no sharp division between magnocellular and parvocellular portions of the red nucleus in Tupaia or any of the Prosimii studied to date. There are however what appear to be some trends. The significance of parvo- and magnocellular divisions have been discussed above (Section V). Of prime importance is the fact that many consider the cells of the magnocellular area to give rise to a crossed descending rubrospinal tract. Others hold the opinion that cells throughout the nucleus give rise to the rubrospinal tract (87). In Galago and Tupaia the distribution of large cells is similar. In the Lorisidae there is a distinct trend for the larger cells to be located in the more caudal portions of the nucleus. In Macaca it has been reported that the large cells are clearly localized in the caudal portion of the nucleus (54). The trend appears to be from tree shrew to Galaginae to Lorisinae. The implications are that the rubrospinal tract, if it does arise from the large cells, is decreasing in importance from Galaginae to Lorisinae. As the magnocellular region of the red nucleus decreases in significance between these subfamilies there is a concurrent increase in the size and differentiation of the motor cortex (42). The corticospinal projections to cord levels as related to function of the forelimb and tail have been emphasized in the literature (78,83). However, the significance of corticospinal fibers projection to lower cord levels as related to locomotor specializations of the lower limb have not been clarified. Results to date on Prosimii (13,47,48,109) indicate that there are trends within the motor component of the corticospinal system related to hindlimb specializations.

Neuroanatomical specializations related to early evolving locomotor patterns are obviously a complex of characteristics. In the basal quadruped (Tupaia) there is limited proprioception from the hindlimb, no basis for a stabilized lower trunk, poorly developed cerebellar cortex (anterior lobe) and cerebellar nuclei, and an even distribution of large, medium and small cells throughout the red nucleus. The first level of locomotor specialization (Galago) shows a distinct increase in hindlimb proprioception, distinct increase in the anterior lobe of the cerebellum, and cerebellar nuclei, but only moderate differentiation of the red nucleus. The dorsal column nuclei are also larger and more differentiated than in Tupaia. It appears that the earliest phylogenetic developments may have been related to a sophistication of the afferent limb of this functioning system followed by a change in the efferent centers. The pattern of fiber projections into spinal cord grey following dorsal rhizotomy, when overlaid with the pattern of corticospinal termination, also prompt this conclusion. Following removal of ipsilateral dorsal root ganglia in both Galago and Tupaia collateral fibers enter the medial base of the ventral horn and terminate in nuclei presumably responsible for innervation of axial and proximal limb musculature (unpublished observations). These same nuclei receive bilateral projections from the motor cortex in Galago (33) while no corticospinal fibers reach medial ventral horn cells in Tupaia (48). In the Lorisinae the deep cerebellar nuclei are larger and there is a further tendency for a segregation of the red nucleus into large and small celled portions. It is also worthy of note that the dorsal column nuclei of Nycticebus are considerably larger than in Galago. This, however, may be partially related to this animal's particular locomotor pattern, its arboreal habit, and its evolution into a specific eco-niche.

The observations of the present report are of a preliminary nature. To adequately answer the questions posed in Section II will entail considerably more work. It has been shown, however, that there are distinct and obvious changes in the nervous system related to evolving locomotor patterns. These presumed specializations can be correlated with behavior, locomotion, and gross anatomical adaptations (43,44,75,76,93,101,102). It is concluded that there is indeed a pool of neuroanatomical characteristics which provided a basis for the divergent locomotor patterns seen in the higher primates. It is acknowledged however, that this pool needs to be filled by a much needed well-spring of research on this problem, the point being that evolving locomotor processes were dynamic and three dimensional and involved more than osteological and myological modifications.

ABBREVIATIONS FOR ALL FIGURES

ALCC - anterior lobe of the corpus cerebelli
as - aqueduct of Sylvius (mesencephalic aqueduct)
Cb - cerebellum
CC - Clarke's column (nucleus dorsalis)
CCr - crus cerebri
CE - cervical enlargement
CNC - central canal
DF - dorsal funiculus
DH - dorsal horn
DL - pars dorsolateralis of nucleus gracilis
DMX - dorsal motor nucleus of the vagus nerve
Dn - Diencephalon
DH - dorsal horn
F - forelimb cortical area
FC - fasciculus cuneatus
FG - fasciculus gracilis
Fl - flocculus
FP - primary fissure of the cerebellum
FPL - posterolateral fissure of the cerebellum
H - hindlimb area of the cortex
h - hilus area of the lateral cerebellar nucleus
Hm - hemisphere of the cerebellum
HPT - habenulopeduncular tract
ICP - inferior cerebellar peduncle

IVN - inferior vestibular nucleus
LCN - lateral cuneate nucleus
LE - lumbar enlargement
LF - lateral funiculus
LFZ - lateral functional zone of cerebellar cortex
Lv - lateral ventricles
Mb - mammillary body
Mes - mesencephalon
Met - metencephalon
Ms - motorsensory cortex
My - myelencephalon
NC - nucleus cuneatus
NCr - neural crest
NF - neural fold
NG - nucleus gracilis
NGr - neural groove
NIA - anterior interposed nucleus of cerebellum
NIP - posterior interposed nucleus of cerebellum
NL - lateral cerebellar nucleus
NM - medial cerebellar nucleus
NP - neural plate
NT - neural tube
NX11 - nucleus of the hypoglossal nerve
OB - olfactory bulb
OT - optic tract
pf - paraflocculus
PFZ - paravermal functional zone
PLCC - posterior lobe of the corpus cerebelli
Pon - pons
Pros - prosencephalon
pt - pituitary
Rhom - rhombencephalon
rn - red nucleus
Sm - sensorymotor cortex
SN&TV - sensory nucleus and tract of the trigeminal nerve
Tel - telencephalon
TS - solitary tract
VF - ventral funiculus
VFZ - vermal functional zone of the cerebellar cortex
VH - ventral horn
VM - pars ventromedialis of the nucleus gracilis
VR - ventral root
3v - third ventricle
4v - fourth ventricle

Roman numerals indicate the lobules of the cerebellum according to the method of Larsell (57).

ACKNOWLEDGMENTS

The majority of this manuscript was completed while one of the authors (DEH) was on the faculty of the Department of Anatomy, Medical College of Virginia. At that Institution the authors are deeply indebted to Dr. William P. Jollie, Professor and Chairman, for his enthusiastic support throughout this and other closely related projects: Mrs. Betsy Harmeling, Ms. Julie Brooks and Ms. Jo Bishop for their technical assistance; Ms. Susan Hall for secretarial help. Some of the animals used in this study were purchased on an A. D. Williams Grant from the Medical College of Virginia. Revisions were completed in the Department of Anatomy, West Virginia University School of Medicine. The authors express their sincere appreciation to Ms. Linda Rogan and Ms. Mary Beth Sustarsic for their patient and skillful typing of the final draft.

REFERENCES CITED

The authors do not intend that this be an exhaustive survey of the literature relative to the question at hand. The references used are considered to be those most pertinent.

1. Albright, B.C. and D. E. Haines. 1973. The morphology of Clarke's column in the lesser bushbaby (*Galago senegalensis*). Brain, Behavior, and Evolution, 8:165-190.

2. Angst, R. P. Mann. 1971. Zur Variabilitat von *Urogale everetti*. Folia primat. 15:148-158.

3. Ariëns Kappers, C., G. C. Huber, and E. C. Crosby. 1936. *The Comparative Anatomy of the Nervous System of Vertebrates Including Man*. The MacMillan Company, New York.

4. Ariëns Kappers, C. U. 1947. *Anatomie Comparée du Systeme Nerveux*. Masson and Cie, Paris.

5. Bischoff, E. 1899. Zur Anatomie der Hinterstrangkerne bei Saugethieren. Jb. Psychiat. Neurol. 18:371-384.

6. Bossy, J. G., and R. Ferratier. 1968. Studies of the spinal cord of *Galago senegalensis*, compared to that of man. J. Comp. Neurol., 132:485-498.

7. Bowsher, D. 1961. The termination of secondary somatosensory neurons within the thalamus of *Macaca mulatta*: an experimental degeneration study. J. Comp. Neurol. 117:213-227.

8. Brauer, K. 1968. Vergleichend-anatomische Untersuchungen am Kleinhirn der Insektivoren. 1. Das Kleinhirn von Erinaceus europaeus. J.f. Hirnforsch. 10:89-100.

9. Brauer, K. 1969. Vergleichend-anatomische Untersuchungen am Kleinhirn der Insektivoren. II. Das Kleinhirn von Sorex araneus und Elephantulus intufi. J.f. Hirnforsch. 11:537-548.

10. Brodal, A. 1969. Neurological Anatomy. Oxford University Press. New York.

11. Buettner-Janusch, J. 1966. Origins of Man. John Wiley and Sons, Inc., New York.

12. Bugge, J. 1972. The cephalic arterial system in the insectivores and primates with special reference to the Macroselidoidea and Tupaioidea and the Insectivore-Primate boundary. Z. Anat. Entwickl-Gesch. 135:279-300.

13. Campbell, C. B. G., D. Yashon, and J. A. Jane. 1966. The origin, course, and termination of the corticospinal fibers in the slow loris, Nycticebus coucang (Boddaert). J. Comp. Neurol. 127:101-112.

14. Carlsson, A. 1922. Uber die Tupaiidae und ihre Beziehungen zu den Insectivora une den Prosimiae. Acta Zool. 3:227-270.

15. Carpenter, M. B. 1956. A study of the red nucleus in the rhesus monkey. Anatomic degenerations and physiologic effects resulting from localized lesions of the red nucleus. J. Comp. Neurol. 105:195-249.

16. Carpenter, M. B., and G. H. Stevens. 1957. Structural and functional relationships between the deep cerebellar nuclei and the brachium conjunctivum in the rhesus monkey. J. Comp. Neur. 107: 109-163.

17. Carpenter, M. B., G. M. Brittin, and J. Pines. 1958. Isolated lesions of the fastigial nuclei in the cat. J. Comp. Neur. 109: 65-89.

18. Cartmill, M. 1972. Arboreal adaptations and the origin of the order Primates. In R. Tuttle (ed.), The Functional and Evolutionary Biology of Primates. Aldine-Atherton, Chicago.

19. Chang. H. T. 1951. Caudal extensions of Clarke's nucleus in the spider monkey. J. Comp. Neurol. 95:43-71.

20. Chang, H.-T., and T. C. Ruch. 1947. Organization of the dorsal columns of the spinal cord and their nuclei in the spider monkey. J. Anat. 81:140-149.

21. Clarke, J. L. 1851. Researches into the structure of the spinal cord. Phil. Trans. Roy. Soc. (London). Part II, 607-621.

22. Clarke, J. L. 1859. Further researches on the gray substance of the spinal cord. Phil. Trans. Roy. Soc. (London). 149:437-467.

23. Courville, J. 1966. Somatotopical organization of the projection from the nucleus interpositus anterior of the cerebellum to the red nucleus. An experimental study in the cat with silver impregnation methods. Exptl. Brain Res. 2:191-215.

24. Crosby, E. C., T. Humphrey, and E. W. Lauer. 1962. Correlative Anatomy of the Nervous System. The MacMillan Company, New York.

25. Crosby, E. C., A. J. Taren, and R. Davis. 1969. The anterior lobe and the lingula of the cerebellum in monkeys and man. Top. Prol. Psychiat. Neurol. 10:22-39.

26. Davis, D. D. 1938. Notes on the anatomy of the tree shrew Dendrogale. Zool. Ser. Field Mus. Nat. Hist. 20:383-404.

27. Eager, R. P. 1963. Efferent cortico-nuclear pathways in the cerebellum of the cat. J. Compt. Neur. 120:81-104.

28. Eager, R. P. 1966. Patterns and mode of termination of cerebellar cortico-nuclear pathways in the monkey (Macaca mulatta). J. Comp. Neurol. 126:551-566.

29. Elliot, O. 1971. Bibliography of the tree shrews 1780-1969. Primates 12:323-414.

30. Ferraro, A., and S. E. Barrera. 1935. Posterior column fibers and their termination in Macacus rhesus. J. Comp. Neurol. 62:507-530.

31. Fulton, J. F., and D. Sheehan. 1935. The uncrossed lateral pyramidal tract in higher primates. J. Anat. 69:181-187.

32. Gerhard, L., and J. Olszewski. 1969. Medulla oblongata and pons. In H. Hofer, A. H. Schultz, and D. Starck (Eds.), Primatologica, Handbook of Primatology. Karger, Basel.

33. Goode, G. E., and D. E. Haines. 1973. Corticospinal fibers in a prosimian primate (Galago senegalensis). Brain Res. 60:477-481.

34. Goodman, D. C., R. C. Hallet, and R. Welch. 1963. Patterns of localization in the cerebellar cortico-nuclear projections of the albino rat. J. Comp. Neurol. 121:51-67.

35. Grant, G. 1962. Spinal course and somatotopically localized termination of the spinocerebellar tracts. An experimental study in the cat. Acta. Physiol. Scand. 56 (Suppl. 193): 1-45.

36. Grundfest, H., and B. Campbell. 1942. Origin, conduction and termination of impulses in the dorsal spino-cerebellar tract of cats. J. Neurophysiol. 5:275-294.

37. Haines, D. E. 1969. The cerebellum of Galago and Tupaia. I. Corpus cerebelli and flocculonodular lobe. Brain, Behavior and Evolution. 2:377-414.

38. Haines, D. E. 1971a. The cerebellum of Galago II. The early postnatal development. Brain, Behavior and Evolution. 4:97-113.

39. Haines, D. E. 1971b. The morphology of the cerebellar nuclei of Galago and Tupaia. Amer. J. Phys. Anthrop. 35:27-42.

40. Haines, D. E., and D. R. Swindler. 1972. Comparative neuro-anatomical evidence and the taxonomy of the tree shrews (Tupaia). J. Hum. Evol. 1:407-420.

41. Haines, D. E. 1973. The cerebellum of some Lorisidae. In R. Martin, G. Doyle, and A. Walker (eds.), Prosimian Biology, Duckworth Co., Ltd., London (in press).

42. Haines, D. E., B. C. Albright, G. E. Goode, and H. M. Murray. 1973. The external morphology of the brain of some Lorisidae. In R. Martin, G. Doyle, and A. Walker (eds.), Prosimian Biology, Duckworth Co. Ltd., London (in press).

43. Hall-Craggs, E. C. B. 1965. An osteometric study of the hind limb of the galagidae. J. Anat. (Lond) 99:119-126.

44. Hall-Craggs, E. C. B. 1966. Rotational movements in the foot of Galago senegalensis. Anat. Rec. 154:287-293.

45. Hatschek, R. 1907. Zur vergleichenden Anatomie der Nucleus Ruber Tegmenti. Arb. Neurol. Inst. Univ. Wien, 15:89-136. Cited by massion, 1967.

46. Huber, G. C., E. C. Crosby, R. T. Woodburne, L. A. Gillilan, J. O. Brown, and B. Tamthai. 1943. The mammalian midbrain and isthmus regions. I. The nuclear pattern. J. Comp. Neurol. 78:129-534.

47. Jane, J. A., C. B. G. Campbell, and D. Yashon. 1965. Pyramidal tract: a comparison of two prosimian primates. Science 147: 153-155.

48. Jane, J. A., C. B. G. Campbell, and D. Yashon. 1969. The origin of the corticospinal tract of the tree shrew (*Tupaia glis*) with observations on its brain stem and spinal terminations. Brain, Behavior and Evolution 2:160-182.

49. Jane, J. A., and D. M. Schroeder. 1971. A comparison of dorsal column nuclei and spinal afferents in the European hedgehog (*Erinacues europeaus*). Exp. Neurol. 30:1-17.

50. Kanagasuntheram, R., and Z. Y. Mahran. 1960. Observations on the nervous system of the lesser bushbaby (*Galago senegalensis senegalensis*). J. Anat. 94:512-527.

51. Kanagasuntheram, R., C. H. Leong, and Z. Y. Mahran. 1966. Observations on some cortical areas of the lesser bushbaby (*Galago senegalensis senegalensis*). J. Anat. 100:317-333.

52. Kaufmann, J. H. 1965. Studies on the behavior of captive tree shrews (*Tupaia glis*). Folia Primat. 3:50-74.

53. Keller, A. D. and W. K. Hare. 1934. The rubrospinal tracts in the monkey: effects of experimental section. Arch. Neurol Psychiat. 32:1253-1272.

54. King, J. A., R. C. Schwyn, and C. A. Fox. 1971. The red nucleus in the monkey (*Macaca mulatta*): a Golgi and an electron microscopic study. J. Comp. Neurol. 143:75-108.

55. Kuypers, H.G.J.M., W. R. Fleming, and J. W. Farinholt. 1962. Subcorticospinal projections in the rhesus monkey. J. Comp. Neurol. 118:107-137.

56. Kuypers, H.G.J.M., and J. Brinkman. 1970. Precentral projections to different parts of the spinal intermediate zone in the rhesus monkey. Brain Res. 24:29-48.

57. Larsell, O., and J. Jansen. 1970. *The Comparative Anatomy and Histology of the Cerebellum from Monotremes through Apes*. University of Minnesota Press, Minneapolis.

58. Larsell, O., and J. Jansen. 1972. The Comparative Anatomy and Histology of the Cerebellum. The Human Cerebellum, Cerebellar Connections, and Cerebellar Cortex. University of Minnesota Press. Minneapolis.

59. Lassek, A. M. 1935. A comparative volumetric study of the gray and white substance of the spinal cord. J. Comp. Neurol. 62:361-376.

60. Lawrence, D. G., and H.G.J.M. Kuypers. 1968. The functional organization of the motor system in the monkey. Brain. 91:1-36.

61. Le Gros Clark, W. E. 1924. On the brain of the tree shrew (Tupaia minor). Proc. Zool. Soc. (London). pp. 1053-1074.

62. Le Gros Clark, W. E. 1926. The anatomy of the pen-tailed tree shrew (Ptilocercus lowii). Proc. Zool. Soc. (London), pp. 1179-1309.

63. Le Gros Clark, W. E. 1959. The Antecedents of Man. Edinburgh University Press, Edinburgh.

64. Le Gros Clark, W. E. 1964. The Fossil Evidence for Human Evolution. University of Chicago Press, Chicago.

65. Lende, R. A. 1969. A comparative approach to the neocortex: localization in monotremes, marsupials and insectivores. In J. M. Petras and C. R. Noback (Eds.), Comparative and Evolutionary Aspects of the Vertebrate Central Nervous System. Ann. N. Y. Acad. Sci. 167:262-276.

66. Lende, R. A. 1970. Cortical localization in the tree shrew (Tupaia). Brain Res. 18:61-75.

67. Liu, C. N., and W. W. Chambers, 1964. An experimental study of the corticospinal system in the monkey (Macaca mulatta). J. Comp. Neurol. 123:257-284.

68. Lundberg, A., and O. Oscarsson. 1956. Functional organization of the dorsal spinocerebellar tract in the cat. Part IV. Synaptic connections of afferents from Golgi tendon organs and muscle spindles. Acta Physiol. Scand. 38:53-73.

69. Manni, E. and R. S. Dow. 1963. Some observations on the effects of cerebellectomy in the rat. J. Comp. Neur. 121:189-194.

70. Martin, R. D. 1968. Reproduction and ontogeny in tree shrews (Tupaia belangeri) with special reference to their general behavior and taxonomic relationships. Zeit. f. Tierpsy. 25: 409-495, 505-532.

71. Martin, G. F., and R. Dom. 1970. The rubro-spinal tract of the opossum (Didelphis virginiana). J. Comp. Neurol. 138: 19-30.

72. Massion, J. 1967. The mammalian red nucleus. Physiol. Rev. 47:383-436.

73. Mehler, W. R., V. G. Vernier, and W.J.H. Nauta. 1958. Efferent projections from dentate and interposate nuclei in primates. Anat. Rec. 130:430-431.

74. Mussen, A. T. 1927. Experimental investigations of the cerebellum. Brain 50:313-349.

75. Napier, J. R. 1967. Evolutionary aspects of primate locomotion. Amer. J. Phys. Anthrop. 27:333-342.

76. Napier, J. R. and A. C. Walker. 1967. Vertical clinging and leaping - A newly recognized category of locomotor behavior of primates. Folia primat. 6:204-219.

77. Noback, C. R., and N. Moskowitz. 1963. The primate nervous system: functional and structural aspects in phylogeny. In J. Buettner-Janusch (Ed.), Evolution and Genetic Biology of Primates, Academic Press, New York. vol. i, pp. 131-177.

78. Noback, C. R., and J. E. Shriver. 1969. Encephalization and the lemniscal systems during phylogeny. In J. M. Petras and C. R. Noback (Eds.), Comparative and Evolutionary Aspects of the Vertebrate Central Nervous System. Ann. N. Y. Acad. Sci. 167:118-128.

79. Noback, C. R., and J. K. Harting. 1971. Spinal cord (spinal medulla). In H. Hofer, A. H. Schultz and D. Starck (Eds.), Primatologica, Handbook of Primatology. Karger, Basel.

80. Orioli, F. L., and F. A. Mettler. 1956. The rubrospinal tract in Macaca mulatta. J. Comp. Neurol. 106:299-318.

81. Oscarsson, O. 1965. Functional organization of the spino- and cuneocerebellar tracts. Physiol. Rev. 45:495-522.

82. Pass, I. J. 1933. Anatomic and functional relationships of the nucleus dorsalis. (Clarke's column). Arch. Neurol. Phychiat. 30:1025-1045.

83. Petras, J. M. 1969. Some efferent connections of the motor and somatosensory cortex of simian primates and felid, canid and procyonid carnivores. In J. M. Petras and C. R. Noback (Eds.), Comparative and Evolutionary Aspects of the Vertebrate Central Nervous System. Ann. N. Y. Acad. Sci. 167:469-505.

84. Peters, M., and A. A. Monjan. 1971. Behavior after cerebellar lesions in cats and monkeys. Physiol. and Behav. 6:205-206.

85. Poirer, L. J., and G. Bouvier. 1966. The red nucleus and its efferent nervous pathways in the monkey. J. Comp. Neurol. 128:223-244.

86. Pompeiano, O. 1957. Analisi degli effecti della stimulazione elettrica del nucleo rosso nel gatto decerebrato. Atti Accad. Naz. Lincei, Mem. Classe Sci. Fis. Mat. Nat., Sez. III. 22: 100-103.

87. Pompeiano, O., and A. Brodal. 1957. Experimental demonstration of a somatotopical origin of rubrospinal fibers in the cat. J. Comp. Neurol. 108:225-252.

88. Preuschoft, H. 1971. Mode of locomotion in subfossil giant lemuroids from Madagascar. Proc. 3rd. Int. Congr. Primat. (Zurich 1970) 1:79-90.

89. Radinsky, L. 1968. A new approach to mammalian cranial analysis, illustrated by examples of prosimian primates. J. Morph. 124:167-180.

90. Radinsky, L. 1970. The fossil evidence of prosimian brain evolution. In C. Noback (ed.) Advances in Primatology. Vol. 1 The Primate Brain. Appleton-Century-Crofts, New York.

91. Rand, R. W. 1954. An anatomical and experimental study of the cerebellar nuclei and their efferent pathways in the monkey. J. Comp. Neurol. 101:167-235.

92. Rexed, B. 1952. The cytoarchitectonic organization of the spinal cord in the cat. J. Comp. Neurol. 96:415-466.

93. Romer, A. S. 1969. Vertebrate history with special reference to factors related to cerebellar evolution. In R. R. Llinas (ed.) Neurobiology of Cerebellar Evolution and Development. A M A Education and Research Foundation, Chicago.

94. Sanides, F., and A. Krishnamurti. 1967. Cytoarchitectonic subdivisions of sensorimotor and prefrontal regions and of bordering insular and limbic fields in slow loris (Nycticebus coucang coucang). J. Hirnforsch. 9:225-252.

95. Schoen, J.H.R. 1964. Comparative aspects of the descending fiber systems in the spinal cord. In J. C. Eccles and J. P. Schade (Eds.), Progress in Brain Research, Elsevier, Amsterdam. vol. 11, pp. 203-222.

96. Schroeder, D. M., and J. A. Jane. 1971. Projection of dorsal column nuclei and spinal cord to brainstem and thalamus in the tree shrew, Tupaia glis. J. Comp. Neurol. 142:309-350.

97. Schultz, A. H., and W. L. Strauss. 1945. The numbers of vertebrae in primates. Proc. Am. Philos. Soc. 89:601-626.

98. Shriver, J. E., and C. R. Noback. 1967. Cortical projections to the lower brain stem and spinal cord in the tree shrew (Tupaia glis). J. Comp. Neurol. 130:25-54.

99. Sorenson, M. W. 1970. Behavior of tree shrews. In L. A. Rosenblum (ed.), Primate Behavior Developments in Field and Laboratory Research. Academic Press, New York.

100. Stern, J. J. 1971. Functional myology of the hip and thigh of cebid monkeys and its implications for the evolution of erect posture. Bibl. Primat. No. 14 pp. 1-318. Karger.

101. Stevens, J. L., V. R. Edgerton, and S. Mitton. 1971. Gross anatomy of the hindlimb skeletal system of the Galago senegalensis. Primates 12:313-321.

102. Stevens, J. L., S. Mitton, and V. R. Edgerton. 1972. Gross anatomy of hindlimb skeletal muscles of the Galago senegalensis. Primates 13:103-109.

103. Straus, W. L. 1949. The riddle of man's ancestry. Quart. Rev. Biol. 24:200-223.

104. Streeter, G. L. 1904. The structure of the spinal cord of the ostrich. Amer. J. Anat. 3:1-27.

105. Szalay, F. S. 1972. Paleobiology of the earliest primates. In R. Tuttle (ed.), The Functional and Evolutionary Biology of Primates. Aldine-Atherton, Chicago.

106. Tilney, F. 1927. The brain of Tarsius. A critical comparison with other primates. J. Comp. Neurol. 43:371-432.

107. Van Valen, L. Tree Shrews, primates, and fossils. Evolution 19:137-151.

108. Verhaart, W. J. C. 1938. The rubrospinal system with monkeys and man. Folia Psychiat. Neurol. Neuochir. Neerl., 42:335-342. Cited by Massion, 1967.

109. Verhaart, W. J. C. 1966. The pyramidal tract of Tupaia, compared to that in other primates. J. Comp. Neurol. 126:43-50.

110. Verma, K. 1965. Notes on the biology and anatomy of the Indian tree-shrew Anathana wroughtoni. Mammalia 29:289-330.

111. von Monakow, C. 1909. Der rote Kern, die Haube und die regio subthalamica bei einigen Saugetieren und beim Menschen. I. Anatomisches und Experimentelles. Arb. Hirnanat. Inst. Zurich 3:51-267. Cited by Massion 1967.

112. von Monakow, C. 1910. Der rote Kern der Saugetiere und des Menschen. Neurol. Zentr., 724-727. Cited by Massion 1967.

113. Wharton, C. H. 1950. Notes on the Philippine tree shrew, Urogale everetti Thomas J. Mamm. 31:352-354.

114. Woollard, H. H. 1925. Anatomy of Tarsius spectrum. Proc. Zool Soc. (London) Part II. 1071-1185.

115. Yoss, R. E. 1952. Studies of the spinal cord. Part I. Topographic localization within the dorsal spino-cerebellar tract in Macaca mulatta. J. Comp. Neurol. 97:5-20.

116. Zervas, N. T. 1969. Paramedial cerebellar nuclear lesions Confin. Neurol. 32:114-117.

NOTE: Edward, S. B. 1972. The ascending and descending projections of the red nucleus in the cat: an experimental study using an autoradiographic tracing method. Brain Res. 48:45-63.
Since this manuscript was prepared, Edwards has reported an absence of the rubrothalamic projections in the cat using radioautography.

OUTLINE OF A PRIMATE VISUAL SYSTEM

D. Max Snodderly, Jr.
Department of Retina Research
Retina Foundation
20 Staniford Street
Boston, Mass., USA, 02114

The visual system of the genus Macaca has been the object of intensive study in recent years. This is partly because the macaques have color vision and visual acuity similar to the visual capabilities of man (DeValois and Jacobs, 1971). Unlike the great apes, they are small enough to be kept in a modest laboratory without massive equipment. The most commonly used species is Macaca mulatta, the rhesus monkey, but some of the neurophysiological and anatomical studies have used Macaca fascicularis. As far as we know, the visual systems of these two species are nearly identical, so I shall refer to both of them simply as macaques.

In this article I sketch out the successive stages of the visual pathways in the macaque. Major emphasis is placed on the response properties of single neurons at different levels of the visual system, with the goal of describing how the eye and brain extract information from the visual world. Our knowledge of this process is still fragmentary and incomplete, but it is sufficient to establish an outline that is both informative and provocative of further study. At the end of the article I try to illustrate how this outline might facilitate the comparison of visual capabilities and visual mechanisms across primate species. Purely motor aspects of eye movement control are beyond the scope of this review.

OPTICAL CHARACTERISTICS OF THE MACAQUE EYE

One of the prerequisites for normal visual function is a clear and undistorted image of the external world on the retina. When accommodation is relaxed in an eye that has no refractive error, the dioptric power of the cornea and the relaxed lens bring far objects into focus on the retina. The act of accommodation increases the curvature and power of the lens to bring nearer objects into focus on the retina. Natural variations in the length of the eye and the curvature of the lens and cornea produce a distribution of refractive errors in the eyes of any population of primates.

Humans between the ages of 5 and 19 and macaques estimated to be less than 7 years old have very similar distributions of refractive errors (Young, 1964). Macaques may have somewhat less astigmatism (distortion of the image due to unequal optical power of different meridions; Young, 1964).

Young animals tend to be slightly out-of-focus in the direction of needing additional plus refracting power when accommodation is relaxed by conventional drugs (Young, 1964). In the normal state they probably overcome this error by active accommodation. As the animals grow older, the refractive error of the relaxed eye shifts gradually until the eye becomes myopic (Young, 1964). This means the relaxed eye is in focus for near objects, so that active accommodation brings the focal plane even nearer, and there is no way for the animal to attain a clear image of distant objects. Confinement in a cage or a restricted visual space hastens this normal aging process for both macaques (Young, 1963, 1964) and chimpanzees (Young et al., 1971). In the chimpanzee the development of myopia seems to be associated with an increase in the axial length of the eye that is caused primarily by an increase in the length of the vitreous chamber (Young, et al., 1971; Young and Farrer, 1970).

The optical components of the eye can have an additional effect on the visual image by filtering out some of the wavelengths of light. This occurs mostly in the blue end of the spectrum. At wavelengths shorter than about 440 nm the lens of the macaque eye begins to exhibit significant absorption of the incident light (Norren, 1972). The amount of absorption increases with decreasing wavelength very quickly so that only $10^{-1.2}$ of the incident photons (Norren) or less (Cooper and Robson, 1969) at 400 nm actually penetrate through the lens. Still shorter wavelengths in the ultraviolet are filtered even more strongly (Cooper and Robson) which may protect the retina from radiation damage. The absorption of human lenses increases with age; macaque lenses that have been studied (from animals 8 years of age or younger) are comparable in absorption characteristics to lenses from children or young adult humans.

Another stage of color filtering is introduced at the retina by the so-called "macular pigment"(Brown and Wald, 1963). In humans the macular pigment is a yellow, light-stable pigment thought to be the carotenoid, xanthophyll. Its absorption curve has a maximum around 450 nm and decreases at both shorter and longer wavelengths, declining to almost zero beyond about 530 nm. The spatial distribution of the pigment in the human is very nonuniform. It has a peak density just outside the center of gaze, the fovea, a lower density in the central fovea, and steadily diminishing density as the measuring locus moves away from the peri-foveal area toward the periphery (P.K. Brown, personal communication). Since the fovea is where the inner layers of the retina thin out, this suggests that at least some of the macular pigment is located in the inner layers where it will modify the spectral distribution of the light before it reaches the receptors. The overall density of the macular pigment in human retinas increases with age up to an absorption as high as 99% at the peak wavelength.

In macaque monkeys the situation is basically similar, except that even the oldest monkeys studied did not develop pigment densities beyond about 0.5 (30% absorption). Thus they were comparable to young or juvenile humans. Exactly equivalent measurements of spatial distribution of the pigment have not been made, because the monkey fovea is smaller and more difficult to scan, but the macular pigment appears to be more dense near the fovea and diminishes to very low levels at 5° or more from the fovea (P.K. Brown, personal communication).

STRUCTURE AND ACTIVITY OF THE RETINA

The precise organization of the macaque visual nervous system already is apparent in the structure of the retina. Cell nuclei and perikarya are neatly arranged into rows, separated by alternating fiber layers with synaptic networks. Characteristic relationships among cells can be identified, and groups of neurons can be combined into modules that are repeated over and over again with some modifications in size of cells and number of connections as one moves laterally across the retina. The connections among the various retinal cell types are diagrammed in Fig. 1. Bear in mind that the retina is "inside out," with the receptors pointing away from the light, toward the outer surface of the eyeball. The terms "inner" and "outer" in retinal anatomy always refer to the spheroidal space of the eyeball, which is often opposite to an inner-outer coordinate system centered in the skull.

The outermost layer of the retina is the pigment epithelium, a monolayer of cells containing dark pigment granules. Light that is not captured by the receptors is absorbed by the pigment epithelium, which prevents scattering and reflections that would degrade the retinal image. In addition to its screening function, the pigment epithelium has an important and still poorly understood metabolic role. If the rest of the retina is detached from the pigment epithelium, the outer segments of the receptors degenerate, and they do not regain normal structure until the retina is reattached and the proximity of the receptors to the pigment epithelium is restored (Kroll and Machemer, 1969).

Two morphological classes of receptors have been distinguished for more than a century on the basis of the distinctive shapes of the outer segments and other structural features. Rods dominate the retinas of nocturnal animals and mediate vision in dim light, where-

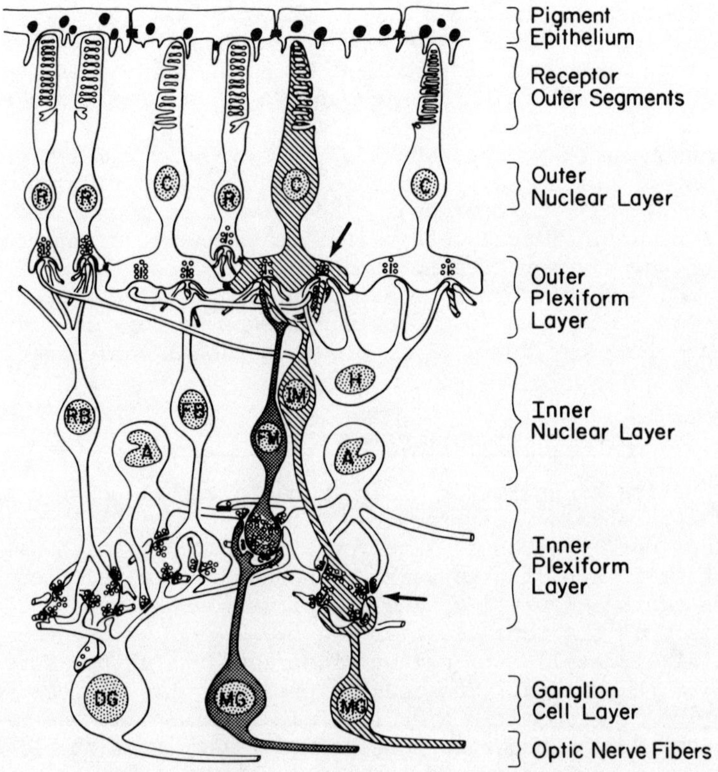

Figure 1. Connections of the macaque retina. Rods, R, end in spherules that receive processes of rod bipolars, RB, and horizontal cells, H, Cones, C, have larger terminals called pedicles that receive processes of horizontal cells as well as three types of bipolars: the flat bipolar, FB, flat midget bipolar, FM, and invaginating midget bipolar, IM. In the inner plexiform layer the amacrine cells, A, contact each other, the bipolars, and the ganglion cells. Midget ganglion cell (MG) dendrites embrace the terminals of a single midget bipolar while diffuse ganglion cells, DG, contact many different bipolars. Arrows indicate the characteristic "triad" synaptic complex of the outer plexiform layer and the "dyad" synaptic pairing at the inner plexiform layer. Adapted from Cohen (1972), Dowling and Boycott (1966), Boycott and Dowling (1969), and Kolb (1970).

as cones are prominent in the retinas of diurnal animals and dominate daylight vision. Both receptor types have outer segments formed from membranous discs stacked on top of each other, that contain the photopigments. In the rods, new discs are continually being added at the base of the outer segment (Young, 1971a) while the oldest discs at the apex are shed by the rod and then phagocytized by the pigment epithelial cells that extend processes into the spaces between receptors (Young, 1971b). The entire outer segment of the rod is thus renewed in 8 to 13 days depending on its length, with about 80 to 90 new discs being assembled each day (Young 1971a).

Cones, on the other hand, are renewed diffusely by synthesis of protein throughout the outer segment (Young, 1971a). Young (1971c) has also suggested that the pointed shape of the cones arises because they do not continue to grow axially as the rods do. When the cone develops, the first discs are smaller than later ones and they are displaced axially as new, larger, discs are added at the base, so that the outer segment takes on the conical, tapered form. Then the cone stops growing and switches to the adult renewal process, unlike the rods, which continue to grow indefinitely, with the result that all discs eventually become the same size.

The difference in renewal processes may be associated with the relationship between the photoreceptor discs and the cell membrane. In rods, the discs are isolated from the extracellular space by the cell membrane, whereas the extracellular fluids have access to many, if not all, of the cone discs (Cohen, 1970). Perhaps this is why the cones dominate the extracellularly recorded early receptor potential, even in areas of the retina where there are many more rods than cones (Goldstein, 1969). The early receptor potential is generated when the visual pigments of the receptors absorb light.

Other electrical signals, the late receptor potentials, can be recorded from the vicinity of primate receptors (Brown et al., 1965) but they are more difficult to interpret. The late receptor potentials attributed to rods have a longer latency and slower rise and fall times than those attributed to the two common types of cones (L, and M, cones, see later; Whitten and Boynton, 1973). Cone late receptor potentials show marked light adaptation when the background intensity is varied while responses are elicited with brief incremental stimuli (Boynton and Whitten, 1970).

It is likely that primate receptors function like those of lower vertebrates, which have been studied more directly. In the rat, the rods and cones generate ionic currents in the dark that flow into the outer segments; when light impinges upon the retina the "dark current" is suppressed (Hagins, 1972). This is probably accomplished by an increase in resistance of the cell membrane, accompanied by a hyperpolarization of the cell (Toyoda et al., 1969). Hagins (1972) has calculated that the dark current of the rat rod is equivalent to turnover of all its cations in 45 sec. Such feverish metabolic activity seems paradoxical, since it means that the receptors are always utilizing large amounts of energy, even when they are not being illuminated. However it may have other important functions, such as maintenance of the "spontaneous" or ongoing activity of the retinal ganglion cells, which in turn influence the resting discharge rate of visual neurons in the brain (Jacobs, 1972).

The currents generated by the receptors are channeled by the relative resistance of the different extracellular pathways. Electron microscopy of the pigment epithelium shows that the adjacent cells are joined by terminal bars that should provide a high resistance barrier to current flow away from the retina. Between the receptor nucleus and the outer segment an "outer limiting membrane" is formed by the processes of glial (Mueller) cells not shown in the drawing. As is appropriate, the junctions of these cells are not sealed and thus should not prevent the currents of the outer segments from exciting the retinal elements (Cohen, 1972). The same Mueller cells extend throughout the thickness of the retina to participate in the inner limiting membrane at the inner surface of the retina adjacent to the optic nerve fibers (i.e., at the very bottom of Fig. 1).

The receptor endings extend into the outer plexiform layer, where they receive processes of horizontal and bipolar cells in characteristic synaptic arrangements. In both the rod spherule and the cone pedicle there is a synaptic ribbon surrounded by synaptic vesicles above an invagination that receives a cluster of processes consisting of lateral elements from the horizontal cells and central elements from the bipolar cells. The synaptic ribbon is an unusual neural structure that could conceivably be an adaptation to the presence of sustained resting currents in the receptors. Rod spherules have only a single invagination that envelopes axon terminals of the horizontal cells as lateral elements and dendrites of two or more rod bipolars as central elements (Kolb, 1970). The bipolar cells that service rods do not contact cones (Boycott and Dowling, 1969).

A separate population of bipolar cells contact the cone pedicles, along with the dendrites of the horizontal cells. Three different types of bipolar cells have been identified as contributing to the synaptic complex (Kolb, 1970; upper arrow in Fig. 1). The most deeply penetrating bipolar process is from the invaginating midget bipolar, which contacts a single cone several times, but does not contact any other receptor. Around this central process are dendrites from a flat midget bipolar, that also contacts only this cone, and dendrites of a flat bipolar that contacts five to six other cones as well (Boycott and Dowling). The flat bipolar also makes contacts with cone pedicles at points outside the invaginating complexes.

Although four types of bipolar cells can be identified, there appears to be only one type of horizontal cell (Boycott and Kolb, 1973). Unlike the situation in other mammals (Dowling et al., 1966) primate horizontal cells have only scattered synaptic vesicles with no evidence of concentration into specialized synaptic loci (Dowling and Boycott, 1966). It is tempting to assume that the identification of the horizontal cell processes in rod spherules as axon terminals (Kolb, 1970) could provide an anatomical basis for the inhibition of rods by cones (see later), but in the absence of interpretable synaptic structure one had best reserve judgement.

Cone pedicles make contacts with rod spherules and with other cone pedicles, but no function can presently be assigned to these contacts (Cohen, 1972).

The bipolar cells carry visual information to the inner plexiform layer, where they end in enlarged axon terminals containing synaptic ribbons like those of the receptor terminals. At each synaptic ribbon the bipolar cell is contacted by a ganglion cell dendrite as well as an amacrine cell process that makes a reciprocal synapse back onto the bipolar cell (as judged by morphological criteria, Dowling and Boycott, 1966). Different types of bipolars tend to terminate at different levels: Rod bipolars end in the inner part of the inner plexiform layer and sometimes contact the ganglion cell somata in a manner that could indicate an electrical synapse (Boycott and Dowling, 1969; Dowling and Boycott, 1966); invaginating midget bipolars also end in the inner third of the inner plexiform layer, but flat midget bipolars usually end in the external third of this layer, even though their perikarya lie lower than those of the invaginating midgets (Kolb, 1970); flat bipolars end in the middle of the inner plexiform layer.

Bipolar terminals are contacted by two broad classes of ganglion cells, midget and diffuse. Midget ganglion cells have compact dendritic branches that mainly embrace the terminals of a single midget bipolar cell. Dendrites of a given midget ganglion cell tend to ramify either in the outer third or the inner third of the inner plexiform layer, but not the middle of the layer (Boycott and Dowling), as might be expected since they contact a flat midget bipolar only, or an invaginating midget bipolar only. Diffuse ganglion cells have more extensive dendritic trees that can make contact with any of the bipolar terminals, including rod bipolars.

Amacrine cells seem to synapse with anything and everything in the inner plexiform layer (Dowling and Boycott). Some of the amacrine and ganglion cells have stratified processes that branch in horizontal strips within the inner plexiform layer. These might make specific connections with particular kinds of bipolar cells that terminate at distinct levels. One sublayer of terminals in the inner plexiform layer of still unidentified origin, but possibly arising in part from amacrine cells, appears to be adrenergic (Ehinger and Falck, 1969). Both amacrine and ganglion cell processes exhibit elaborate variations in geometrical form (Boycott and Dowling) but is not yet clear what functional significance these may have.

The pattern of connections can be remembered more easily by grouping the neurons into horizontal and vertical pathways through the retina. Rods have only a diffuse vertical pathway through the rod bipolars to diffuse ganglion cells. Cones also have diffuse pathways through the flat bipolars and diffuse ganglion cells, but they can have in addition discrete vertical pathways via the two types of midget bipolars and the midget ganglion cells. Horizontal cells provide horizontal pathways at the outer plexiform layer, whereas amacrine cells mediate lateral interactions in the inner plexiform layer.

A natural unit is a module consisting of a single ganglion cell and its immediate input elements. This includes: (1) bipolars that contact the ganglion cell; (2) receptors that contact the bipolars; (3) horizontal cells that contact the receptors and bipolars; and (4) amacrine cells that synapse with the ganglion cell and its bipolars.

The module of vertical and horizontal elements undergoes systematic modifications as one moves from the center of the retina to the periphery. At the center of gaze, the retina thins out and forms a pit, named the fovea, in its inner surface by displacing laterally the cells of the inner nuclear layer and the ganglion cells. The flattened central region of the fovea, the foveola, contains only cone receptors that are long, narrow, and close-packed into a regular array (see Polyak, 1957 for illustration) that achieves the highest density of cones of any part of the retina. The number of cones per unit area declines monotonically as one moves along either the nasal or temporal meridian into peripheral retina (Rolls and Cowey, 1970). Rods have a different distribution, being absent in the fovea, then rising to a maximum density near the boundary of the central retina and declining in number in the periphery (Polyak, 1957; Young, 1971a).

The central retina or area centralis is usually defined as the area of retina where the ganglion cell layer is more than one cell deep. Since the ganglion cells that subserve the fovea are displaced laterally, they accumulate in as many as 6 to 8 tiers just outside the fovea (Polyak), but like the cones their number declines rapidly toward the periphery (Rolls and Cowey; Van Buren, 1963).

Concurrent with the changes in cell populations are changes in cell sizes and connections. The inner segments of the cones and the diameters of the cone pedicles are both larger in the peripheral retina (Polyak; Boycott and Dowling). Cone pedicles in the periphery have more invaginations (Kolb) and horizontal cells are larger and contact more cones (Boycott and Kolb). In the fovea, a horizontal cell may contact 6 to 9 cones, but in the far periphery the number is as high as 30 to 40 cones. Bipolar and amacrine cells both seem to have large variations in branching patterns, but more precise studies than are presently available will be necessary to determine how these might change with retinal position (Boycott and Dowling).

There is agreement that ganglion cells often have more compact dendritic trees near the fovea than further peripheral (Boycott and Dowling); Polyak (1957) even states that the majority of the ganglion cells in the central retina are midget ganglion cells. Nevertheless, too few ganglion cells exist in the central retina for every cone to be serviced by two discrete pathways through two midget ganglion cells (Rolls and Cowey), so some variations in the patterns of connections of midget bipolars to ganglion cells might be expected.

The overall picture is one of a fine mosaic of neural modules in the fovea and central retina that becomes gradually coarser in the periphery. This is probably associated with the decline in visual acuity that occurs (in man) when the test stimulus is moved from the fovea progressively greater distances into the periphery (e.g., see Fig. 2 of Alpern (1962)). Similarly detailed psychophysical information is not available for macaques, but from lesion data it has been inferred that extrafoveal acuity is poorer than foveal acuity (Cowey and Ellis, 1969).

OPTIC NERVE

Axons of the retinal ganglion cells course along the surface of the retina to the optic disc, where they all become invested with myelin sheaths and leave the eyeball to form the optic nerve. Total counts of the optic nerve fibers of two rhesus monkeys gave values of 1.5 and 1.8 million fibers, somewhat more than the 1.2 and 1.3 million found in two human optic nerves (Potts et al., 1972a). As would be expected from the range of ganglion cell sizes, the optic nerve fibers have a wide range of diameters from $1/4\mu$ or less to several microns (not including the myelin sheath). The mode of the distribution is 0.5μ for both macaque and human (Potts et al., 1972b). Immediately upon leaving the eye, the fibers from the central retina are located in the temporal portion of the optic nerve but then gradually move into the center of the nerve as it courses proximally. Most of these fibers are very small, with diameters of 0.75μ or less in the macaque (Potts, et al., 1972c).

The conduction velocities of the optic nerve fibers are related to the fiber diameter according to the equation $V=3.2D$, where V is velocity in meters/second and D is fiber diameter (including myelin) in microns (Ogden and Miller, 1966). Since the distribution of fiber sizes is continuous and unimodal, the conduction velocities also range along a continuum between the limits of 1.3 and 20 meters/sec. For the short portion of their travel within the eyeball toward the optic disc, the optic nerve fibers are unmyelinated and they have slower conduction velocities, ranging from 0.45 to 1 meter/sec (Ogden and Miller).

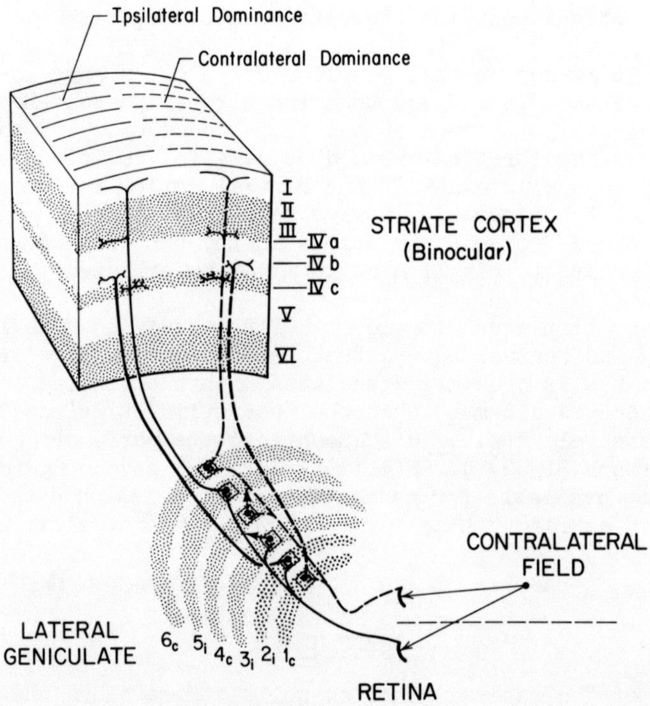

Figure 2. Geniculostriate afferent pathway of the macaque monkey. Retina: The temporal retina of the ipsilateral eye and the nasal retina of the contralateral eye send axons to the lateral geniculate. This can be remembered by simply noting that each hemiretina projects to the geniculate closest to it. Lateral Geniculate: The dotted regions in this schematic coronal section through the nucleus indicate the cell laminae, the clear areas between them being fiber layers. Cell laminae traditionally have been numbered 1-6 starting with the most ventral layer; here the number is followed by a "c" or an "i" to denote that the layer is connected to the contralateral or the ipsilateral eye respectively. The four (parvocellular) dorsal layers have smaller cells than the two (magnocellular) ventral layers. Two additional thin intercalated laminae, one on each side of layer 1 have been described recently (Guillery and Colonnier, 1970) and the ventral-most one is innervated by the central retina of the ipsilateral eye (Campos-Ortega and Hayhow, 1970c). Another small lamina, called the "intermediate cell group," lies medial to the main laminae and receives segregated inputs from both eyes (Campos-Ortega and Hayhow, 1971). Nothing is known about the physiology of these minor laminae, nor are their anatomical connections clearly specified, so they are not

Many people have wondered whether the primate retina receives efferent inputs originating in the brain. Some silver-stained preparations do show fibers branching on the inner surface of the retina and entering the retina to travel along the inner nuclear layer (Honrubia and Elliott 1970). Physiological evidence for efferent control of retinal activity in mammals has so far been severely criticized (Brindley, 1970).

GENICULO-CORTICAL PATHWAY

The major thalamic receiving station for the optic nerve fibers is the lateral geniculate nucleus (LGN, Fig. 2). The caudal half of the nucleus, which receives input from the central retina, is the most highly differentiated part, with 6 laminae separated by fiber layers. The four dorsal layers (3-6) are composed of a fairly uniform population of small cells, whereas the two ventral layers (1 and 2) have larger cells and different physiological properties (see later). In the rostral half of the nucleus, which receives input from the peripheral retina, the dorsal layers coalesce to form only two layers (Kaas et al., 1972).

Although synapses occur in both the cell and the fiber layers (Guillery and Colonnier, 1970), there still is no evidence for binocular interaction in this nucleus (unlike the situation in the cat; Sanderson et al., 1971 and their cited references). Instead, the input from each eye seems to be segregated into separate cell layers, with the topographic retinal maps in the different layers brought into register; a point in the contralateral visual field is thus represented in a line of projection through the nucleus (See reviews by Brindley, 1970; Polyak, 1957). The correspondence between the retina and the lateral geniculate is so precise that the optic disc, where the optic nerve fibers exit and form a "hole" in the cellular layers of the retina, is represented by a "hole" in the cellular laminae of the geniculate (Kaas et al., 1973).

Neurons in a given line of projection send their axons to a

---

included in this drawing. Striate Cortex: A section perpendicular to the cortex shows a horizontal layering as schematically depicted here; the numbering system is that of Hubel and Wiesel (1972). Layers II, III, IVa, IVc, and VI are densely filled with cell bodies, while the intervening layers contain rather sparsely distributed cells. The stria of Gennari, a plexus of myelinated fibers arising mostly from cortical cells, occupies layer IVb. A second horizontal axon plexus is found in layer V. The axon endings of the lateral genculate cells are located in layers IV and I as described in detail in the text. For a summary of numbering systems of cortical layers see Garey (1971).

common region of the striate cortex, which also preserves the topographic map of the visual field (Talbot and Marshall, 1941; Daniel and Whitteridge, 1961). Only striate cortex (sometimes called V1) receives fibers from the LGN in the macaque, although several additional cortical areas receive geniculate inputs in the cat (Hubel and Wiesel, 1972; Wilson and Cragg, 1967). Axons of geniculate neurons in the dorsal parvocellular layers branch and terminate most densely in layer IVc but give a second less extensive input to layer IVa and a still sparser fiber contribution to layer I (Hubel and Wiesel, 1972). The branches in layer IVc and IVa may arise from the same axons, but those in layer I probably come from a different set of LGN cells (Lund, 1973). Fibers from the ventral magnocellular layers of the geniculate terminate in a separate region confined primarily to the lower part of layer IVb (Hubel and Wiesel, 1972).

The segregation of input from each eye is continued in the termination pattern of geniculate fibers in layer IV. Long bands of terminals about 0.5 to one mm wide from one eye alternate with bands of terminals from the other eye. The long axis of these stripes or "columns" would be perpendicular to the plane of the paper in Fig. 2, with the bands in IVa lying in register above those in IVc. Physiological results correlate well with the anatomy, for most cells in layer IV are driven by one eye only.

Rather subtle differences in the neuronal populations of the cortical layers are probably also associated with the pattern of geniculate inputs. Traditionally, two types of cortical cells have been distinguished - stellate and pyramidal. Most stellate cells have fairly symmetrical dendritic branching patterns, while pyramidal cells have one vertical (apical) dendrite that is markedly longer than the rest (Lund, 1973). Both stellate and pyramidal cells can have dendritic spines that are the sites of synaptic contacts; some stellate cells (presumably a different population) have few or no spines (Garey, 1971; Lund, 1973). Where the largest number of geniculate fibers terminate, in layers IVa and IVc, spinous stellate cells dominate the neuropil (Lund, 1973). Cell bodies of pyramidal neurons are not found in lamina IVc nor do the dendrites of pyramidal cells passing through this layer show their usual density of spines (Lund, 1973). Since most of the geniculate fibers terminate on dendritic spines, (Garey and Powell, 1971) it is likely that their principal targets are the spiny stellate cells.

In cortical layers other than number IV, neurons are driven more strongly by the eye giving input to the nearest part of layer IV, but they also respond to stimulation of the other eye (Hubel and Wiesel, 1968; Poggio, 1972). This presumably requires some horizontal or diagonal connections within the cortex, which could include fibers from the stellate cells that travel within the horizontal plexus known as the stria of Gennari in layer IVb (Fisken et al., 1973).

There is little evidence at present for structural differentiation of the striate cortex related to the topographic heterogeneity of the retina. The only exception is a 50% increase in the thickness of the cortex in the foveal projection area (Dow and Gouras, 1973).

Smaller cell groups around the lateral geniculate named the parageniculate and pregeniculate also receive optic nerve fiber inputs (Hendrickson et al., 1970), but their projections to other visual structures have not been studied with modern techniques. However, physiological findings on the pregeniculate indicate that it has a very different role from the lateral geniculate. For example some of the neurons are binocular and can be excited from both eyes (DeValois, 1960). In addition, many pregeniculate cells discharge in relation to saccadic eye movements, whereas 95% of the lateral geniculate neurons are unaffected by eye movements (Buttner and Fuchs, 1973). The saccade-related activity of pregeniculate cells has a curious time course, for it often does not begin until after an eye movement has been initiated, and it seems to be independent of the duration of the saccade.

Receptive Field Hierarchy

Our knowledge of the receptive fields of visual neurons is based primarily on the results of extracellular single-unit recording (Snodderly, 1973). This technique utilizes a very fine electrode with a tip of about 1-3μ positioned just outside the cell so that the action potentials and the firing patterns of the cell can be monitored. Usually a small region of the visual field (or of the retina) can be identified where the presentation of restricted visual stimuli modulates the rate at which the cell fires impulses; this is the receptive field.

Most retinal ganglion cells have circular receptive fields containing two mutually antagonistic response mechanisms. An increment in light intensity is translated into an increased firing rate (excitation) by one mechanism but a decreased firing rate (inhibition) by the other mechanism. When an incremental stimulus is turned off, or when a decrement from the adaptation level is used, the two response mechanisms reverse their roles. Thus one response mechanism is on-excitatory, off-inhibitory, while the other is on-inhibitory, off-excitatory. For simplicity's sake the first is usually referred to as the excitatory mechanism and the second as the inhibitory mechanism.

The two response mechanisms are concentrically distributed within the circular receptive field, and are approximately radially symmetric (Fig. 3). One is more sensitive than the other and peaks

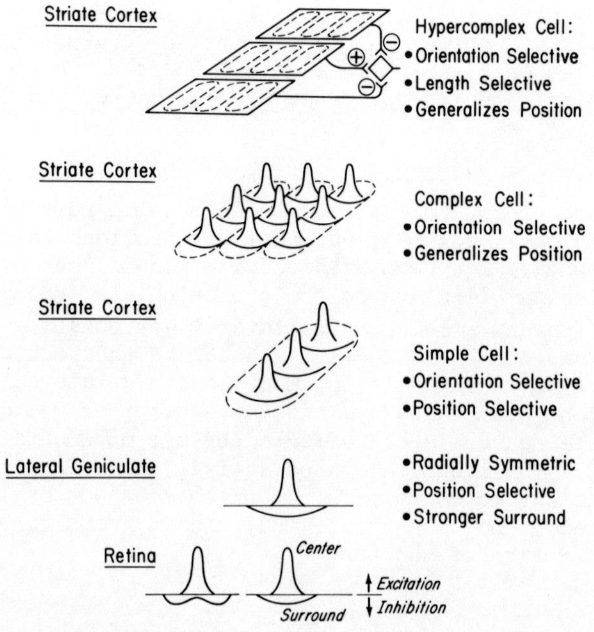

Figure 3. Receptive Field Hierarchy.
Retina: Schematic profile along the radius of a receptive field demonstrating possible distribution of excitation and inhibition in the circular receptive fields of retinal ganglion cells. Response sensitivity on vertical axis, distance on horizontal. This illustration is for an excitatory center, with two different types of inhibitory surround, but the converse organization with inhibitory center, excitatory surround, is also frequently found. It looks like this drawing upside-down. A three dimensional representation would be generated by spinning these curves around the central vertical axis through the center mechanism. Lateral Geniculate: The surround of the non-color-coded lateral geniculate cells is stronger than that of retinal ganglion cells, but the exact profile is not yet known. This representation is a relatively simple one that is likely to be valid for some, but perhaps not all, cells. Striate Cortex: Cells in the striate cortex acquire the ability to respond selectively to lines of particular orientations. Simple cells require that the stimulus be positioned precisely, but complex and hypercomplex cells generalize over a greater region of space. Hypercomplex cells are inhibited, though, if the stimulus extends past the activating region into the inhibitory flanking fields. The illustration given here is for cortical cells that prefer light slits, but a comparable hierarchy

in the center of the field; it is called the center mechanism. The less sensitive mechanism has a broader spatial distribution extending further laterally than the center mechanism, and is therefore known as the surround mechanism. These distributions can be interpreted as the magnitude of the response from each mechanism that would be expected for a small stimulus as a function of its position within the receptive field. Alternatively, the distributions might indicate the reciprocal of the threshold for each mechanism at different receptive field loci. Note particularly that the surround has a high threshold (low sensitivity) for small spots.

By proper choice of stimuli the response elicited from the cell can be made characteristic of the center or of the surround (Fig. 4). A small spot centered on the field gives principally a center-type response, whereas an annulus with an inner diameter large enough to exclude most of the center mechanism will evoke a surround-type response (Hubel and Wiesel, 1960).

When both the center and the surround are stimulated together, the antagonistic inputs sum (in a still unspecified, though possibly algebraic way) to give the net response. As a result, the retinal ganglion cells are "tuned" to spots or lines of sizes that achieve the best balance between the opposing center and surround influences. When the entire field is uniformly covered by a large figure, the center mechanism usually dominates, but the response is weakened by the surround input (Fig. 4).

If the center mechanism is excitatory, the surround will be inhibitory as diagrammed in Fig. 3, and the cell will fire most vigorously to an incremental spot of intermediate size centered on the receptive field. On the other hand, the center mechanism can also be inhibitory, with an excitatory surround (turn Fig. 3 upside down). Such a cell is tuned to decremental stimuli in the same way that a cell with an excitatory center is turned to incremental ones. It will fire most vigorously to a decremental (i.e., black) spot of intermediate size centered on the receptive field.[1]

[1] This precise experiment has not been reported, but the result can be inferred from our studies on lateral geniculate cells and from the literature on the cat retina.

---

built around retinal receptive fields with an inhibitory center, excitatory surround (just turn the figure upside down) would respond to dark bars. Presumably a combination putting together excitatory-center fields with inhibitory-center fields could respond best to a light-dark contrast border. Try your hand.

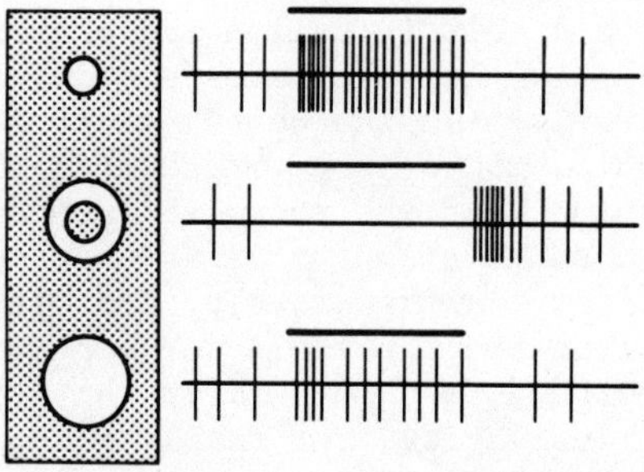

Figure 4. Idealized responses to incremental stimuli of a retinal ganglion cell with an excitatory center and an inhibitory surround. The stimulus is on during the period indicated by the horizontal bar. Top: On-response to a central spot. Middle: On-inhibition and off-excitation to an annulus. Bottom: Weak center-type (on) response to diffuse or large-spot illumination

We still do not know the exact spatial distributions of the center and the surround mechanisms within retinal receptive fields, so the diagrams of Fig. 3 should be considered only schematic. As indicated in the figure, more than one type of surround distribution could be consistent with the available data. The common feature remains the differential distribution of the excitatory and inhibitory influences, giving rise to a "spatially opponent" organization in which appropriately chosen spatial stimuli can give opposing responses.

It has been suggested that the center and the surround mechanisms of retinal receptive fields arise from separate and specific neural pathways (Dowling and Boycott, 1966). The center mechanism could be mediated by the vertical pathways of the retina from receptors to bipolars to ganglion cells. Surround influences, on the other hand, would more likely be transmitted by the horizontal pathways--the amacrine cells and possibly horizontal cells. Known cellular dimensions and sizes of receptive fields are at least consistent with this idea.

A similar concentric receptive field organization also exists in the lateral geniculate nucleus (Wiesel and Hubel, 1966) but there the effectiveness of the surround mechanism (in non-color-coded cells) in antagonizing the influence of the center is increased (Marrocco,

1972b). How this occurs is still a matter of controversy, as several different models have recently been proposed for the same phenomenon in the more intensively studied LGN of the cat (e.g., Levick et al., 1972; Maffei and Fiorentini, 1972; Singer and Creutzfeldt, 1970). The diameter of the center mechanism of retinal ganglion cells and lateral geniculate cells ranges from about 2' of arc to 2° and there is a loose positive correlation between center size and the distance of the field from the fovea (Hubel and Wiesel, 1960; Wiesel and Hubel, 1966).

One important consequence of the center-surround organization of the lateral geniculate receptive fields can be demonstrated by using large stimulus figures. Consider, for example, an excitatory-center cell, which is tested with a large white contrast figure on a gray background. If the figure is centered on the receptive field, a weak response is elicited because both the center and surround mechanisms are stimulated. But if the same figure is positioned so that the contrast border falls just to one side of the receptive field center, the surround mechanism is not stimulated as strongly, and the response is increased. This is known as border enhancement and it is exhibited by both color-coded and non-color-coded cells when they are presented with large excitatory luminence stimuli (DeValois and Pease, 1971).

It has long been thought that the excitatory-inhibitory interactions within the receptive field provide a basis for coding brightness contrast. They probably do begin that process, but most likely do not complete it, for the geniculate receptive fields are too small. Changing the luminence of the surround around a 2° stimulus figure can make the figure appear black or white by simultaneous contrast, while little affecting the firing of a geniculate cell in the center of the figure (DeValois and Pease, 1971). Cells whose receptive fields are located nearer the borders will respond to such surround shifts, and their signals must be carrying the information about the contrast of large figures. It remains for the cortex or other parts of the visual system to somehow interpret the border information as coding brightness for the whole enclosed area.

In the lateral geniculate of the squirrel monkey (Jacobs and Yolton, L968) the center and the surround mechanisms of non-color-coded fields are known to overlap, but the exact profiles, like those of retinal ganglion cells, remain to be specified. It is difficult to determine the properties of the surround because the more sensitive center mechanism contaminates surround responses under most conditions. In color-coded receptive fields of the macaque, chromatic adaptation has been used to suppress the center mechanism and show that the surround mechanism peaks at the center of the receptive field as shown in Fig. 3 and Fig. 7 (Mead, 1967; for illustration see DeValois, in press).

Receptive fields of a kind radically different from the retina are first generated in the striate cortex (Hubel and Wiesel, 1968). There many cells are maximally activated by an elongated stimulus: a bright line, a dark bar, or a contrast border between a light and a dark field. The cortical neurons are commonly classified according to a scheme devised by Hubel and Wiesel (1962, 1965) on the basis of their studies of the cat, which were later extended to the monkey. One group, called simple cells, respond best to a stimulus of a particular orientation and width in a specific position. Their receptive fields frequently have an elongated central region that is either inhibitory or excitatory, flanked by parallel regions of opposite sign. The excitatory and inhibitory influences sum to determine the output of the cell as described above. Simple cells are thought to receive inputs from lateral geniculate cells whose receptive fields are arranged along a line as depicted in Fig. 3.

Another group, called complex cells, have receptive fields not easily separated into distinct excitatory and inhibitory regions. They respond best to an elongated stimulus correctly oriented, but over a wider range of positions. Still, the optimum width of the stimulus can remain only a fraction of the width of the receptive field. Complex receptive fields are thought to be generated by the convergence of inputs from simple cells whose receptive fields are juxtaposed side-by-side (Fig. 3).

The most complicated neurons of the striate cortex, the hypercomplex cells, are selectively tuned to one more stimulus feature--the length. A line of the proper orientation evokes a stronger response as it is lengthened until the end of the line encroaches on inhibitory flanking areas, and then the firing of the cell declines or ceases. This behavior could result from the convergence of complex cells on the hypercomplex cell (Fig. 3), with a central complex cell providing an excitatory input, and cells with flanking fields of the same orientation specificity giving inhibitory inputs. At eccentricities of 1 to 4° from the fovea, simple cells have fields ranging from 1/4° X 1/4° to 1/2° X 3/4°, while the fields of complex and hypercomplex cells are 1 1/2 to 2 times as large in linear dimensions (Hubel and Wiesel, 1968). Another estimate gives field sizes of 0.1 to 0.4° near the center of the fovea and up to 0.5 to 1.0° at 3 to 6° parafoveal (Poggio, 1972).

The chief strength of this hierarchical scheme is the insight it provides into the way the cortex combines inputs to analyze visual stimuli for specific spatial features, such as width, length, and orientation. Nevertheless, it is undoubtedly oversimplified (Hoffman and Stone, 1971; Stone, 1972, and discussion following his paper) and further work will probably bring some modification.

For example, one of the pieces of evidence for the serial construction of a hierarchical arrangement was the finding that simple cells were most frequent in the layer IV of the cortex whereas complex and hypercomplex cells predominated in the other layers, which presumably receive much of their input from layer IV (Hubel and Wiesel, 1968). A recent report (Poggio, 1972) presents a somewhat different picture, with complex cells distributed almost evenly throughout the cortical layers. The general picture presented in that paper is one of less specificity for the form of the stimulus than was formerly thought, with only half the neurons of foveal striate cortex and about 65% of parafoveal neurons showing orientation selectivity.

Receptive field maps of cortical neurons have also been constructed by automated scanning of the field with a moving disc (Spinelli et al., 1970). However, it has never been demonstrated that the maps derived from this procedure are adequate predictors of the response of the cell when it is stimulated by a different shape or size of stimulus or with a different temporal pattern of presentation. Until this necessary control experiment is done, the results of such automated mapping are uninterpretable.

Wurtz (1969a) has shown that an additional distinction can be made within the hierarchy by comparing the responses of the cells to stationary and to moving stimuli. He trained awake monkeys to fixate steadily while he studied the receptive fields of neurons in the parafoveal projection area of striate cortex. In response to an appropriate stationary stimulus some neurons responded with a brief phasic discharge, and others responded with sustained firing.[2] Both "phasic" and "sustained" cells were found among neurons of all spatial categories. When tested with a moving stimulus traveling perpendicular to its long axis the phasic cells (except for those with-

---

[2] A phasic discharge begins with a burst of spikes at the onset of the stimulus but then rapidly decays back to the prestimulus firing rate, even though the stimulus remains on. A sustained discharge begins with an initial transient burst of spikes, followed by a decline in firing rate to a sustained level that is higher than the prestimulus level and is maintained for the duration of the stimulus.

In the interest of simplicity I have imposed a single terminology where three separate ones exist. What I call phasic/sustained, Wurtz labels adapting/nonadapting, Cleland designates transient/sustained, and Gouras calls phasic/tonic. I prefer the phasic/sustained choice because it cannot be confused with longer-term adaptation, the initial transient of a longer response, or the tonic retinal discharge in the absence of light, as the other labels can. Both phasic and sustained fibers of the non-color-coded variety can be identified in the monkey optic nerve (Marrocco, 1972a) so this distinction cannot be used to discriminate between color-coded and non-color-coded cells.

out orientation selectivity) responded more vigorously than they had to a stationary stimulus, with a preference for a limited range of velocities. Some of the phasic cells also showed directional selectivity, increasing their firing when the stimulus moved in one direction, and inhibiting or remaining unaffected by movement in the opposite direction. Sustained neurons, on the other hand, generally were not directionally selective and showed only small differences in responses when the stimulus was moved in opposite directions. Statistics are not available for behaving animals, but in animals paralyzed with neuromuscular blocking agents, 50% of complex parafoveal neurons are said to be directionally selective (Hubel and Wiesel, 1968; Poggio, 1972) while only 30% of foveal complex cells are directionally selective. (Poggio, 1972).

It would be interesting to know how these different responses to movement are related to the presumed serial convergence of the response hierarchy. Recent work on the cat visual system indicates that a phasic/sustained dichotomy already exists in the retina, where cells with fast-conducting axons are phasic, and those with slower conducting axons are sustained (Cleland et al., 1971). This separation is retained all the way to the visual cortex, since lateral geniculate cells of the cat can also be classed as phasic or sustained. Thus the visual cortex could easily build parallel hierarchies, one from sustained, and the other from phasic inputs. But in the monkey, no counterparts for the phasic cells that exist in the retina (Gouras, 1969) have been described for the lateral geniculate. So we are not sure whether the cortex receives phasic inputs. This is a problem for further research.

Wurtz (1969b,c) also described another effect of stimulus movement that may indicate the need for more detailed subdivisions of receptive field categories before the behavior of cortical neurons can be predicted under a wide range of stimulus conditions. He first identified neurons that were excited by a stationary or slowly moving stimulus, and then moved the same stimulus across the receptive field at a high velocity. One group of the cells continued to excite, but a second group now gave no response, and a third group was inhibited by the rapidly moving stimulus. The same transformation was ovserved when a rapid eye movement displaced the receptive field across a stationary stimulus: the response of the cortical neuron was identical as long as eye movement or stimulus movement produced the same motion of the retinal image. Thus these neurons could not reliably distinguish between motions of objects in the external world and movements of the eyes. However, when the eyes execute slow pursuit movements instead of rapid saccadic movements, some cortical cells respond slightly more to a stimulus moving slowly with respect to a background than to a stationary stimulus (Bridgeman, 1972). Since the reported effects are still very small, brain structures other than the visual cortex may be of greater importance in signaling the difference between object movement and eye movement.

Columnar Organization

In the striate cortex, cells with a common orientation preference tend to be clustered together (Hubel and Wiesel, 1968). Such a grouping is called a "column," since it is demonstrated most easily in recordings made from a series of cells while an electrode advances perpendicularly through the thickness of the cortex. Earlier studies on the cat (Hubel and Wiesel, 1963), as well as the results of oblique penetrations in the monkey, indicate that the columns in some areas are ordered in a rotational sequence as shown in Fig. 5. There may also be other less exact arrangements and variations, but the extraordinary precision displayed by the cortical architecture in this case is quite stunning. The columns appear to be only one, or a few, cells wide, which makes a whole 180° sequence small enough to receive input from the same restricted area of the retina. Each portion of the retinal image is therefore subjected to an intensive geometrical analysis by the array of cortical columns. This is presumably one reason why small regions of the retina are serviced by so much larger areas of cortex (Talbot and Marshall, 1941; Daniel and Whitteridge, 1961).

Ocular Dominance and Binocular Interaction

An independent set of columns is based on "ocular dominance," and this second set overlaps or includes the orientation columns. Ocular dominance is a rating of the relative effectiveness of the two eyes in exciting a cortical neuron. Cells dominated (i.e., more effectively driven) by the contralateral eye are clustered into separate columns from those with ipsilateral dominance (Hubel and Wiesel, 1968). Like the orientation columns, the ocular dominance columns are also shaped into parallel slabs, although somewhat larger ones. They are perpendicular to the plane of the page in Fig. 2, while the orientation columns might lie parallel to the plane of the page (Hubel and Wiesel, 1972). Presumably the alternating bands of fiber projections from the geniculate layers provide the anatomical basis for the ocular dominance columns.

The implications for binocular vision of this elaborate dominance organization still have not been clarified. It is much easier to appreciate what happens at the next stage of the visual pathway, in cortical area 18 or V2 (a part of the circumstriate belt, see Fig. 8). Here, almost half the cells give only a weak response or no response to stimulation of either eye separately, but strong responses when the two eyes are stimulated together in an appropriate manner. These were termed "binocular depth cells" by Hubel and Wiesel (1970), because some of them respond best when the images on the two retinae fall on disparate points: The receptive field in one eye is displaced from the corresponding position in the other eye at right angles to the optimal stimulus orientation.

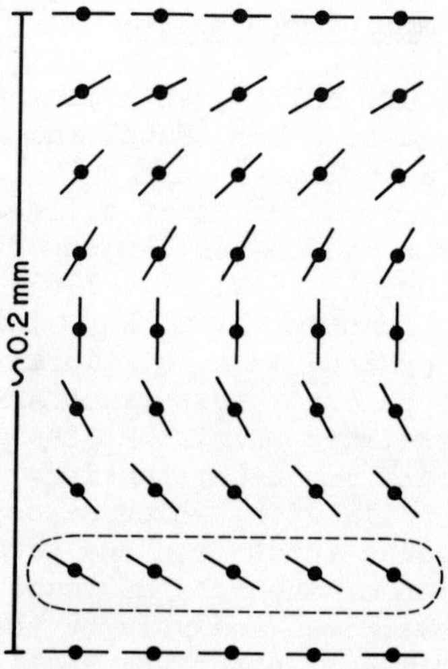

Figure 5. Idealized arrangement of orientation columns in monkey striate cortex, as viewed from the outer surface of the cortex. A column (one is outlined by the dashed line) is a set of cells with the same orientation preference, symbolized here by an appropriately tilted line. The columns are ordered in a continuous sequence according to the inclination of their optimum stimulus. This diagram is not meant to imply anything about the exact number of discriminable angles included in a 180° rotational sequence. The columns of the cortical cube in Fig. 2 might look like this if viewed from the top surface.

The horizontal component of this disparity is just the condition that should give rise to stereopsis, or binocular depth perception (Graham, 1965). Since the detection of small retinal disparities in anesthetized animals is difficult (Barlow et al., 1967; Nikara et al., 1968) it is conceivable that such cells exist in striate cortex (V1) as well as in V2. The fact that Hubel and Wiesel (1970) find more of the binocular depth cells at increasing retinal eccentricities, where the receptive field disparities should be large (Joshua and Bishop, 1970) underlines this possibility.

Again, there appear to be selective groupings of the cells into columns, this time associating those cells that represent a particular stereoscopic depth relative to the plane of fixation. As more work is done on the visual cortex, still more columns are likely to be described, and eventually this intricate jigsaw puzzle should begin to fit together.

COLOR CODING

Cones

The fact that color sensations require the activation of the cones has focused a great deal of attention on measuring the spectral absorption characteristics of these receptors. With recent refinements in spectrophotometric techniques, it is now possible, although still difficult, to obtain difference spectra from single cones of primates. These measurements indicate that there are three types of cones, with absorption peaks in the short (S), middle (M), and long-wavelength (L) portions of the spectrum. The total number of receptors whose curves have been published is still quite small—8 cones from macaque monkeys (an unspecified number from *Macaca mulatta* and the rest from *Macaca nemestrina*) and 6 cones from humans (Marks, Dobelle, and MacNichol, 1964; Brown and Wald, 1964). Even these data show a good deal of variation from receptor to receptor, which might be due to variation between subjects, variability within the populations of receptors, or measurement error. However, partial bleaching studies of large foveal populations of cones give rather similar results, and support the view that the cones of the macaque and those of man have quite similar pigments (Brown and Wald, 1963). For our purposes, the primate cones will be considered to peak near the wavelengths 445 nm (S), 535 nm (M), and 570 nm (L) (Marks, Dobelle, and MacNichol). These values agree well with other indirect estimates of the cone properties and they have been used in much of the theorizing about central color mechanisms (DeValois, in press).

Lateral Geniculate Cells

Since the most detailed studies of chromatic mechanisms have been conducted on the lateral geniculate nucleus (LGN), it is convenient to begin with a summary of that work first, and then refer back to the retina. Of the cells in the dorsal, parvocellular layers of the LGN, approximately 70-80% excite to some wavelengths of diffuse light, and inhibit to other wavelengths (DeValois et al., 1966; Wiesel and Hubel, 1966). These are called chromatically opponent cells, to distinguish them from the 20-30% of the cells that either excite to diffuse incremental stimuli of any wavelength, or inhibit to increments of all wavelengths.

The chromatically opponent cells can easily be further subdivided into one subpopulation that excites to long wavelengths and inhibits to short wavelengths, and another subpopulation of mirror image counterparts that excite to short wavelengths and inhibit to long wavelengths. However, within the group of cells that excited to long wavelengths, DeValois et al. (1966) were able to demonstrate a bimodal distribution of crosspoints, the wavelength at which the response changed from excitation to inhibition. (For example the cross-

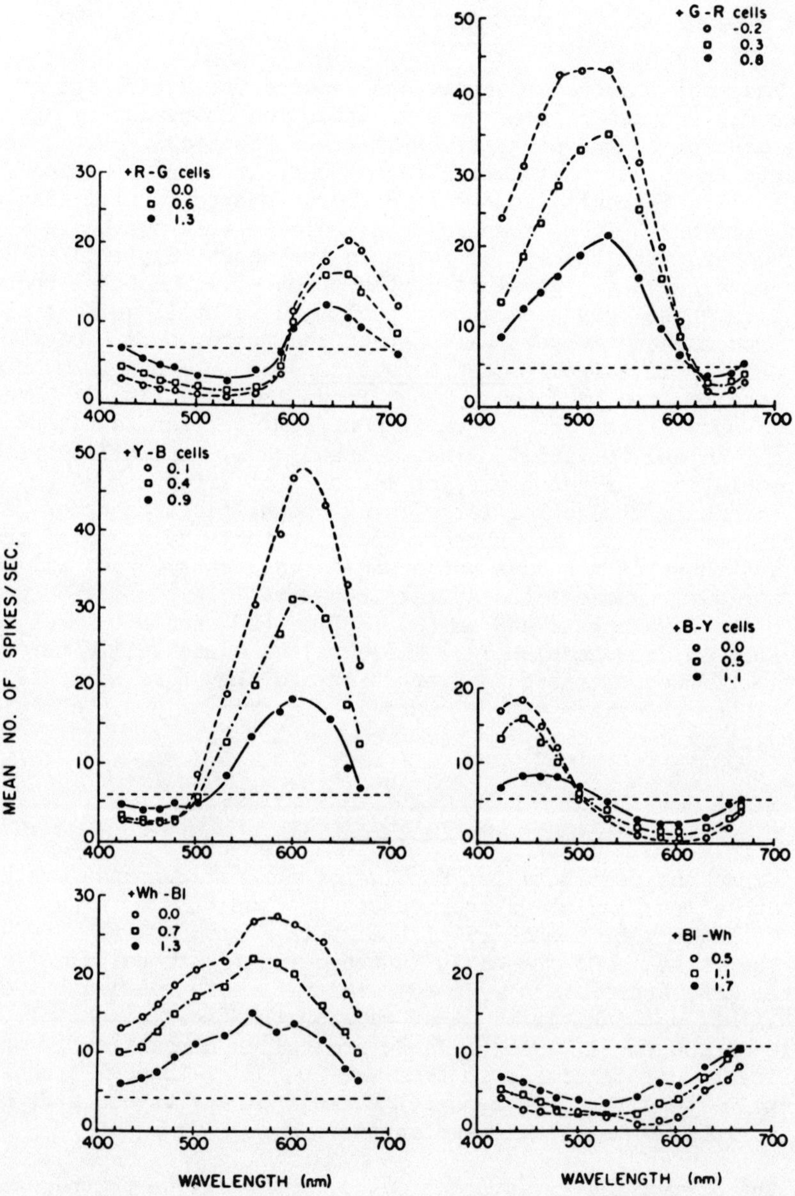

Figure 6. Average responses of the six different types of lateral geniculate nucleus cells. The average number of action potentials during a one-second presentation of the stimulus is plotted on the vertical axis versus wavelength on the horizontal axis. Responses were obtained to an equal-energy series of spatially diffuse increments of monochro-

point in the upper left panel of Fig. 6 is about 600 nm). This suggested that two different classes were represented, so they separated the long wavelength excitatory cells into two groups, one with crosspoints below 560 nm, and the other with crosspoints above 560 nm. Averaging across the cells within these two classes produced the curves shown in the upper left (+R-G cells) and middle left (+Y-B cells) panels of Fig. 6. On the basis that the situation should be symmetrical, they also separated the short-wavelength excitatory cells into two populations with crosspoints above 560 nm and below 560 nm and averaged within these populations. This gave the graphs of the upper right (+G-R cells) and middle right (+B-Y cells) panels.

The symmetry argument was justified in two ways: (1) If equal-response spectral sensitivity curves were constructed from the average data, the short wavelength peak of the +G-R cells fell at the same point as the short wavelength valley of the +R-G cells, with a similar result for the YB pair. Thus the original assumption of symmetry is borne out by the averaged data. (2) Chromatic adaptation experiments to separate the excitatory and inhibitory components (DeValois, 1965, 1970, details still not published), show that the excitatory component for the +G-R cells has the same spectral sensitivity as the inhibitory component for the +R-G cells, and the same relationship holds true for excitatory and inhibitory components of the YB pair.

This classification is also consistent with wavelength discrimination experiments, which demonstrate that the macaque, like the human (DeValois and Jacobs, 1968, 1971), performs best in the spectral regions near the crosspoints of the RG and YB cell classes. Since the slopes of the cell response curves are greatest in these regions, small shifts in wavelengths here produce the greatest change in firing, and hence the lowest thresholds for discrimination (DeValois et al., 1967).

In addition to the chromatically opponent cells, we must include in our scheme the spectrally nonopponent excitatory (+Wh-Bl) and inhibitory (+Bl-Wh) cells (Fig. 6, lowest panel). This gives a total of six chromatic classes of LGN cells. For simplicity's sake, we usually refer to them according to the excitatory response component: Left-to-right and top-to-bottom in Fig. 6 they would be red, green, yellow, blue, white and black cells. These color names are

---

matic lights at each of three intensity levels as indicated by the symbols, which denote the $\log_{10}$ attenuation from the maximum intensity available. The dotted line is the average spontaneous firing rate of the cells. Points above that line are excitatory responses, those below it are inhibitory. From DeValois (in press). See text for further details.

used deliberately instead of words with less subjective connotations, because the cell response components have shapes similar to the frequency curves humans generate when asked to assign these four colors to monochromatic flashes of light (DeValois et al., 1966).

Cone Connections

Given that we have only three types of cones as inputs, it is clear that the six classes of geniculate cells must be derived from combinations of the different cone types. The available evidence indicates that the green-red system subtracts the outputs of the L and M cones, green cells getting excitation from the M cones and inhibition from the L cones, while the red cells have the mirror image arrangement, getting excitation from the L cones and inhibition from the M cones (DeValois, 1965, 1970; Abramov, 1968). Unfortunately specifying the cone inputs to the yellow-blue system has been more difficult. It was originally thought that an opponent pairing of the L cone with the S cone should provide the basis for the basis for the yellow-blue system (DeValois, 1967, 1970; Abramov, 1968), but this is clearly impossible. The peak response of the yellow cells is at 600 nm, a longer wavelength than the peak of the L cone (570 nm). But subtracting the S cone from the L cone could not produce this peak shift (compare curves in Abramov, 1968). Therefore there must be some M cone input to the yellow-blue system. Since the curves of Fig. 6 are averages of populations of cells chosen on the basis of the crosspoint criterion one cannot distinguish with certainty between two possibilities: (1) The yellow-blue cells receive inputs from all three cone types. (2) Cells classified as yellow or blue cells on the basis of a crosspoint below 560 nm represent two distinct groups, one pairing L cones with M cones and the other pairing S cones with M cones. On the basis of other evidence (Paul Pease, personal communication; Wiesel and Hubel, 1966; Gouras, 1968), I believe the latter case is more likely to be correct.

Unlike the chromatically opponent cells, the achromatic WhBl system both sums and differences inputs from more than one cone type (Type III cells of Wiesel and Hubel, 1966). From studies on similar kinds of retinal ganglion cells we can infer that these cells should be connected to at least the L and M cones (Gouras, 1968). Within the center mechanism inputs from both cones sum, and within the surround mechanism the same two cones add. But since the center and the surround are mutually antagonistic, the spatial opponency subtracts outputs from the same two cones without generating chromatic opponency.

With this background, it will hopefully be clear to the reader why cones should not be given color names (DeValois, in press). The L cone, for example, is connected to some cells that are maximally responsive to red, but others that are not even color-coded, and still others that may be coding yellow. Color is determined by the organization of the nervous system, not the cone type; the cones are necessary, but not sufficient, for color vision.

A few of the chromatically opponent cells have triphasic response curves with two spectral areas of excitation separated by a spectral region of inhibition or vice-versa (DeValois, et al., 1966). These cells have not been well studied and might easily have three cone inputs.

Rod Inputs

No examples of cells with only rod inputs have yet been reported. Some chromatically opponent cells have both rod and cone inputs, and some have only cone inputs (Wiesel and Hubel, 1966; DeValois, 1967), even in the parafoveal regions where rods are numerous. The nonopponent white-black system also can have either pure cone inputs or both rod and cone inputs (Wiesel and Hubel). In all cases the rods seem to have the same center-surround excitatory-inhibitory arrangement that cones do.

Retinal Ganglion Cells

Rod-cone interaction has been studied more specifically at the level of the retinal ganglion cells (Gouras and Link, 1966; Gouras, 1967). Here it has been found that the rod and cone signals can be played against each other in the dark-adapted state. Whichever signal arrives at the ganglion cell first leaves a transitory refractoriness or inhibition. Under normal conditions of light adaptation, however, the cone signals have a shorter latency and will be transmitted at the expense of the rod signals. Since these experiments utilized only brief (8-12 msec) flashes of light, it is difficult to predict what the results would be with stimuli of longer duration, especially in the mesopic range.

So far, the chromatically opponent retinal ganglion cells do not appear to differ in their chromatic properties from the lateral geniculate cells in any striking way (Marrocco, 1972a). It looks as if the chromatic interactions are well established already in the retina.

Receptive Fields

My earlier discussion of receptive fields implicitly considered only their spatial properties (Fig. 3). However, we can extend that description to include spectral parameters if we specify in addition the spectral sensitivities of the center and the surround mechanisms. The two mechanisms can be studied separately by determining thresholds with small spots of light centered on the receptive field to stimulate the center mechanism selectively, or by using annuli to activate selectively the surround mechanism. The response of the surround mechanism can also be isolated by suppressing the center mechanism with either punctiform or diffuse chromatic adaptation (Wiesel and Hubel, 1966; Gouras, 1968).

In the simplest case, the achromatic or WhBl cells (Type III of Wiesel and Hubel), the center and surround have the same spectral sensitivity, which matches closely the CIE photopic luminosity curve (DeValois et al., 1966; Gouras, 1968). The only known difference between the receptive fields of the achromatic retinal ganglion cells and the corresponding LGN cells is the enhanced strength of the surround in geniculate cells (Marrocco,1972b). Since even this slight receptive field transformation has not been demonstrated in the chromatically opponent system, I will assume for present purposes that the color-coded receptive fields of retinal ganglion cells and lateral geniculate cells can be described by the same schema.

Basically, the novel feature of the color-coded receptive fields is the distinctly different spectral sensitivity of the center and the surround (Wiesel and Hubel, 1966; Gouras, 1968). The best-studied receptive fields of cells in the dorsal, parvocellular layers of the LGN have a center mechanism fed by the L cones with the surround mechanism fed by the M cones or vice-versa (Wiesel and Hubel, 1966).

Our recent experiments suggest that these receptive field types can be related to the chromatic classifications based on crosspoints (Snodderly et al., in preparation). We find that green cells, for example, usually have excitatory receptive field centers. Since they also have excitatory inputs from the M cones, they must have M cones feeding an excitatory green center mechanism with L cones feeding an inhibitory red surround. This is diagrammed in Fig. 7. Yellow cells also have excitatory centers, which must be fed by the L cones, and some of them must have inhibitory surrounds fed by the M cones.

This leaves us the cells with inhibitory centers to describe. Most red cells probably have inhibitory green centers fed by the M cones with excitatory red surrounds fed by the L cones. Some blue cells probably have inhibitory yellow centers fed by L cones with excitatory blue surrounds fed by M cones. Other blue cells most likely have excitatory blue centers fed by S cones and inhibitory surrounds fed by M cones (Wiesel and Hubel, 1966: Gouras, 1968). Perhaps there are some other minor combinations as well. This type of organization is consistent with a number of interesting observations from human visual psychophysics that will be discussed elsewhere (Snodderly et al., in preparation).

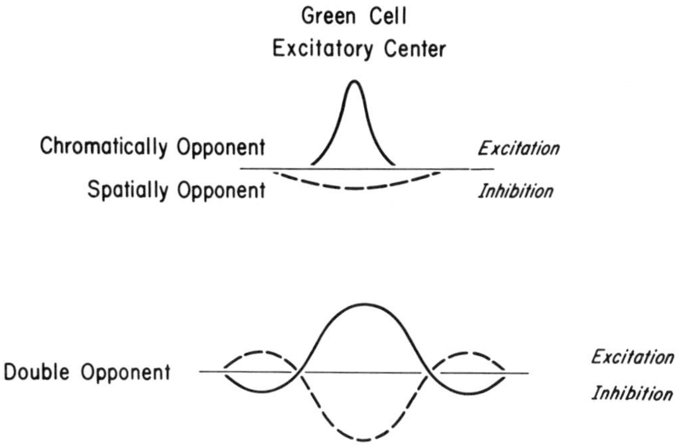

Figure 7. Examples of the color-coded receptive fields of the GR system. Upper Panel: Chromatically opponent lateral geniculate cell and retinal ganglion cell receptive fields. The solid lines are the spatial sensitivity profiles for the green mechanism and the dashed lines are the sensitivity profiles of the red mechanism.
Other conventions same as Fig. 3. The surround distribution is only schematic and may in fact have several forms.
Lower Panel. Profile through a green-red double opponent receptive field. The relative heights and detailed shapes of the various peaks are not yet known and may be rather different from this diagram.

A small fraction of the geniculate cells appear to have coextensive excitatory and inhibitory mechanisms such that the less sensitive "surround" mechanism does not have a greater spatial extent than the more sensitive "center" mechanism (Type II of Wiesel and Hubel, 1966). However, it is possible that chromatic aberration of stimuli designed to elicit the surround response has smeared the stimulus and caused an error in estimation of the extent of the center mechanism in some of these neurons.

Wiesel and Hubel (1966) also described an unusual cell type (Type IV) found only in the ventral, magnocellular layers of the geniculate, that has an excitatory center of broad spectral sensitivity and an inhibitory surround whose spectral sensitivity is shifted to longer wavelengths. The ventral layers contain, in addition, Type III, or white-black cells.

As in the case of achromatic cells, the antagonistic center-surround arrangement causes most chromatically opponent cells to be selectively tuned to spatial stimuli that produce the best balance between the excitatory and inhibitory inputs. This will depend upon the size, shape, and spectral composition of the stimulus, in a manner that can be roughly predicted from diagrams like Fig. 7. More precise predictions will only be possible when we know much more about the receptive field profiles, and particularly about the profile of the surround mechanism, which is shown only schematically in the figure. At least some geniculate cells appear to have surround mechanisms maximally sensitive in the center of the receptive field as drawn here (Mead, 1967), but others may not.

Although the chromatically opponent cells are clearly involved in color discriminations, they remain responsive to white and black contrast figures (DeValois, 1972; Gouras, 1972; Snodderly et al., in preparation). Recent data suggest that most cells with excitatory centers excite to white figures and those with inhibitory centers excite to black figures regardless of the chromatic type of the cell (Snodderly et al., in preparation). Furthermore the spatial opponency of the receptive field gives rise to maximal responses to achromatic stimuli of intermediate size in the same manner as occurs in white and black cells. Luminance flicker drives the chromatically opponent cells, too, and they give temporal tuning curves much like those of the achromatic cells (Spekreisje et al., 1971). It seems as if the chromatically opponent system can do most of the things that the nonopponent system does, and then some. This feat is accomplished at the cost of a certain ambiguity in the message, however--the nervous system cannot read off the response of a chromatically opponent cell and know whether it is being stimulated by color or by an appropriate white-black contrast figure. Presumably other cells, such as cortical neurons, must make this distinction.

Further convergence within the chromatic system in the striate cortex certainly does occur, and gives rise to new receptive field types, whose properties are only beginning to be understood. The most striking example is the double opponent cell (Hubel and Wiesel, 1968), which is not thought to be common at the geniculate level (Daw, 1972). Double opponent receptive fields are radially symmetric with both excitation and inhibition mediated by each color mechanism as shown in Fig. 7 (lower panel). Cortical neurons in layer IV apparently build up double opponent fields from geniculate cell inputs (Michael, 1972). The double opponent cells are thought to mediate color contrast effects, since they should respond best to a spot of one color surrounded by its opponent color (a green spot in a red surround for the example of Fig. 7). This is exactly the stimulus situation that gives the most saturated color sensations. Color induction in an achromatic central stimulus area caused by a colored annulus is probably also related to this receptive field organization.

Other transformations in color processing appear to be associated with the concurrent transformation in spatial properties of cortical neurons. It is likely that separate hierarchies as illustrated in Fig. 3 are constructed from different subtypes of geniculate receptive fields. For instance, several examples of simple cells with elongated excitatory regions of highest sensitivity to long wavelengths, flanked on either side by inhibitory regions more sensitive to shorter wavelengths, have been described (Hubel and Wiesel, 1968). These might well be constructed from yellow excitatory center geniculate cells. A few complex and hypercomplex cells give responses to colored stimuli over very restricted portions of the spectrum, or responses to colored stimuli and not white, and these presumably are also related to a hierachy based on the chromatically opponent cells. Nevertheless, many cortical cells with precise spatial requirements seem to be much like the achromatic geniculate cells in their disregard for color (Dow and Gouras, 1973).

As more research is conducted in this area, the percentages of cells recognized to be concerned with color processing will probably increase. Indeed, the estimates of percentages of cortical cells handling color information have already risen from about 10% (Hubel and Wiesel, 1968) to above 50% (Gouras, 1972). The fact that many cortical neurons receive clear inputs from all three cone types (Gouras, 1970) suggests that some convergence pattern employing more than one chromatic type of geniculate cell may also be operating at the striate cortex.

FEEDBACK OF CORTICAL INFLUENCE TO THE LATERAL GENICULATE

When a local region of the macaque visual cortex is cooled sufficiently to suppress activity of the cortical neurons, the responses of some of the lateral geniculate cells are modified (Hull, 1968). This indicates that a corticofugal feedback onto the LGN cells has been removed. Under these conditions, some LGN cells give increased responses to flashes of diffuse light, while others give decreased responses. Most of the neurons that altered their properties during cortical cooling were yellow or green cells, which we would expect to have excitatory receptive field centers. Either the neurons with inhibitory receptive field centers are less readily influenced by cortical connections, or perhaps their response alterations are more difficult to detect.

The anatomical pathway of the centrifugal influences is still obscure. Recent autoradiographic studies suggest the presence of direct connections from macaque visual cortex (V2) to the LGN (Hendrickson, 1972). However, other authors using degeneration techniques in conjunction with electron microscopy could find no evidence for a corticogeniculate pathway in the macaque (Campos-Ortega et al., 1970a). In the squirrel monkey, such a pathway does exist (Wong-Riley, 1972), and the corticofugal axons also originate in secondary cortical areas outside the striate cortex (Spatz et al., 1970).

BEYOND THE STRIATE CORTEX

The initial electrophysiological studies of retinal projections to the cortex of the monkey (Talbot and Marshall, 1941; Daniel and Whitteridge, 1961) provided no evidence for visual activity past the limits of the striate cortex, V1 (defined as the region containing the stria of Gennari; Duke-Elder, 1961; Polyak, 1957). But more recent work shows that there are detailed topographic projections from striate cortex to the surrounding area, the circumstriate belt or prestriate cortex (Fig. 8). The vertical meridian of the visual field is represented along the boundary between striate and circumstriate cortex, and the first projection V2 occurs in mirror-image fashion across the boundary (Cowey, 1964). This is presumably the region designated area 18 by Hubel and Wiesel in their study of binocular depth cells.

A second topographically ordered projection from striate cortex to the circumstriate belt has also been described by anatomical means (Cragg, 1969; Zeki, 1969). Called V3, it lies adjacent to the first striate projection, and is contained in large part within the folds of the lunate sulcus and the inferior occipital sulcus. The macular (i.e., central retinal) representations in V1, V2, and V3 are all adjacent, and occupy roughly the area indicated by the dashed lines in Fig. 8. Both V2 and V3 project slightly further anterior to areas within the lunate and inferior occipital sulci (V4 and V4a), as well as regions extending onto the prelunate gyrus (Zeki, 1971b; which we shall call V5) in a manner that appears to have less precise topographic specification. By the time the visual projections reach the posterior bank of the superior temporal sulcus, there is extensive convergence from V1, V2, and V3 coming from different locations within the visual field, with either a loss or a dramatic transformation of the visual map (Zeki, 1971a; see Fig. 10 for block diagram).

The properties of single neurons in the posterior bank of the superior temporal sulcus reflect this anatomical convergence, for the cells have very large receptive fields up to 21° x 16° in extent. They also seem to have lost some of the specificity that striate neurons have for stimulus form, being principally responsive to slow movement of a figure across the receptive field. Shape or size of the figure, and whether it is black or white, appear to be relatively less important for most of these neurons (Dubner and Zeki, 1971). Over 70% of the cells studied, however, were directionally selective, and the majority of these had a preferred direction with a horizontal component away from the fovea. This should remind us that true specificity along any one stimulus dimension can only be achieved by generalizing along other dimensions, and extensive convergence may be necessary for the capacity to generalize. The authors suggest that these neurons may play a role in slow, smooth-pursuit eye movements.

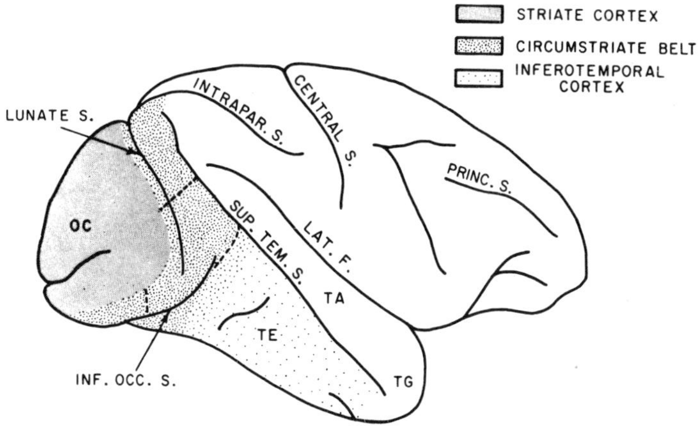

Figure 8. Lateral view of the right hemisphere of a macaque brain with visual cortical areas indicated. Symbols are: OC, striate cortex; Lunate S., lunate sulcus; Sup Temp. S., superior temporal sulcus; TE, inferotemporal cortex. The prelunate gyrus is the area between the lunate sulcus and the superior temporal sulcus. The dashed lines indicate the region of projection of the central retina or macula. (From Gross, 1973).

The little else that is known about the properties of the circumstriate belt is derived mainly from anatomical and lesion studies. The circumstriate area receives extensive inputs from the corpus callosum (Fig. 9), particularly in the areas that represent the vertical meridian or midline (Cragg, 1969; Zeki, 1970). Some of these callosal fibers also penetrate slightly across the boundary into the striate cortex (Zeki, 1970). The callosal connections provide the first opportunity in the visual pathway for combining of the contralateral and ipsilateral hemifields, and this is probably why the receptive fields of neurons in the superior temporal sulcus can cross several degrees over the midline into the ipsilateral visual hemifield (Dubner and Zeki, 1971).

Another "first" for the circumstriate cortex is its specific involvement in visual learning. Lesions that destroy the large macular projection area (Fig. 8) markedly reduce the ability of monkeys to learn and retain a discrimination between paired visual stimuli, although striate cortex lesions do not (Cowey and Gross, 1970). This may also be related to the function of the pulvinar (a thalamic nucleus near the lateral geniculate), which provides yet another input to this area (Cragg, 1969). Outputs from the circumstriate belt, though not from areas V2 and V3 (Zeki, 1971b) go to the inferotemporal cortex (Kuypers et al., 1967).

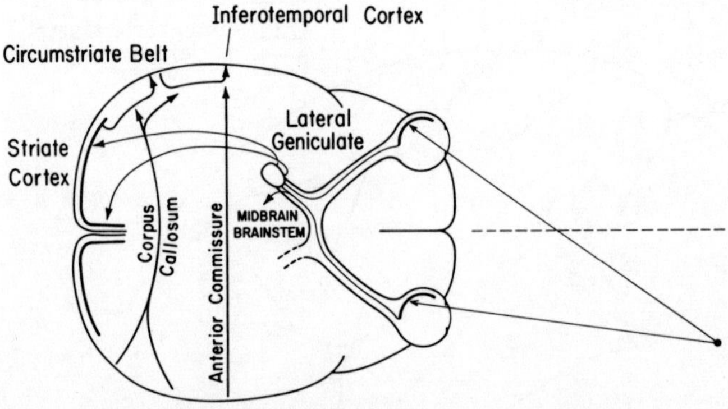

Figure 9. Geniculocortical Afferent Pathway. Afferent fibers from the lateral geniculate end in the striate cortex, which projects in turn to the circumstriate belt. Neurons in the circumstriate belt send their axons to the inferotemporal cortex. Fibers from the corpus callosum enter both the circumstriate belt and the inferotemporal cortex; the anterior commissure also projects to the inferotemporal cortex. Another thalamic nucleus that is immediately adjacent to the lateral geniculate (the pulvinar, not shown here) provides additional inputs to both the circumstriate and inferotemporal areas.

Similar but more complicated visual learning deficits occur when the inferotemporal cortex (Fig. 8) is bilaterally destoyed (For review, see Gross, 1973). Monkeys so treated do poorly on difficult visual tasks, especially when they have to learn several problems concurrently (Cowey and Gross, 1970). Yet they appear to have normal sensory functions for visual acuity, critical flicker frequency, increment thresholds, or tasks involving other sensory modalities such as audition (Gross, 1973, and in press).

The mechanisms underlying these fascinating phenomena are still not known. All indications are that the inferotemporal cortex is yet another locus for convergence of inputs (Fig. 9). It receives fibers from the opposite hemisphere via the corpus callosum (Pandya, et al., 1971; Zeki, 1970), and the anterior commissure (Fox et al., 1948; Whitlock and Nauta, 1956), and from the thalamus via the pulvinar (Cowey and Gross, 1970; Chow, 1950). As a result of the commissural connections, the neuronal receptive fields can extend well into the ipsilateral hemifield and some fields are even totally ipsilateral (Gross et al., 1972; Bender et al., 1972). A further expansion of the foveal influence occurs to such an extent that the receptive fields of all neurons studied include the fovea.

The inferotemporal region has been subdivided into two different anatomical areas on the basis of layering patterns of the cortical cell laminae (Von Bonin and Bailey, 1947; Gross et al., 1972). Area OA extends anterior to the small sulcus to the left of the letters TE on Fig. 8, and area TE extends from there to the temporal pole. Single unit recordings have been made only as far anterior as half the distance between the inferior occipital sulcus and the temporal pole.

In area OA receptive fields are large (median area 69 deg^2) and many units respond to both light and dark slits or bars moving across the receptive field. Most cells are direction selective, and some require complex stimulus figures for strong activation. Within the large receptive field, the foveal area is the most responsive, although its preferred stimulus is the same as the rest of the field. Some of these units may be the same as those identified with the posterior bank of the superior temporal sulcus (Dubner and Zeki, 1971).

In area TE receptive fields are even larger (median area 409 deg^2) and more of the cells require complex dark stimuli for adequate stimulation. One striking example was a unit that responded best to a silhouette shaped like the shadow of a monkey hand. In keeping with the greater complexity of the TE units, many of them require longer interstimulus intervals to avoid habituation of the response than do OA units. Movement of the stimulus is usually effective in eliciting a response, and about half the cells are directionally selective.

The complicated phenomenology of the inferotemporal neurons probably arises from several sources that need to be identified one by one. One surprising result is that lesions of the ventral and caudal portions of the pulvinar increase the receptive field sizes of inferotemporal neurons to greater than $64° \times 64°$ without changing the stimulus specificity of the cells (Gross, in press).

The pulvinar receives a weak input from the optic nerve (Campos-Ortega et al., 1970b), but probably also has additional inputs from other visual nuclei, possibly including the lateral geniculate (Allman et al., 1972). In the squirrel monkey the inferior pulvinar as well as the lateral geniculate nucleus project to the hippocampal region where neurons of the limbic system fire in a sustained fashion during illumination of the eye (MacLean and Creswell, 1970; MacLean et al., 1968). The detailed characteristics of the visual responses in the pathways involving the pulvinar and the limbic system are still unexplored.

RETINO-HYPOTHALAMIC PROJECTION

It has long been known that light-dark cycles affect neuroendocrine mechanisms that probably are regulated by the hypothalamus. Recent anatomical studies using autoradiography have demonstrated a projection from the optic nerve to the two suprachiasmatic nuclei that lie just above the optic chiasm in the hypothalamus on either side of the third ventricle (Hendrickson et al., 1972; Moore, 1973). No other hypothalamic nuclei appear to have such a projection in the macaque, or in other mammals that have been studied. The contralateral eye contributes the greater input, but ipsilateral fibers also terminate in each nucleus. No physiological data on the suprachiasmatic nuclei are yet available.

RETINO-TECTAL SYSTEM

A substantial population of the optic nerve fibers terminate in and around the brainstem region known as the tectum. Several small nuclei, such as the lateral terminal nucleus of the accessory optic tract, the pretectal nucleus, and the nucleus olivaris, as well as other unnamed small regions, receive optic nerve inputs (Hendrickson et al., 1970), but nothing is know about their physiology in the macaque. The major visual structures of the brainstem are the two superior colliculi, which receive inputs from both the optic nerve fibers and the striate cortex V1. The contralateral half of the visual field is mapped topographically onto each colliculus, with the central part of the hemifield represented rostrally and the periphery caudally (Cynader and Berman, 1972). As in other visual structures the foveal and perifoveal regions occupy a disproportionately large area. However the optic nerve fibers only supply peripheral inputs beyond the central 5° of the visual field, whereas the ipsilateral striate cortex projects to the entire colliculus including the area representing the central 5° of vision (Wilson and Toyne, 1970; Hendrickson, 1972). What this division of labor accomplishes is still not clear.

Superficial Layers, Superior Colliculus

Like most of the visual centers, the superior colliculus has a layered structure of alternating strata of high and low cell density. Functional differences between neurons are correlated with the layer of the colliculus in which the cells are found. Neurons in the superficial gray and optic layers (stratum griseum superficiale and stratum opticum) have visual receptive fields with properties rather different from those of the geniculostriate system. Most cells respond well to flashing spots of light with bursts of spikes at both the onset and the offset; some units also fire tonically during the light. The receptive fields as mapped with small spots are roughly circular or elliptical and they range in size from a few tenths of a degree near the fovea to nearly a quadrant of the visual field in the periphery. There are gradients in sensitivity within the receptive fields; in those near the midline the steepest spatial decline in sensitivity often occurs on the side of the receptive field nearest the center of gaze (Goldberg and Wurtz, 1972a).

Moving stimuli of almost any shape activate these cells, and the contrast of the stimulus, whether white or black, seems to make little difference (Cynader and Berman, 1972). The optimal stimulus usually is smaller than the diameter of the receptive field, however, and the response can be virtually eliminated when the stimulus encroaches upon the area around the central activating region. This suppressive surround does not evoke responses itself but reduces those elicited via the activating region. The suppressive field may also underlie the activating region and participate in the determination of the optimal stimulus size, which is said to be usually less than 5° in diameter (Schiller and Koerner, 1971).

A small fraction (10%) of the neurons in the superficial gray layer were more demanding in their stimulus requirements. They did not respond to flashed stationary stimuli but were most vigorously excited by a moving edge oriented perpendicular to the preferred direction of motion. Stimulus movement in the opposite direction elicited no response or inhibited the cell (Goldberg and Wurtz, 1972a).

Approximately 20% of the cells in the superficial layers appeared to lack the suppressive surround and responded well to large stimulus figures and diffuse light (Goldberg and Wurtz, 1972a). All cells within the area of binocular overlap are driven by both eyes from approximately corresponding points on the two retinae (Cynader and Berman, 1972). When the monkey made spontaneous eye movements in the dark, over half the cells tested were inhibited during the eye movements (Goldberg and Wurtz, 1972a).

Intermediate and Deep Layers

In animals trained to fixate during the experiment, the parameters of stimulus movement tested had no dramatic effects (Goldberg and Wurtz, 1972a). However, when the eyes were immobilized surgically or with neuromuscular blocking agents, the visual stimulus had to be moved in a jerky manner in order to activate cells of the intermediate layers (stratum griseum intermediale, stratum album intermediale; Schiller and Koerner, Schiller, and Stryker, Cynader and Berman). This type of motion is probably provided by involuntary microsaccadic eye movements in the fixating animal. Receptive fields of cells in the intermediate layers are larger than those of the superficial layers, and in anesthetized animals these neurons tended to habituate and give variable responses to repeated stimulation (Cynader and Berman).

This trend continues in the deep layers (stratum griseum profundum and stratum album profundum) where receptive fields are still larger, and neurons habituate even more dramatically. Some cells respond to non-visual stimuli in the auditory, tactile, or vibratory modalities in a manner that interacts with visual responses (Cynader and Berman). Color coding has not been identified in the macaque superior colliculus although it appears to be present in the squirrel monkey colliculus (Kadoya et al., 1971).

With awake animals — either spontaneously looking about or fixating stimuli in specified locations it is clear that many neurons in the intermediate layers fire in relation to saccadic eye movements (Schiller and Koerner; Wurtz and Goldberg, 1972a). They do not appear to discharge in relation to smooth pursuit eye movement (Wurtz and Goldberg, 1972a). Most of the movement-related cells begin a burst of firing from 20 to 30 msec before a saccadic eye movement. The saccade can be spontaneous, elicited by a visual stimulus, induced by a vestibular stimulus or even performed in total darkness so long as it is in a particular direction and of a specific length. The initial position of the eye in the orbit has little influence; the important feature is the movement accomplished by the fovea, not the absolute direction of gaze. Many of the saccade-related units also have visual receptive fields that lie in the general area toward which the saccadic eye movement shifts the fovea. This led Schiller and Koerner to propose that the collicular neurons are part of an eye-centering system that enables the animal to shift the fovea onto stimuli that initially fall upon extrafoveal retina.

Electrical stimulation of local regions of the superior colliculus produces eye movements that are consistent with the "foveation hypothesis" (Robinson, 1972; Schiller and Stryker, 1972). Brief pulse trains delivered to a particular site elicit a saccade (conjugate, with both eyes) that directs the foveas approximately toward the point in the visual field corresponding to the locus of brain stimulation. The lowest thresholds for movement were found in the lower levels of the colliculus, where many of the cells discharge only in relation to eye movements and have no visual receptive fields. This is consistent with the idea that visual neurons in the upper layers of the colliculus signal in a rather non-specific way the presence of a stimulus, and movement neurons in the lower levels output to motor areas that direct the fovea toward the stimulus for closer examination. The latency of eye movements elicited by collicular stimulation is at least 20 msec, which indicates that several other synaptic stages must be traversed before the final motor act is triggered.

An alternative hypothesis for collicular function has been advanced by Wurtz and Goldberg. By having the monkey perform saccades to different points in the visual field, they identified a set of points with which presaccadic discharges were associated. This was called the "movement field" of the cell. When the cell also had a visual receptive field, the two fields overlapped but one was frequently larger than the other and the two were not coterminous. Nevertheless, the region of maximal movement-associated activity always lay in the area with maximal visual response (Wurtz and Goldberg, 1972a). The basic doubt about the foveation hypothesis was whether the collicular neurons could control the eye movement with sufficient accuracy, since the visual and motor fields of single cells are large compared to the small errors committed by trained monkeys executing voluntary saccades.

Wurtz and Goldberg (1972b) also showed that the superior colliculus was not necessary to achieve accurate saccades. They lesioned the part of the colliculus representing the location to which the fovea was to move, or even much larger areas of the colliculus. The only deficit found after this drastic treatment was an increase of 150 to 300 msec in latency of the saccade. No decrease in accuracy was observable.

They proposed that the colliculus performs a more basic orienting function than would be required by the foveation hypothesis. Instead they suggest that the colliculus participates in shifting attention and facilitating eye and head movements toward a stimulus that appears in the visual field. This is supported by their demonstration that the response to a visual stimulus of about half the neurons in the superficial layers is enhanced if the monkey realizes he is to attend to the stimulus and make a saccade to it (Goldberg and Wurtz, 1972b). The enhancement consisted of a stronger on burst and an augmented late response to a flashed stimulus. Although some enhancement continued even if the monkey no longer shifted gaze to the stimulus, the cell's response habituated slowly over a number of trials. It was therefore not tightly locked to the eye movement per se, but to the significance of the stimulus in drawing the attention of the animal.

This is probably still not the whole story, since the deepest layers of the colliculus have not been studied thoroughly. Some cells there are different from the neurons of the upper layers in that they may code eye position in the orbit and not just movements relative to the fovea (Wurtz and Goldberg, 1972a).

OVERVIEW OF THE MACAQUE VISUAL SYSTEM

Reflected light from objects in the visual world is imaged on the retina by the dioptric power of the cornea and the lens. Macaque eyes have about the same distribution of refractive errors and the same spectral filtering characteristics as those of young humans. Ultraviolet light and some of the energy in the blue end of the spectrum are filtered out by the lens. More of the blue light is removed by the macular pigment before it reaches the receptors within 5° of the fovea.

The receptors have a high rate of metabolic activity that is dependent on the adjacent pigment epithelium. They initiate the transfer of excitation through the retina by both discrete and diffuse vertical pathways. Discrete vertical pathways go from single cones to midget bipolar cells and then to midget ganglion cells. Diffuse vertical pathways go from several cones through flat bipolar cells to diffuse ganglion cells, or from several rods through rod bipolars to diffuse ganglion cells. These vertical pathways may mediate the center mechanism of retinal ganglion cell receptive fields. Horizontal pathways through amacrine cells and possibly horizontal cells probably mediate the surround mechanisms of the receptive fields.

The greatest cone density, the largest number of midget ganglion cells, and the highest proportion of receptive fields with small centers are all associated with the fovea and the central retina. This is also the region of greatest visual acuity.

Axons of the retinal ganglion cells, from 1.5 to 1.8 million of them, leave the eye to form the optic nerve. They are distributed to several areas of the brain, including the hypothalamus, the thalamus, and the brainstem (Fig. 10).

The most complicated central pathway for visual information is the geniculocortical pathway. In the lateral geniculate, as in the retina, the center-surround organization of the receptive field causes the cells to be tuned to spots or lines of particular sizes. Most neurons excite to only one contrast (white or black) whether they are color-coded or not, and most color-coded neurons excite to only one color. No selectivity for direction of stimulus movement or for orientation has been reported.

Lateral geniculate cells project to the striate cortex, where at least half of the neurons respond best to elongated stimuli of a particular orientation. Some of the cortical complex cells also respond selectively to movement in only one direction. Concurrent with this increased form and motion specificity there appears a limited ability to generalize across position and contrast: Complex cells respond to their preferred stimulus over a range of retinal positions, and some cells are activated by both black and white stimuli.

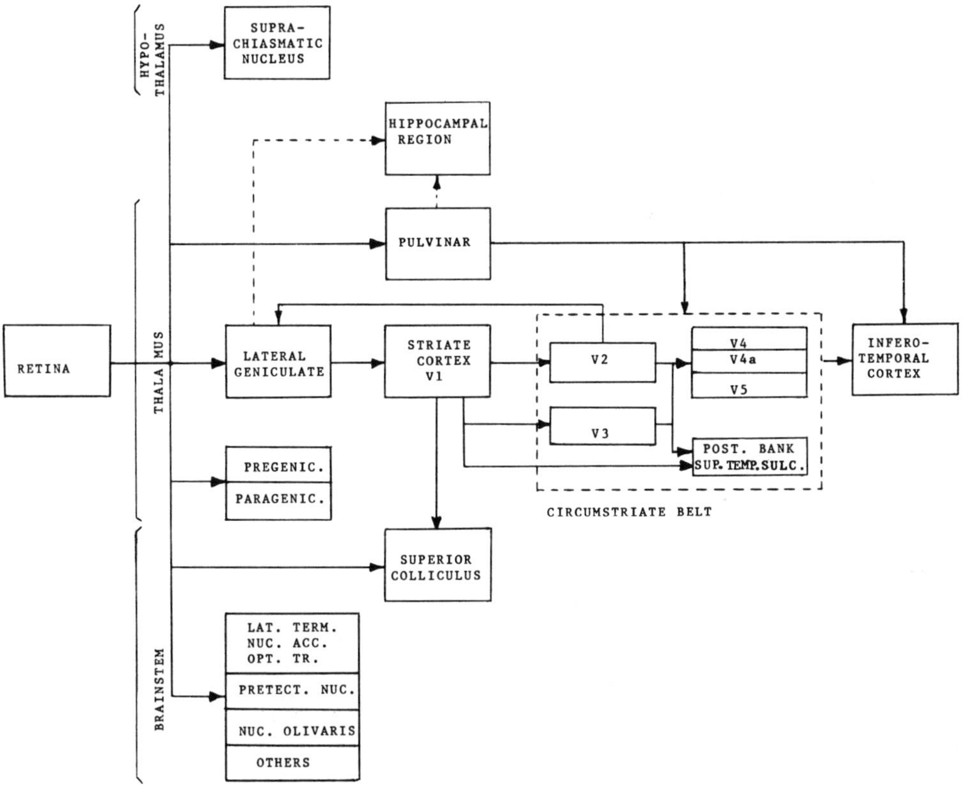

Figure 10. Major pathways of the macaque visual system. The dashed lines from the lateral geniculate and inferior pulvinar represent connections inferred from work on the squirrel monkey. Abbreviations are: Pregenic., pregeniculate nucleus; Paragenic., parageniculate nucleus; Post. Bank. Sup. Temp. Sulc., posterior bank of the superior temporal sulcus; Lat. Term. Nuc. Acc. Opt. Tr., lateral terminal nucleus of the accessory optic tract; Pretect. Nuc., pretectal nucleus; Nuc. Olivaris, nucleus olivaris.

The trend toward contrast generalization continues in the circumstriate cortex (posterior bank of the superior temporal sulcus) where direction of contrast makes little difference to many cells, and in the inferotemporal cortex, where about half the cells respond to both dark and light stimuli. Appropriate to the dual development of generalization and specificity, some neurons in the inferotemporal cortex have very complicated stimulus requirements for forms that approach the shapes of natural objects rather than simple geometrical figures such as spots and slits of light. The results of lesion studies suggest that this region of the brain is critical for normal visual learning, and it may participate in storage of visual memories.

The pattern of repeated neuronal relays with successive stages of convergence seems to be continued all the way from the retina through the striate cortex and circumstriate belt into the temporal lobe. Receptive fields that have centers of 1° or less in the lateral geniculate are combined and recombined to produce fields covering hundreds of square degrees. As a result of this process single visual neurons acquire the ability to respond selectively to some features of the retinal images while generalizing across others. This is apparently the brain's approach to form vision, an approach that is utilized also by the neural structures specifically associated with visual learning.

The other major visual pathway, to the superior colliculus, has rather different physiological properties. Here single neurons in the superficial layers respond rather non-specifically to the appearance of a wide variety of visual stimuli within their receptive fields. Neurons in the lower layers fire in association with saccadic eye movements that would be appropriate to direct the center of gaze in the general direction of the object that is stimulating visually responsive cells of the superficial layers. This seems to be a mechanism for orienting the eyes toward an object in the visual field and shifting attention to it. Such a function is nicely complementary to the pattern recognition and visual learning activity of the geniculocortical pathway.

One must guard against oversimplifying these results, however. Lesions of the striate cortex do not eliminate pattern vision, but they do raise visual thresholds (Weiskrantz, 1972). Likewise, lesions of the superior colliculus do not eliminate precisely targeted saccadic eye movements, but they do increase their latency. There are probably many functional interconnections not shown in Fig.10 that allow a given neural structure to play more than one role, particularly when the nervous system is damaged. Except for cases involving the most peripheral sensory or motor elements, I am not aware of a clear demonstration that any behavior important to the survival of the animal is exclusively mediated by a single neural structure.

COMPARISONS WITH OTHER PRIMATES

Man

A great deal of research on the primate visual system has been motivated by the desire to establish an animal model for the human visual system. In many respects the macaque monkey is a suitable choice. It has color vision and visual acuity that are very similar to human visual functions measured under identical conditions (DeValois and Jacobs, 1971; Cavonius and Robbins, 1973). Some of the other nonhuman primates have poorer visual acuity (for summary see DeValois and Jacobs, 1971) and therefore poorer acuity than man. This is important because it means that fine details apparent to humans will not be visible to many primates. On the other hand, it should be reassuring to field workers to know that their subjects are not able to use visual cues that humans cannot see. They are therefore spared the necessity for using special instrumentation to detect sensory stimuli of importance to the animal.

This is not true of the auditory sense modality. Many primates and prosimians detect auditory signals of higher frequencies than humans can hear (Stebbins, 1971). Thus a student of primate vocal behavior really requires physical measurements to determine whether his subjects are communicating in a frequency range that he himself can not detect.

Since the visual acuity of many species of primates has not been studied under controlled conditions, one can not assume that man's vision is as acute as that of all other primates. Each case must be considered separately.

There is a tendency to conclude too quickly that nonhuman primates and man have similar or identical visual mechanisms. For example, it has been said that the extraocular muscles that move the eye of the macaque have "the same anatomical distribution and organization as is found in man" (Young and Farrer, 1970). This is an oversimplification. The macaque has an additional small muscle, the accessory lateral rectus, and a slightly different locus of insertion of the inferior oblique onto orbital bone (Bast, 1933, Wojtowicz et al., 1969). The accessory lateral rectus was also omitted from a recent atlas (Szebenyi, 1969). Such details may not be trivial, because *Macaca speciosa* can make eye movements of higher velocity than those of humans (Fuchs, 1967).

With the greater development of the cerebral cortex in man, there must be important differences in central visual mechanisms. At present it is not clear how dramatic these differences really are, nor how they will be manifested in normal and abnormal visual function (see e.g., Weiskrantz, 1972).

Saimiri

Another useful example that has been documented in some detail is the comparison between the visual capabilities of the macaque and the squirrel monkey, *Saimiri*. Under identical conditions, the squirrel monkey has slightly poorer visual acuity (0.74 min of arc minimum separable) than the macaque (0.65 min), in a free viewing situation (Cowey and Ellis, 1967). Several factors may account for this. First of all, the squirrel monkey has a smaller eye, which means that the size of the retinal image is smaller. It is estimated that one degree of the visual field occupies 0.18 mm of the retina in the squrrel monkey (Rolls and Cowey, 1970), whereas one degree covers 0.2 (Wilson and Toyne, 1970) to 0.25 mm (Rolls and Cowey) of the macaque retina. Given a retinal mosaic of finite dimensions, the macaque has an optical advantage because fine details actually cover more retinal elements.

Saimiri partially compensates for its optical disadvantage by achieving a greater number of cones per square mm in the central retina, thus reducing the spacing of one part of the retinal mosaic. However, there may be other limitations on acuity that involve higher stages of the visual pathway, particularly the striate cortex. The smaller brain of *Saimiri* has only about 1/2 as much area devoted to striate cortex as the macaque (Cowey and Ellis, 1969). The central one degree of the field, nevertheless, is given about the same amount of cortex in *Saimiri* as in *Macaca*. This probably helps to maintain reasonably good acuity as long as the animals can use the fovea. But if the foveal region of striate cortex is removed from both macaques and squirrel monkeys, the squirrel monkey's visual acuity is more severely impaired (Cowey and Ellis, 1969). Presumably this reflects the fact that extrafoveal regions of the visual field are being processed by much smaller regions of striate cortex in the squirrel monkey (Rolls and Cowey, 1970). Thus, the extrafoveal "cortical mosaic" of *Saimiri* may be too coarse to maintain acute vision beyond the center of gaze.

In addition to having a smaller spatial extent of its visual projection area, *Saimiri* may have fewer synaptic relays in its geniculo-cortical pathway. Striate cortex projects twice, to areas V2 and V3, in the macaque circumstriate belt (Fig. 10), but only once in the same region in *Saimiri* (Spatz and Tigges, 1972). Learning the functional consequences of this difference may provide another insight into the way the smaller brain of *Saimiri* analyzes the same visual stimuli that larger primates deal with. The utilization of a large area of striate cortex for the foveal projection, at the expense of a reduced cortical area for the peripheral field, suggests that quite subtle mechanisms are at work. Similar kinds of compromises should be expected in the information processing conducted by the synaptic relays.

In the realm of color vision, *Macaca* and *Saimiri* provide another interesting comparison. Although behavioral tests show that the macaque's color vision is quite similar to man (DeValois and Jacobs, 1971), the squirrel monkey has less sensitivity to long wavelengths and more sensitivity to short wavelengths. The reduced sensitivity at long wavelengths may be due to a shift of the L cone pigment to a shorter peak wavelength in *Saimiri* (DeValois and Jacobs, 1968). But the S cone pigment is thought to be the same in both species. One obvious question is what role the macular pigment might play here.

If Saimiri had much less macular pigment it would tend to raise the visual sensitivity to short wavelengths.

On two basic tests of color discrimination the squirrel monkey is inferior to the macaque. It has generally poorer wavelength discrimination, which is especially bad at long wavelengths. And it requires more energy in a monochromatic light in order to distinguish it from an achromatic white. This may be due to the much smaller population of color-coded cells in the nervous system. Only 20% of the LGN cells of the squirrel monkey are chromatically opponent, as compared with 70-80% in the macaque. Furthermore most of the opponent cells belong to the yellow-blue system and relatively few red-green cells are found. Even then the green cells respond much more weakly in Saimiri than in the macaque, presumably because of the displaced L pigment that creates a closer balance between excitation and inhibition (DeValois and Jacobs, 1968).

To understand why Saimiri's visual system evolved in this way, we will need to know a great deal about the visual stimuli important for the survival of the animal. Perhaps posing the problem will stimulate the necessary investigations.

It would be interesting to extend the species comparisons to nocturnal primates and prosimians with rod-dominated retinas and poor visual acuity (Ordy and Samorajski, 1968; Hamasaki, 1967). What mechanisms do they use, how are they specialized, and what benefit is derived? More data will certainly be needed to answer these questions. I hope the present outline of the visual system of the macaque will be useful as a reference framework when considering these problems.

ACKNOWLEDGMENTS

Preparation of this paper was supported in part by PHS Research Grant EY 00899-01 and in part by the Gustavus and Louise Pfeiffer Research Foundation.

REFERENCES

1. Abramov, I. (1968) Further analysis of the Response of LGN Cells. J. Opt. Soc. Amer. 58: 574-579.

2. Allman, J. M., Kaas, J. H., Lane, R. H. and Miezin F. M. (1972). A representation of the visual field in the inferior nucleus of the pulvinar in the owl monkey (Aotus trivirgatus). Brain Res. 40: 291-302.

3. Alpern, M.,(1962) Introduction to Movements of the Eyes, in the Eye, Vol. III ed by H. Davson, Academic Press, N. Y.

4. Barlow, H.B., Blackmore, C., and Pettigrew, J. D. (1967) The Neural Mechanism of Binocular Depth Discrimination. J. Physiol. 193: 327-342.

5. Bast, T. H. (1933) The Eye and The Ear. in The Anatomy of the Rhesus Monkey., ed. by C. G. Hartman and W. L. Straus. Hafner, New York.

6. Bender, D. B., Rocha-Miranda, C. E., Gross, C. G., Volman, S. and Mishkin, M. (1972) Effects of striate lesions and commissure section on the visual response of neurons in inferotemporal cortex. Physiologist 15: 84 (Abstract only).

7. von Bonin, G., and Bailey, P. (1947) The Neocortex of *Macaca mulatta*. University of Illinois Press. Urbana, Illinois.

8. Boycott, B. B. and Dowling, J. E. (1969) Organization of the Primate Retina: Light Microscopy. Phil.Trans. Roy. Soc. Lond. B. 255: 109-184.

9. Boycott, B.B. and Kolb, H. (1973) The horizontal cells of the rhesus monkey retina. J. Comp. Neur. 148: 115-140.

10. Boynton, R. M. and Whitten, D. N. (1970) Visual Adaptation in Monkey Cones: Recordings of Late Receptor Potentials. Science 170: 1423-1426.

11. Bridgeman, B. (1972) Visual receptive fields sensitive to absolute and relative motion during tracking. Science 178:1106-1108.

12. Brindley, G. S. (1970) Physiology of the Retina and Visual Pathway. 2nd ed. Williams and Wilkins, Baltimore.

13. Brown, K. T., Watanabe, K., and Murakami, M. (1965) The Early and Late Receptor Potentials of Monkey Cones and Rods. Cold Spring Harbor Symp. Quant. Biol. 30: 457-482.

14. Brown, P. K., and Wald, G. (1963) Visual Pigments in Human and Monkey Retinas. Nature 200:37-43.

15. Brown, P. K., and Wald, G. (1964) Visual Pigments in Single Rods and Cones of the Human Retina. Science 144: 45-52.

16. Büttner, U., and Fuchs, A. F. (1973) Influence of Saccadic Eye Movements on Unit Activity in Simian Lateral Geniculate and Pregeniculate Nuclei. J. Neurophysiol. 36: 127-141.

17. Campos-Ortega, J. A., Hayhow, W. R., and de V. Clüver, P.F., (1970a) The Descending Projections from the Cortical Visual Fields of *Macaca mulatta* with Particular Reference to the Question of a Cortico-Lateral Geniculate Pathway. Brain Behav. Evol. 3: 368-414.

18. Campos-Ortega, J. A., Hayhow, W. R. and de V. Clüver, P. F. (1970b) A note on the problem of retinal projections to the inferior pulvinar nucleus of primates. Brain Res. 22: 126-130.

19. Campos-Ortega, J.A., and Hayhow, W. R. (1970c) A new lamination pattern in the lateral geniculate nucleus of primates. Brain Research 20: 335-339.

20. Campos-Ortega,, J.A. and Hayhow, W. R. (1971) A note on the connexions and possible significance of Minkowski's "intermediare Zellgruppe" in the lateral geniculate body of cercopithecid primates. Brain Research 26: 177-183.

21. Cavonius, C. R., and Robbins, D.O. (1973) Relationships between luminance and visual acuity in the rhesus monkey. J. Physiol., 232: 239-246.

22. Chow, K. L. (1950) A Retrograde Cell Degeneration Study of the Cortical Projection Field of the Pulvinar in the Monkey. J. Comp. Neurol. 93: 313-340.

23. Cohen, A. I. (1970) Further studies of the question of the patency of saccules in outer segments of vertebrate photoreceptors. Vision Res. 10: 445-453.

24. Cohen, A. I. (1972) Rods and Cones, in Handbook of Sensory Physiology VII/2, Physiology of Photoreceptor Organs. Ed by M.G.F. Fuortes, Springer, Berlin.

25. Cooper, G. F., and Robson, J. G. (1969) The Yellow Colour of the Lens of Man and Other Primates. J. Physiol. 203: 411-417.

26. Cowey, A. (1964) Projection of the Retina on to Striate and Prestriate Cortex in the Squirrel Monkey, Saimiri Sciureus J. Neurophysiol. 27: 366-393.

27. Cowey, A., and Ellis, C. M. (1967) Visual acuity of rhesus and squirrel monkeys. J. Comp. Physiol. Psych. 64: 80-84.

28. Cowey, A., and Ellis, C. M. (1969) The cortical representation of the retina in squirrel and rhesus monkeys and its relation to visual acuity. Exp. Neurol. 24: 374-385.

29. Cowey, A., and Gross, C. G. (1970) Effects of Foveal Prestriate and Inferotemporal Lesions on Visual Discrimination by Rhesus Monkeys. Exp. Brain Res. 11: 128-144.

30. Cragg, B.G. (1969) The Topography of the Afferent Projections in the Circumstriate Visual Cortex of the Monkey Studied by the Nauta Method. Vision Res. 9: 733-747.

31. Cynader, M., and Berman, N. (1972) Receptive-Field Organization of Monkey Superior Colliculus. J. Neurophysiol. 35: 187-201.

32. Daniel, P.M., and Whitteridge, D. (1961) The Representation of the Visual Field on the Cerebral Cortex in Monkeys. J. Physiol. 159: 203-221.

33. Daw, N. W. (1972) Color-Coded Cells in Goldfish, Cat, and Rhesus Monkey. Invest. Ophthalmol. 11: 411-417.

34. DeValois, R. L. (1960) Color Vision Mechanisms in the Monkey. J. Gen. Physiol. 43: 115-128.

35. DeValois, R. L. (1965) Analysis and Coding of Color Vision in the Primate Visual System. Cold Spring Harbor Symp. Quant. Biol. 30: 567-579.

36. DeValois, R. L., Abramov, I., and Jacobs, G. H., (1966). Analysis of Response Patterns of LGN Cells. J. Opt. Soc. Amer. 56: 966-977.

37. DeValois, R. L., Abramov, I., and Mead, W. R. (1967) Single Cell Analysis of Wavelength Discrimination at the Lateral Geniculate Nucleus in the Macaque. J. Neurophysiol. 30: 415-433.

38. DeValois, R. L. and Jacobs, G. H. (1968) Primate Color Vision. Science. 162: 533-540.

39. DeValois R. L. (1970) Physiological Basis of Color Vision. In Tagungsbericht Internationale Farbtagung Color 69. Stockholm.

40. DeValois, R. L. (1971) Processing of Intensity and Wavelength Information by the Visual System. Invest. Ophthalmol. 11: 417-427.

41. DeValois, R. L. and Jacobs, G. H. (1971) Vision, in Behavior of Nonhuman Primates, Vol. 3, ed. by A. M. Schrier and F. Stollnitz. Academic Press, New York.

42. DeValois, R. L., and Pease, P. L. (1971) Contours and Contrast: Responses of Monkey Lateral Geniculate Nucleus Cells to Luminance and Color Figures. Science 171: 694-696.

43. DeValois, R. L. (in press) Central Mechanisms of Color Vision. In Handbook of Sensory Physiology, Vol. VII/3, pt. A, ed. by R. Jung, Springer, Berlin.

44. Dow, B. M., and Gouras, P. (1973) Color and Spatial Specificity of Single Units in Rhesus Monkey Foveal Striate Cortex. J. Neurophysiol. 36: 79-100.

45. Dowling, J. E., and Boycott, B. B. (1966) Organization of the

Primate Retina: electron microscopy. Proc. Roy. Soc. (Lon.) B. 166: 80-111.

46. Dubner, R., and Zeki, S. M. (1971) Response Properties and Receptive Fields of Cells in an Anatomically Defined Region of the Superior Temporal Sulcus in the Monkey. Brain Research 35: 528-532.

47. Duke-Elder, S. (1961) System of Ophthalmology. Vol. II. The Anatomy of the Visual System. C. V. Mosby, St. Louis.

48. Ehinger, B., and Falck, B. (1969) Morphological and Pharmacohistochemical Characteristics of Adrenergic Retinal Neurons of Some Mammals. Albrecht von Graefes Arch. Klin. Ophthalmol. 178: 295-305.

49. Fisken, R. A., Garey, L. J., and Powell, T. P. S. (1973) Patterns of Degeneration after Intrinsic Lesions of the Visual Cortex (area 17) of the Monkey. Brain Res. 53: 208-213.

50. Fox, C. A., Fisher, R. R. and DeSalva, S. J. (1948) The Distribution of the Anterior Commissure in the Monkey (Macaca mulatta) J. Comp. Neurol. 89: 245-278.

51. Fuchs, A. F. (1967) Saccadic and Smooth Pursuit Eye Movements in the Monkey. J. Physiol. 191: 609-631.

52. Garey, L. J. (1971) A Light and Electron Microscopic Study of the Termination of the Lateral Geniculo - Cortical Pathway in the Cat and Monkey. Proc. R. Soc. Lond. B. 179: 21-40.

53. Garey, L. J. and Powell, T. P. S. (1971) An Experimental Study of the Termination of the Lateral Geniculo - Cortical Pathway in the Cat and Monkey. Proc. R. Soc. Lond. B. 179: 41-63.

54. Goldberg, M. E., and Wurtz, R. H. (1972a) Activity of Superior Colliculus in Behaving Monkey. I. Visual Receptive Fields of Single Neurons. J. Neurophysiol. 35: 542-559.

55. Goldberg, M.E., and Wurtz, R. H. (1972b) Activity of Superior Colliculus in Behaving Monkey. II. Effect of Attention on Neuronal Responses. J. Neurophysiol. 35: 560-574.

56. Goldstein, E. B. (1969) Contribution of Cones to the Early Receptor Potential in the Rhesus Monkey. Nature 222: 1273-74.

57. Gouras, P., and Link, K. (1966) Rod and Cone Interaction in Dark-Adapted Monkey Ganglion Cells. J. Physiol. 184: 499-510.

58. Gouras, P. (1967) The Effects of Light-Adaptation on Rod and Cone Receptive Field Organization of Monkey Ganglion Cells. Physiol. 192: 747-760.

59. Gouras, P., (1968) Identification of Cone Mechanisms in Monkey Ganglion Cells. J. Physiol. 199: 533-547.

60. Gouras, P. (1969) Antidromic Responses of Orthodromically Identified Ganglion Cells in Monkey Retina. J. Physiol. 204: 407-419.

61. Gouras, P. (1970) Trichromatic Mechanisms in Single Cortical Neurons. Science 168: 489-492.

62. Gouras, P. (1972) Color opponency from fovea to striate cortex. Invest. Ophthalmol. 11: 427-434.

63. Graham, C. H. (1965) Visual Space Perception, in Vision and Visual Perception, ed. by C. H. Graham, Wiley, New York.

64. Gross, C. G., Rocha-Miranda, C. E., and Bender, D. B. (1972) Visual Properties of Neurons in Inferotemporal Cortex of the Macaque, J. Neurophysiol. 35: 96-111.

65. Gross, C. G., (1973) Visual Functions of Inferotemporal Cortex, in Handbook of Sensory Physiology, Vol. 7, Part 3B, ed. by R. Jung. Springer, Berlin.

66. Gross, C. G. (in press) Inferotemporal Cortex and Vision, in Progress in Physiological Psychology, Vol. V.

67. Guillery, R. W., and Colonnier, M. (1970) Synaptic Patterns in the Dorsal Lateral Geniculate Nucleus of the Monkey. Z. Zellforsch. 103: 90-108.

68. Hagins, W. A. (1972) The Visual Process: Excitatory Mechanisms in the Primary Receptor Cells. Ann. Rev. Biophys. Bioeng. 1: 131-158.

69. Hamasaki, D. I. (1967) An anatomical and electrophysiological study of the retina of the owl monkey, *Aotes trivirgatus*. J. Comp. Neurol. 130: 163-174.

70. Hendrickson, A., Wilson, M. E., and Toyne, M. J. (1970) The distribution of optic nerve fibers in *Macaca mulatta*. Brain Res. 23: 425-427.

71. Hendrickson, A. E., Wagoner, N., and Cowan, W. M. (1972) An Autoradiographic and Electron Microscopic Study of Retino-Hypothalamic Connections. Z. Zellforsch. 135: 1-26.

72. Hendrickson, A. E. (1972) Paper presented at annual meeting, Association for Research in Vision and Ophthalmology, Sarasota, Florida.

73. Hoffman, K. P., and Stone, J. (1971) Conduction velocity of afferents to cat visual cortex: a correlation with cortical receptive field properties. Brain Res. 32: 460-466.

74. Hogan, M. J., Alvarado, J.A., and Weddell, J. E. (1971) Histology of the Human Eye. Saunders, Philadelphia.

75. Honrubia, F. M., and Elliott, J. H. (1970) Efferent Innervation of the Retina II. Morphologic study of the monkey retina. Invest. Ophthalmol. 9: 971-976.

76. Hubel, D. H, and Wiesel, T. N. (1960) Receptive Fields of Optic Nerve Fibers in the Spider Monkey. J. Physiol. 154: 572-580.

77. Hubel, D. H., and Wiesel, T. N. (1962) Receptive Fields, Binocular Interaction, and Functional Architecture in the Cat's Visual Cortex. J. Physiol. 160: 106-154.

78. Hubel, D. H., and Wiesel, T.N. (1963) Shape and Arrangement of Columns in Cat's Striate Cortex. J. Physiol. 165: 559-568.

79. Hubel, D. H., and Wiesel, T. N. (1965) Receptive Fields and Functional Architecture in Two Nonstriate Visual Areas (18 and 19) of the Cat. J. Neurophysiol. 28: 230-289.

80. Hubel, D. H., and Wiesel, T. N. (1968) Receptive Fields and Functional Architecture of Monkey Striate Cortex. J. Physiol. 195: 215-243.

81. Hubel, D. H., and Wiesel, T. N. (1969) Anatomical Demonstration of Columns in the Monkey Striate Cortex. Nature, 221: 747-750.

82. Hubel, D. H., and Wiesel, T. N. (1970) Stereoscopic Vision in Macaque Monkey. Nature 225: 41-42.

83. Hubel, D. H., and Wiesel, T.N. (1972) Laminar and Columnar Distribution of Geniculo-cortical Fibers in the Macaque Monkey. J. Comp. Neurol. 146: 421-450.

84. Hull, E. M., (1968) Corticofugal influence in the macaque lateral geniculate nucleus. Vision Res. 8: 1285-1298.

85. Jacobs, G.H. (1972) Spontaneous Activity in Visual Systems. Am. J. Optom. 49: 905-921.

86. Jacobs, G. H., and Yolton, R. L. (1968) Distribution of Excitation and Inhibition in Receptive Fields of Lateral Geniculate Neurons. Nature 217: 187-188.

87. Joshua, D. E. and Bishop, P.O. (1970) Binocular Single Vision and Depth Discrimination. Receptive Field Disparities for Central and Peripheral Vision and Binocular Interaction on Peripheral Single Units in Cat Striate Cortex. Exp. Brain Res. 10: 389-416.

88. Kaas, J. H., Guillery, R. W., and Allman, J. M. (1972) Some Principles of Organization in the Dorsal Lateral Geniculate Nucleus. Brain, Behav. Evol. 6: 253-299.

89. Kaas, J. H, Guillery, R. W., and Allman, J. M. (1973) Discontinuities in the dorsal lateral geniculate nucleus corresponding to the optic disc: a comparative study. J. Comp. Neurol. 147: 163-180.

90. Kadoya, S., Wolin, L. R., and Massopust, L.C. (1971) Collicular unit responses to monochromatic stimulation in squirrel monkey. Brain Res. 32: 251-254.

91. Kolb, H. (1970) Organization of the outer plexiform layer of the primate retina: electron microscopy of Golgi-impregnated cells. Phil. Trans. Roy. Soc. Lond. B. 258: 261-283.

92. Kroll, A. J., and Machemer, R. (1969) Experimental retinal detachment and reattachment in the rhesus monkey. Am. J. Ophthalmol. 68: 58-77.

93. Kuypers, H.G.J.M., Szwarcbart, M. K., Mishkin, M., and Rosvold H. E., (1965) Occipitotemporal corticocortical connections in the rhesus monkey. Exp. Neurol. 11: 245-262.

94. Levick, W. R., Cleland, B. G., and Dubin, M. W., (1972) Lateral Geniculate Neurons of Cat: Retinal Inputs and Physiology. Invest. Ophthalmol. 11: 302-311.

95. Lund, J. S. (1973) Organization of Neurons in the Visual Cortex, Area 17, of the Monkey (*Macaca mulatta*). J. Comp. Neurol. 147: 455-496.

96. MacLean, P. D., Yokota, T., and Kinnard, M. A. (1968) Photically sustained On-responses of units in posterior hippocampal gyrus of awake monkey. J. Neurophysiol. 31: 870-883.

97. MacLean, P.D., and Creswell, G. (1970) Anatomical connections of visual system with limbic cortex of monkey. J. Comp. Neurol. 138: 265-278.

98. Maffei, L., and Fiorentini, A. (1972) Retinogeniculate Convergence and Analysis of Contrast. J. Neurophysiol. 35: 65-72.

99. Marks, W. B., Dobelle, W. H., and MacNichol, E. F. (1964) Visual Pigments of Single Primate Cones. Science 143: 1181-1183.

100. Marrocco, R. T. (1972a) Responses of Monkey Optic Tract Fibers to Monochromatic Lights. Vision Res. 12: 1167-1174.

101. Marrocco, R. T. (1972b) Maintained Activity of Monkey Optic Tract Fibers and Lateral Geniculate Nucleus Cells. Vision Res. 12: 1175-1181.

102. Mead, W. R. (1967) Analysis of the Receptive Field Organization of Macaque Lateral Geniculate Nucleus Cells. Unpublished Ph. D. Thesis. Indiana University, Bloomington, Indiana.

103. Michael, C. R. (1971) Dual Opponent-Color Cells in the Lateral Geniculate Nucleus of the Ground Squirrel. J. Gen. Physiol. 57: 254 (Abstract only).

104. Michael, C. R. (1972) Double Opponent-Color Cells in the Primate Striate Cortex. Physiologist 15: 216.

105. Moore, R. Y. (1973) Retinohypothalamic projection in mammals: a comparative study. Brain Res. 49: 403-409.

106. Nikara, T., Bishop, P. O. and Pettigrew, J. D. (1968) Analysis of Retinal Correspondence by Studying Receptive Fields of Binocular Single Units in Cat Striate Cortex. Exp. Brain Res. 6: 353-372.

107. Norren, D. V. (1972) Macaque lens absorption in vivo. Invest. Ophthalmol. 11: 177-181.

108. Ogden, T. E., and Miller, R. F. (1966) Studies of the Optic Nerve of the Rhesus Monkey: Nerve Fiber Spectrum and Physiological Properties. Vision Res. 6: 485-506.

109. Ordy, J. M., and Samorajski, T. (1968) Visual acuity and ERG-CFF in relation to the morphologic organization of the retina among diurnal and nocturnal primates. Vision Res. 8: 1205-1225.

110. Pandya, D. N., Karol, E. A., and Heilbronn, D. (1971) The topographical distribution of the interhemispheric projections in the corpus callosum of the rhesus monkey. Brain Res. 32: 31-43.

111. Pirenne, M. H. (1967) Vision and the Eye, 2nd. ed. Chapman and Hall, London.

112. Poggio, G. F. (1972) Spatial properties of neurons in striate cortex of unanesthetized macaque monkey. Invest. Ophthalmol. 11: 368-377.

113. Polyak, S. (1957) The Vertebrate Visual System, University of Chicago Press, Chicago.

114. Potts, A. M., Hodges, D., Shelman, C.B., Fritz, K.J., Levy, N.S., and Mangnall, Y. (1972a) Morphology of the primate optic nerve I. Method and total fiber count. Invest. Ophthalmol. 11: 980-988.

115. Potts, A. M., Hodges, D., Shelman, C. B., Fritz, K. J., Levy, N. S., and Mangnall, Y. (1972b) Morphology of the primate optic nerve II. Total fiber size distribution and fiber density distribution. Invest. Ophthalmol. 11: 989-1003.

116. Potts, A. M, Hodges, D., Shelman, C. B., Fritz, K. J., Levy, N. S., and Mangnall, Y. (1972c) Morphology of the primate optic nerve III. Fiber characteristics of the foveal outflow. Invest. Ophthalmol. 11: 1004-1016.

117. Robinson, D. A. (1972) Eye Movements Evoked by Collicular Stimulation in the Alert Monkey. Vision Res. 12:1795-1808.

118. Rolls, E. T., and Cowey, A. (1970) Topography of the Retina and Striate Cortex and its Relationship to Visual Acuity in Rhesus Monkeys and Squirrel Monkeys. Exp. Brain Res. 10: 298-310.

119. Sanderson, K. J., Bishop, P.O., and Darian-Smith, I. (1971) The Properties of the Binocular Receptive Fields of Lateral Geniculate Neurons. Exp. Brain Res. 13: 178-207.

120. Schiller, P. H., and Koerner, F. (1971) Discharge Characteristics of Single Units in Superior Colliculus of the Alert Rhesus Monkey. J. Neurophysiol. 34: 920-936.

121. Schiller, P.H., and Stryker, M. (1972) Single-Unit Recording and Stimulation in Superior Colliculus of the Alert Rhesus Monkey. J. Neurophysiol. 35: 915-924.

122. Singer, W., and Creutzfeldt, O.D. (1970) Reciprocal Lateral Inhibition of On-and Off-Center Neurons in the Lateral Geniculate Body of the Cat. Exp. Brain. Res. 10: 311-330.

123. Snodderly, D. M. (1973) Extracellular Single Unit Recording, in Methods in Physiological Psychology, Vol. 1. Recording of Bioelectric Activity, ed. by R. F. Thompson and M. M. Patterson. Academic Press, New York.

124. Spatz, W. B., and Tigges, J. (1972) Species difference between Old World and New World monkeys in the organization of the striate-prestriate association. Brain Res. 43: 591-594.

125. Spekreijse, H., van Norren, D., and van den Berg, T.J.T.P. (1971) Flicker Responses in Monkey Lateral Geniculate Nucleus and Human Perception of Flicker. Proc. Nat. Acad. Sci. USA 68: 2802-2805.

126. Spinelli, D. N., Pribram, K.H., and Bridgeman, B. (1970) Visual Receptive Field Organization of Single Units in the Visual Cortex of Monkey. Intern. J. Neuroscience 1: 67-74.

127. Stebbins, W. C. (1971) Hearing. in Behavior of Nonhuman Primates. ed by Schrier, A. M., and Stollnitz, F., Academic Press, N.Y.

128. Stone, J. (1972) Morphology and Physiology of the geniculocortical synapse in the cat: The question of parallel input to the striate cortex. Invest. Ophthalmol. 11: 338-346.

129. Szebenyi, E.S. Atlas of Macaca mulatta. Fairleigh Dickinson University Press. (1969) 307 pp.

130. Talbot, S. A. and Marshall, W. H. (1941) Physiological Studies on Neural Mechanisms of Visual Localization and Discrimination. Amer. J. Ophthalmol. 24: 1255-1264.

131. Toyoda, J., Nosaki, H., and Tomita, T. (1969) Light-induced resistance changes in single photoreceptors of Necturus and Gekko. Vision Res. 9: 453-463.

132. Van Buren, J. M. (1963) The Retinal Ganglion Cell Layer. Thomas, Springfield, Ill.

133. Weiskrantz, L. (1972) Behavioral analysis of the monkey's visual nervous system. Proc. R. Soc. Lond. B. 182: 427-455.

134. Whitlock, D. G., and Nauta, W.J.H. (1956) Subcortical projections from the temporal neocortex in Macaca mulatta. J. Comp. Neurol. 106: 183-212.

135. Whitten, D. N., and Boynton, R. M. (1973) The time courses of late receptor potentials from monkey cones and rods. Vision Res. 13: 107-135.

136. Wiesel, T.N., and Hubel, D. H. Spatial and Chromatic Interactions in the Lateral Geniculate Body of the Rhesus Monkey, J. Neurophysiol. 29: 1115-1156 (1966)

137. Wilson, M. E., and Cragg, B. G. (1967) Projections from the lateral geniculate nucleus in the cat and monkey. J. Anat. 101: 677-692.

138. Wilson, M. E., and Toyne, M. J. (1970) Retino-Tectal and Cortico-Tectal Projections in Macaca Mulatta. Brain Res. 24: 395-406.

139. Wojtowicz, Z., Sadowski, T., and Kurek, D. (1969) The external ocular muscles in Macacus rhesus. Folia Morph. (Warszawa) 28: 235-240.

140. Wong-Riley, M.T.T. (1972) Changes in the dorsal lateral geniculate nucleus of the sqirrel monkey after unilateral ablation of the visual cortex. J. Comp. Neurol. 146: 519-548.

141. Wurtz, R. H. (1969a) Visual Receptive Fields of Striate Cortex Neurons in Awake Monkeys. J. Neurophysiol. 32: 727-742.

142. Wurtz, R. H. (1969b) Response of Striate Cortex Neurons to Stimuli During Rapid Eye Movements in the Monkey. J. Neurophysiol. 32: 975-986.

143. Wurtz, R. H. (1969c) Comparison of Effects of Eye Movements and Stimulus Movements on Striate Cortex Neurons of the Monkey. J. Neurophysiol. 32: 987-994.

144. Wurtz, R. H., and Goldberg, M. E. (1972a) Activity of Superior Colliculus in Behaving Monkey. III. Cells Discharging Before Eye Movements. J. Neurophysiol. 35: 575-586.

145. Wurtz, R. H., and Goldberg, M. E. (1972b) Activity of Superior Colliculus in Behaving Monkey. IV. Effects of Lesions on Eye Movements. J. Neurophysiol. 35: 587-596.

146. Young, F. A. (1963) The effect of restricted visual space on the refractive error of the young monkey eye. Invest. Ophthalmol. 2: 571-577.

147. Young, F. A. (1964) The distribution of refractive errors in monkeys. Exp. Eye. Res. 3: 230-238.

148. Young, F. A., and Farrer, D. N. (1970) Visual Similarities of Nonhuman and Human Primates. In Medical Primatology, ed. by E. I. Goldsmith and J. Moor-Jankowski. Karger, Basel.

149. Young, F. A., Leary, G. A., and Farrer, D. N. (1971) Four Years of Annual Studies of Chimpanzee Vision. Am. J. Optom. 48:407-416.

150. Young, R. W. (1971a) The renewal of rod and cone outer segments in the rhesus monkey. J. Cell. Biol. 49: 303-318.

151. Young, R. W. (1971b) Shedding of Discs from Rod Outer Segments in the Rhesus Monkey. J. Ultrastruct. Res. 34: 190-203.

152. Young, R. W. (1971c) An hypothesis to account for a basis distinction between rods and cones. Vision Res. 11: 1-5.

153. Zeki, S. M. (1969) Representation of Central Visual Fields in Prestriate Cortex of Monkey. Brain Research 14: 271-291.

154. Zeki, S. M. (1970) Interhemispheric Connections of Prestriate Cortex in Monkey. Brain Research 19: 63-75.

155. Zeki, S. M., (1971a) Convergent input from the striate cortex (area 17) to the cortex of the superior temporal sulcus in the rhesus monkey. Brain Research 28: 338-340.

156. Zeki, S. M. (1971b) Cortical Projections from Two Prestriate Areas in the Monkey. Brain Research 34: 19-35.

THE STUDY OF CHROMOSOMES

B.Chiarelli
Institute of Anthropology
University of Turin
Turin, Italy

THE IMPORTANCE OF CHROMOSOMES IN TAXONOMIC AND PHYLOGENETIC STUDY

In many different animal groups, recently acquired knowledge on the number and morphology of chromosomes has furnished important data for a reconstruction of their phyletic evolution and has provided one more criterion for taxonomic organization (IO).

The knowledge of chromosomes is important for the study of phyletic evolution and taxonomy because these structures are the direct carriers of genetic information which has a stable pattern of organization on the chromosomes. For each species of Eukaryota the structural affinity between homologous chromosomes is consistently controlled by their pairing at meiosis in each individual. Incomplete chromosome pairing at meiosis infact suggests the presence of structural changes in the organization of genetic information which is linearly distributed along the chromosomes. If such structure changes are not restricted and limited, they are no longer compatible with the functional organization of genetic information. The result of this is the inability of homologous chromosomes to pair at meiosis and consequently the failure of the meiotic process, making it impossible to complete the formation of gametes (Fig.I).

In all organisms, homologous chromosomes pairing at meiosis plays the role of a filter, through which only a functionally patterned genetic system can pass. Furthermore in native populations meiosis represents a barrier which prevents exchange between diverging genetic systems. It has therefore to be considered as a basic mechanism for separation in diversifying populations into new species.

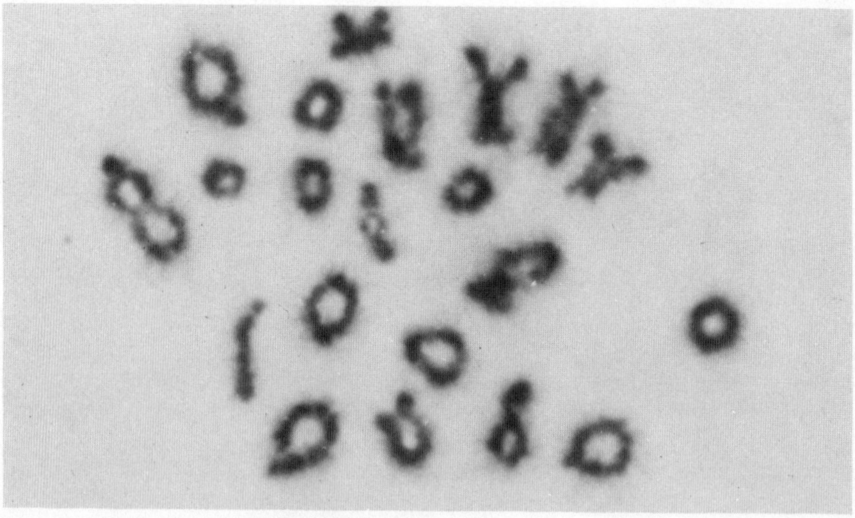

Fig.I. Meiotic chromosomes: a diakinesis of *Macaca mulatta*.

The constancy of the features of the karyotype for each species (which we can appreciate at the level of each chromosome) is therefore a consequence of the meiotic filter. On the other hand, the existing difference between karyotypes of different species is due to the existing variation in the somatic chromosomes of the germinal cell line which could filter through the meiotic sieve and establish themselves if having some advantage.

By studying and hypothesizing the possible steps by which the chromosomes of two related species differentiated, one can construct a phyletic line connecting them and establish their taxonomic affinities.

However, before one can establish chromosome identity between individuals of different species, it is important to be sure of their homologous genetic content. The only control we can have at the moment is the observation of chromosome behaviour during meiosis (especially diakinesis) in F_I individuals derived from cross-breeding between two related species. In cytotaxonomic research therefore, hybridological data deserves close consideration.

HOW CHROMOSOMES MAY CHANGE IN NUMBER AND IN MORPHOLOGY

Coming back to the possible steps by which the chromosomes of two related species differentiated, we know now a number of mechanisms capable of altering the karyotype of individuals or of groups of individuals to the point of isolating them reproductively from other representatives of their species. Among numerical variation, polyploidy has certainly played a very important role in the evolution of vegetal species and of many lower animals. In the case of mammals, and therefore of the Primates, this mechanism may be disregarded because of the difficulty interposed by the presence of well differentiated sex chromosomes. Single chromosomes may be increased in number by meiotic or mitotic nondisjunction. Another theoretical means of increasing the number of chromosomes is the possibility of some of them reduplicating asynchronously in a cell. Such a duplication, if occurring during the first mitotic division of the germinal line could have some importance in the evolution of the species. Another mechanism which can lead to an increase in the number of chromosomes, without any variation in the genetic material, is the transverse misdivision of the centromeres. In this case the region of the centromere divides transversely rather than longitudinally during mitosis, so that the chromosome divides into two chromosomes, each with a functioning centromere.

The chromosome number can be not only increased but also reduced. The mechanism which leads to a reduction in the number of chromosomes without causing severe damage to the genome is that of centric fusion. This occurs between chromosomes having the centromere in the terminal position (acrocentric) and consists in the fusion of these terminal regions.in two chromosomes. In other words, the phenomenon is the opposite of "misdivision" of the centromere.

The reduction of the number of chromosomes by means of centric fusion has been demonstrated in the most diverse organisms. The number of chromosomes undoubtedly has evolutionary importance, and yet it is difficult to determine the reason. Probably it corresponds to particular requisites of an ecological nature. In fact it is possible that a low chromosome number could represent an advantage for a very specialized animal. Such a condition would create blocks of genes which would tend to be grouped together during meiosis and thus reduce the possibility of harmful deviations in the descendants.

Chromosomes may vary not only in number but also in their morphology. There are a number of ways in which these variations may come into being. In figure 2. these chromosome changes are visualized and their possibility of survival evaluated.

Several mechanisms may intervene at the same time or one mechanism may have several effects in the differentiation of the karyotype of a species, and therefore it is not always easy to distinguish the action of one of these.

THE KARYOTYPE OF PRIMATES

Chromosomal variations, concomitant with special ecological isolation and with the succession of generations, have certainly played an important role in the diaspora of the primate species and in human evolution. Similarities between the chromosome complements of diffe-

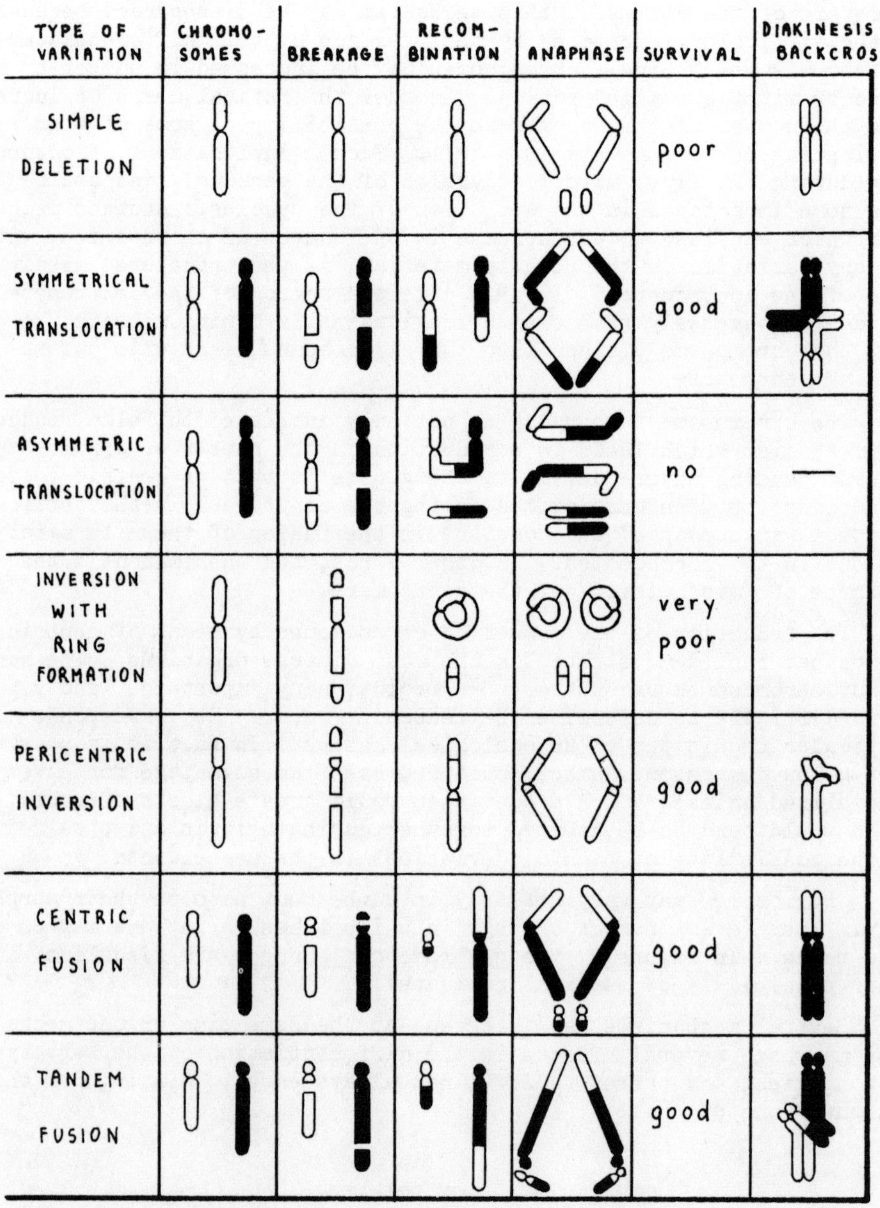
TYPE OF VARIATION
CHROMOSOMES
BREAKAGE
RECOMBINATION
ANAPHASE
SURVIVAL
DIAKINESIS IN BACKCROSS
SIMPLE DELETION
poor
SYMMETRICAL TRANSLOCATION
good
ASYMMETRIC TRANSLOCATION
no
INVERSION WITH RING FORMATION
very poor
PERICENTRIC INVERSION
good
CENTRIC FUSION
good
TANDEM FUSION
good

Fig.2. Graphical representation of the mechanisms of chromosomal variations in number and in morphology.

rent but related species can be taken as indication of a common origin while the differences make practicable a rough reconstruction of the stages by which they differentiated. In this manner although inspection of the karyotype of the ancestor of living species is unfeasible, the study of the karyotype of the latter has provided many examples of the stages through which karyotypes have passed during their evolution.

It is essential, however, to point out that a likeness between the chromosomes of related species does not necessarily imply that they are homologous. Only the study of the meiotic chromosomes of interspecific hybrids will permit the establishment of possible homologies between apparently similar chromosome complements.

Despite many limitations, the data so far collected on Primates chromosomes can reliably be used for taxonomic and phylogenetic purposes and will certainly serve as a basis for the new available techniques of chromosome banding or identification of DNA fractions on the chromosomes.

THE CHROMOSOMES OF THE PROSIMIANS

The group of Prosimiae, so heterogeneous for their remote evolutionary history and for their geographic isolation, seems particularly suitable for a karyological approach to their phylogeny.

The phylogenetic organization of the species currently assigned to this group has been in fact the object of continuous discussion, notably with respect to the classificatory order at a supra-familial level.

A synthesis of the available information for the chromosomes of the species which have been studied to date is presented in Table I. Data on the DNA content per nucleus is reported in Table 2.
From this table the following can really be seen:

I. The lack of any karyological information for many important genera: Dendrogale, Ptilocercus.
2. The incompleteness of data for many of other genera in which only poor and preliminary information are available (Propithecus, Microcebus, Cheirogaleus, Phaner, Lepilemur, Daubentonia, Avahi and Indri)
3. The variability existing in the number of chromosomes in all the genera more extensively studied as Galago and Lemur.

Disregarding, at the moment, this intergeneric chromosomal variation mentioned in point 3., which will be discussed later, a more general question may be asked: can the phylogenetic relationship between families be established upon variations in the chromosome number between genera belonging to each family in this group of Primates?

Much information on the diploid number of somatic chromosomes overlooks one of the most frequently easily occurring chromosome mutations, namely : centric fusion. This difficulty can partially be

Table I. Numerical data on the chromosomes of Tupaioidea, Lorisoidea, Lemuroidea and Tarsioidea. (For references on single data see J. Human Evol. 2(4): 279-281.)

Taxa*	2*n*	S–M	A	X	Y
1. Tupaiidae					
1.1. *Tupaia glis*	60–62	14–12	44–48	M–S	A
1.2. *T. montana*	52–68			M	A
1.5. *T. minor*	66				
3.1. *Urogale everetti*	44				
2. Lorisidae					
1.1. *Loris tardigradus*	62	34–38	26–22	S	S–A
2.1. *Nycticebus coucang*	50–52	48		S	S
2.2. *N. pygmaeus*	50				
3.1. *Arctocebus calabarensis*	52	50		S	
4.1. *Perodicticus potto*	62	24	36	S	A
3. Galagidae					
1.1. *Galago senegalensis*	36–38	22–24–30	14–12–6	S	S–A
1.2. *G. crassicaudatus*	62	6–30	54–30	S	A–S
1.3. *G. alleni*	40				
1.4. *G. demidovii*	58				
4. Lemuridae					
1.1. *Microcebus murinus*	66		64	A	A
2.1. *Cheirogaleus major*	66		64	A	A
2.2. *C. medius*	66		64	S–M	?
3.1. *Phaner furcifer*	48				
4.1. *Hapalemur griseus*	54–58	10–6	42–50	A	A
4.2. *H. simus*	60	4	54	M	A
5.1. *Lemur catta*	56	10–14	44–50	S–A	A
5.2. *L. variegatus*	46	18	26	S	A
5.3. *L. macaco*	44–48–52–58–60	20–16–4	22–30–52–54	A	A
5.5. *L. mongoz*	58–60	4	52–54	A	A
6.1. *Lepilemur mustelinus*	22–38				
5. Indridae					
1.1. *Propithecus diadema*	48				
1.2. *P. verreauxi*	48				
2.1. *Avahi laniger*	64				
3.1. *Indri indri*	44				
6. Daubentoniidae					
1.1. *Daubentonia madagascariensis*	30				
7. Tarsiidae					
1.2. *Tarsius syrichta*	80				
1.3. *T. bancanus*	80	14	66		

* The code numbers for the species are the ones used in A. B. Chiarelli (1972). *Taxonomic Atlas of Living Primates*. London: Academic Press.

Table 2. Nuclear DNA content and area of Primate lymphocytes (from M.G. Manfredi-Romanini, 1972)

Species	DNA ± S.D. (arbitrary units)	DNA (μg)
Lemur catta	12.18±0.11	5.91
Galago senegalensis	13.97±0.09	6.78
Tarsius syrichta	18.95±0.77	9.19
Cebuella pygmaea	12.31±0.12	5.97
Cebus albifrons	14.07±0.13	6.83
Saimiri sciureus	11.66±0.13	5.65
Alouatta palliata	12.82±0.70	6.22
Ateles geoffroy	11.48±0.12	5.57
Cercopithecus aethiops	10.42±0.35	5.05
Cercopithecus cephus	12.49±0.57	6.06
Cercocebus galeritus	16.72±0.29	8.11
Macaca mulatta	11.11±0.26	5.39
Papio hamadryas	12.48±0.48	6.05
Colobus polykomos	12.75±0.13	6.18
Nasalis larvatus	15.27±0.19	7.41
Hylobates lar	10.35±0.16	5.02
Symphalangus syndactylus	10.54±0.22	5.11
Pongo pygmaeus	14.50±8.40	7.03
Pan troglodytes	13.60±4.60	6.60
Gorilla gorilla	12.62±7.30	6.12
Homo sapiens	12.36±0.11	6.00

S.D. = standard deviation

overcome by resorting to Matthey's fundamental number (N.F.) since it takes into consideration solely the number of arms. The use of the fundamental number is moreover based on the assumption that centric fusion (or misdivision of the centromere) is one of the more successful mutations to pass the sieve of meiosis. Such mutation in fact would not interfere with the organization of the genetic information of the chromosomes. The only change which would result would be the reduction or increase in the random distribution of the genetic information in the offspring (+)

The reduction or increase of chromosome units certainly presents an advantage to an organism; reducing or increasing the potential variability in a population. Moreover, a relation between chromosome morphology and the chiasma frequently exists, although for the moment being we do not have enough information to define the exact relationship.

These type of data are shown in the fourth column of Table 3.

The karyological data elaborated in this way lend themselves more safely to analysis and yield more reliable results on taxonomic grounds.

Matthey's fundamental number is 70 to 84 in Tupaioidea; 87 to I02 in Lorisidae; 6I to 94 in Galagidae; 62 to 70 in Lemuridae, 94 in Tarsidae and 54 in Daubentoniidae.

No problems exists with regard to the taxonomic, and hence phylogenetic, separation between Tupaioidea with respect to Lorisoidea and Lemuroidea. However, difficulties do arise for the last two superfamilies, concerning which more controversial opinions exist. A ready explanation is provided by the data on the fundamental number available to us. Lorisidae can, in fact, be sharply distinguished from Lemuridae because their fundamental numbers are different and are in no way superimposable (87-I02, 64-70).

The taxonomic position of the Galagidae which for their fundamental number represents in some way a bridge between the Lemuridae and Lorisidae (having a F.N. varying between 6I and 94) is particularly interesting.

The striking large variation in the diploid chromosome in the two Galago species previously studied (Galago senegalensis with 2n=38 and G. crassicaudatus with 2n=62) was an attractive data to envisage the possibility of a polyploid mechanism in the origin of this variation. However, recent finding :

a) of chromosomal polymorphism in the chromosome number of Galago senegalensis due to a mechanism of centric fusion (2I)

(+) The calculation of the fundamental number (N.F.) is made by counting metacentric and submetacentric as 2 and acrocentric and subacrocentric as I in a general karyotype.

b) of a diploid chromosomal number of 58 in *Galago demidovii* and of 40 in *G. alleni* (I)
c) and finally the establishing of identical DNA content in the two *Galago* species (7.54±0.I7 in *G. senegalensis* and 7.26±0.09 in *G. crassicaudatus* in a.u.) by Manfredi Romanini *et al.* (I8)

definitely eliminated the hypothesis of polyploids. These data, moreover lend specific support to a closer taxonomic relation of the Galagidae to the Lorisidae and open the field to an extensive speculation on the adaptive advantages for such a chromosomal polymorphism in the Galagidae.

Tab. 3. Karyological information on Prosimiae

Genera	Known/Studied	2n	Fundamental number		
Tupaia	II/6	52-60-62-66-69	70-72-74-76	70	
Dendrogale	2/0				
Urogale	I/I	44	84-80		
Ptilocercus	I/0			84	
Loris	I/I	62	93-IOI	87	
Nycticebus	2/I	50	I00		
Arctocebus	I/I/	52	I02		
Perodicticus	I/I	62	87	I02	
Galago	6/2	38-62	6I-64-69-75-94	6I	94
Microcebus	2/I/	66	68	62	
Cheirogaleus	2/I	66	68		
Phaner	I/0	48	62		
Hapalemur	2/2	54-58	64		
Lemur	6/6/	44-46-48-52-56-58-60	62-64-66-70		
Lepilemur	I/0	22-38	42-36	70	
Propithecus	2/I	48			
Avahi	I/0	64			
Indri	I/0	44			
Daubentonia	I/0	30	54	54	
Tarsius	3/2	80	94	94	

The fact that also in the genus Lemur (see Tab. 3) exists an extensive polymorphism due to a centric fusion mechanism with the production of karyological subspecies as it has been recently underlined by Rumpler et al. (I9) enormously increases the interest in the adaptative advantage of these variations and in such a peculiar mechanism of speciation.

THE CHROMOSOMES OF THE NEW WORLD MONKEYS

The diploid number of chromosomes of the platyrrhine monkeys studied to date varies from 2n=20 to 2n=62 (see Tab.4)

The various species studied within the family Callithricidae present a natural uniformity in the chromosome number (from 44 to 46) and morphology. Therefore they represent a particularly closeknit group of species. The variations which have led to the diversification of the number of chromosomes can all be brought back to mechanisms of centric fusion.

The family Cebidae includes ten genera, six of which have been studied in detail.

In general, the relationship between the total number of chromosomes and the number of acrocentric chromosomes is valid for the group of the Cebidae as well. This fact demonstrates that the Robertsonian mechanism for the reduction of the chromosome number has played an important role in the evolution and differentiation of the Cebidae.

Furthermore it seems that there exists in this group a direct relationship between the number of chromosomes and the degree of specialization to arboreal life. In fact the less specialized species, like those belonging to the genus Cebus, have in general a greater number of chromosomes and especially of acrocentric chromosomes, while those with a smaller number of chromosomes and few acrocentric ones, as for example Ateles, are better adapted to life in the forest.

Callimico goeldi presents 48 chromosomes.

In general the chromosomes of Cebidae differ greatly in number and morphology from the chromosomes of the Callithricidae. Callimico goeldi provides an example of a karyotype having characteristics intermediate between the two groups.

The taxonomic position of Callimico is still much discussed and there are doubts as to whether it should be classified among the Callithricidae or Cebidae. Its limbs and overall appearance would tend to classify it as a Saguinus, but its dental formula would place it among the Cebidae. From a karyological point of view Callimico, in general; appears to be more similar to a Callithrix although still presenting many resemblances to the karyotypes of Cebus, Callicebus, Aotes and Cacajao.

An important aspect to which the study of chromosomes will probably contribute a solution is the origin of the South American monkeys.

Tab. 4. Numerical data on the chromosomes of the Platyrrhine primates.
(For references on single data see Chiarelli I973)

Taxa	2*n*	M	S	A	X	Y
Callithricidae						
Callithrix chrysoleuca	46	4	26	14	S	S
C. jacchus	46	4	28	12	S	A
C. argentata	44	4	28	10	S	M
C. humeralifer	44	4	28	10	S	A
Cebuella pygmaea	44	4	28	10	S	A
Saguinus oedipus	46	4	26	14	S	M
S. fuscicollis	46	4	26	14	S	M
S. nigricollis	46	4	26	14	S	M
S. mystax	46	4	26	14	S	M
S. leucopus	46	–	–	-	–	–
S. tamarin	46	–	–	-	–	–
Leontideus rosalia	46	4	28	12	S	M
Callimiconidae						
Callimico goeldii	48	4	24	18	S	A
Cebidae						
Alouatta seniculus	44	4	12	26	A	S
A. villosa	53 (?)	–	–	–	–	–
A. caraya	52	4	16	30	S	A
Aotus trivirgatus	54 (52)	4	16	32	S	S
Cebus albifrons	54	4	16	32	S	A
C. capucinus	54	4	14	34	S	A
C. apella	54	4	20	28	A	A
Callicebus moloch	46	4	16	24	S	A,M
Callicebus torquatus	20	–	10	10	–	–
Cacajao rubicundus	46	4	16	24	S	–
Pithecia pithecia	46	–	–	–	–	–
Saimiri sciureus	44	4	26	12	S	A
Saimiri madeirae	44	4	28	10	S	A
Saimiri boliviensis	44	4	28	10	S	A
Lagothrix ubericola	62	–	–	–	–	–
Ateles arachnoides	34	–	–	–	–	–
A. paniscus	34	–	30	2	S	A
A. belzebuth	34	–	30	2	S	A
A. geoffroyi	34	–	30	2	S	A,S

According to the traditional theory, the South American Monkeys originated from a group of existing North American Prosiminae (Omomydae) whose representatives migrated from North to South America. However, recent reevaluation of the theory of the Continental Drift makes it possible to postulate a direct migration of some early stock of Primates directly from Africa to South America as late as the end of Eocene.

THE CHROMOSOMES OF THE OLD WORLD MONKEYS

The number of chromosomes in the Old World Primates varies from 2n=42 to 2n=72 (Tab.5)

All the species of the genus *Macaca*, *Papio*, *Theropithecus* and *Cercocebus* present a diploid number of chromosomes equal to 42. The differences between the karyotypes of the different species are of little importance and can be reconciled with structural rearrangements of the inversion or translocation types.

In all the species of the genera *Macaca*, *Papio*, *Cercocebus* and *Theropithecus* the X chromosome appears to be of average size with the centromere more or less in the middle. The Y chromosome is very small and, when it is possible to distinguish the centromere, it seems to be metacentric.

A common characteristic to all these species, moreover, is that of having one chromosome of medium size with an achromatic region on one arm surmounted by a linear satellite.

This common feature of chromosomes lends support to the theory that these four genera of the family Cercopithecidae had a common origin. Later geographical and ecological isolation would have led to the differentiation of the karyotypes and to the formation of groups of individuals reproductively isolated from the others, who were to become the ancestors of so many species. If a comparative analysis is carried out on the morphology of each single chromosome, it is found that a closer similarity exists between the different species of these genera. The resemblance is closer between both *Macaca* and *Papio* than between *Macaca* and *Cercocebus*. A classification based on these likenesses corresponds perfectly to the system based on general external morphology.

These karyological similarities are thus to some extent validated by the idioplasmatic continuity still evident among the various species of this genus. It is noteworthy that hybridological data bring to light a remarkable frequency of chance hybrids between individuals of different species of this group, whereas this does not happen among the others (2).

The species of the genus *Cercopithecus* (among which *Erythrocebus* must also be included) present a variable number of chromosomes ranging between 48 to 72 and several different chromosome numbers may

Table 5. Numerical data on the chromosomes of the Old World Primates. (For references on single data see J. Human Evol. 2(4): 297-300.)

Taxa*	2*n*	S–M	A	X	Y
1.1. Macaca					
1. *M. silenus*	42	40	—	S	S
2. *M. nigra*	42	40	—	S	S
3. *M. sylvana*	42	40	—	S	S
4. *M. arctoides*	42	40	—	S	S
5. *M. maura*	42	40	—	S	S
6. *M. sinica*	42	40	—	S	S
7. *M. radiata*	42	40	—	S	S
8. *M. cyclopis*	42	40	—	S	S
9. *M. mulatta*	42	40	—	S	S
10. *M. fuscata*	42	40	—	S	S
11. *M. nemestrina*	42	40	—	S	S
12. *M. fascicularis*	42	40	—	S	S
13. *M. assamensis*	42	40	—	S	S
1.2. Papio					
1. *P. hamadryas*	42	40	—	S	S
2. *P. ursinus*	42	40	—	S	S
3. *P. anubis*	42	40	—	S	S
4. *P. cynocephalus*	42	40	—	S	S
5. *P. papio*	42	40	—	S	S
6. *P. sphinx*	42	40	—	S	S
7. *P. lecuophaeus*	42	40	—	S	S
1.3. Theropithecus					
1. *T. gelada*	42	40	—	S	S
1.4. Cercocebus					
1. *C. galeritus*	42	40	—	S	S
3. *C. torquatus*	42	40	—	S	S
4. *C. aterrimus*	42	40	—	S	S
5. *C. albigena*	42	40	—	S	S
1.5. Cercopithecus					
1. *C. aethiops*	60	34	24	S	A
2. *C. cynosuros*	60	36	22	S	A
3. *C. sabaeus*	60				
4. *C. cephus*	6o	46	18	S	A
5. *C. diana*	58	42	14	S	A
6. *C. l'hoesti*	58–60	48	14–16	S	A
7. *C. preussi*	66				
9. *C. mona*	66–68	46	18–20	S	S, A
10. *C. campbelli*	66–68	46	18–20	S	S, A
13. *C. denti*	66				
14. *C. petaurista*	66				
15. *C. neglectus*	58–62	44–46	12–14	S	S
16. *C. nictitans*	66–70	46	18–22	S	A
17. *C. ascanius*	66				
20. *C. mitis*	72	44–52	22	S	A, S
21. *C. nigroviridis*	60	36	22	S	A
22. *C. talapoin*	54	38	14	S	A
23. *C. hamlyni*	64	50	12	S	—
1.6. Erytrocebus					
1. *E. patas*	54	36	16	S	A, S
2.1. Presbytis					
1. *P. entellus*	44	40	2	S	A
2. *P. senex*	44	40	2	S	A
10. *P. obscurus*	44	40	2	S	A
2.2. Pygathrix					
1. *P. nemaeus*	44	42	—	S	A
2.5. Nasalis					
1. *N. larvatus*	48	46	—	S	?
2.6. Colobus					
1. *C. polykomos*	44	42	—	S	?
4. *C. badius*	44	42			
5. *C. kirkii*	44	42	—	S	A
3.1. Hylobates					
1. *H. lar*	44	42	—	S	S
2. *H. agilis*	44	42	—	S	S
3. *H. moloch*	44	42	—	S	S
4. *H. concolor*	52	44	6	S	A
5. *H. hoolock*	44	42	—	S	S
3.2. Symphalangus					
1. *S. syndactylus*	50	46	2	S	S
4.1. Pongo					
1. *P. pygmaeus*	48	26	20	S	S
4.2. Pan					
1. *P. troglodytes*	48	34	12	S	A
2. *P. paniscus*	48	34	12	S	A
4.3. Gorilla					
1. *G. gorilla*	48	30	16	S	S
5.1. Homo					
1. *H. sapiens*	46	34	10	S	A

* The code numbers for the species are the ones used in A. B. Chiarelli (1972). *Taxonomic Atlas of Living Primates.* London: Academic Press.

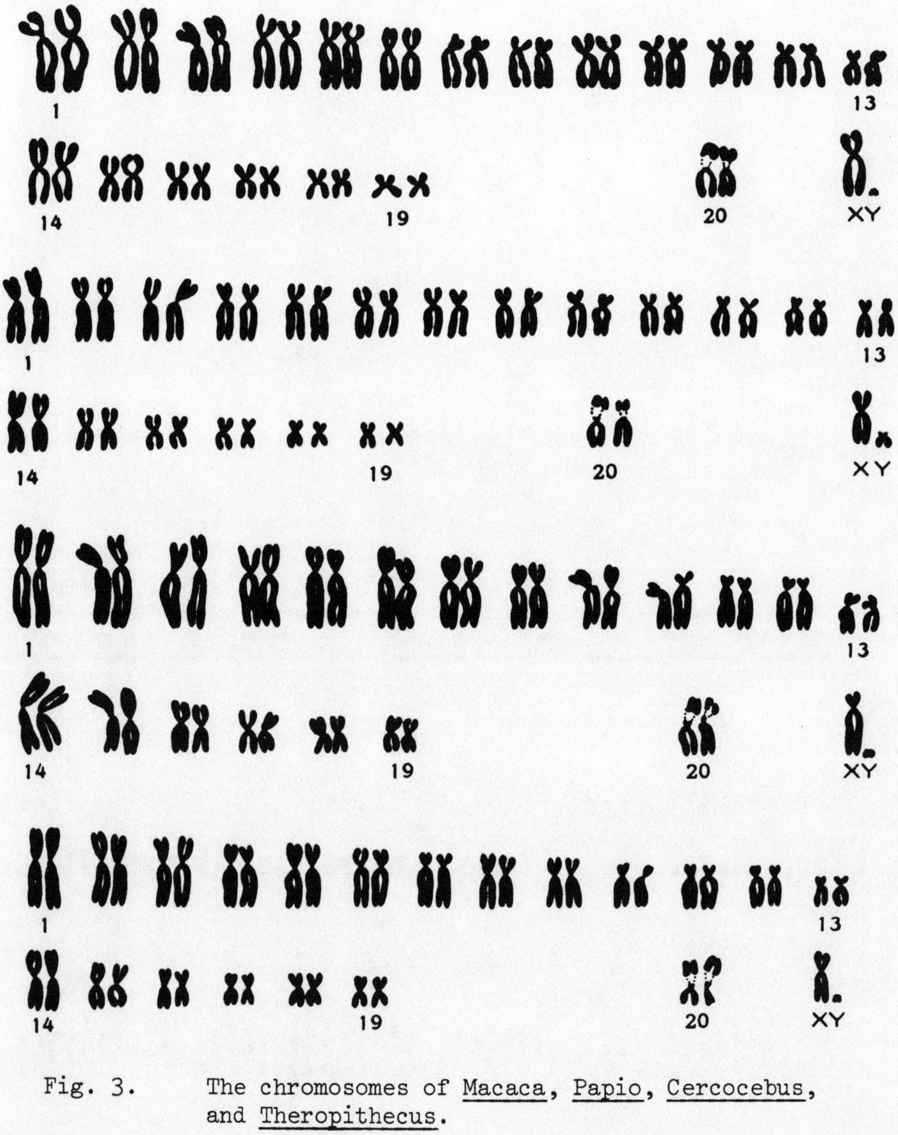

Fig. 3. The chromosomes of <u>Macaca</u>, <u>Papio</u>, <u>Cercocebus</u>, and <u>Theropithecus</u>.

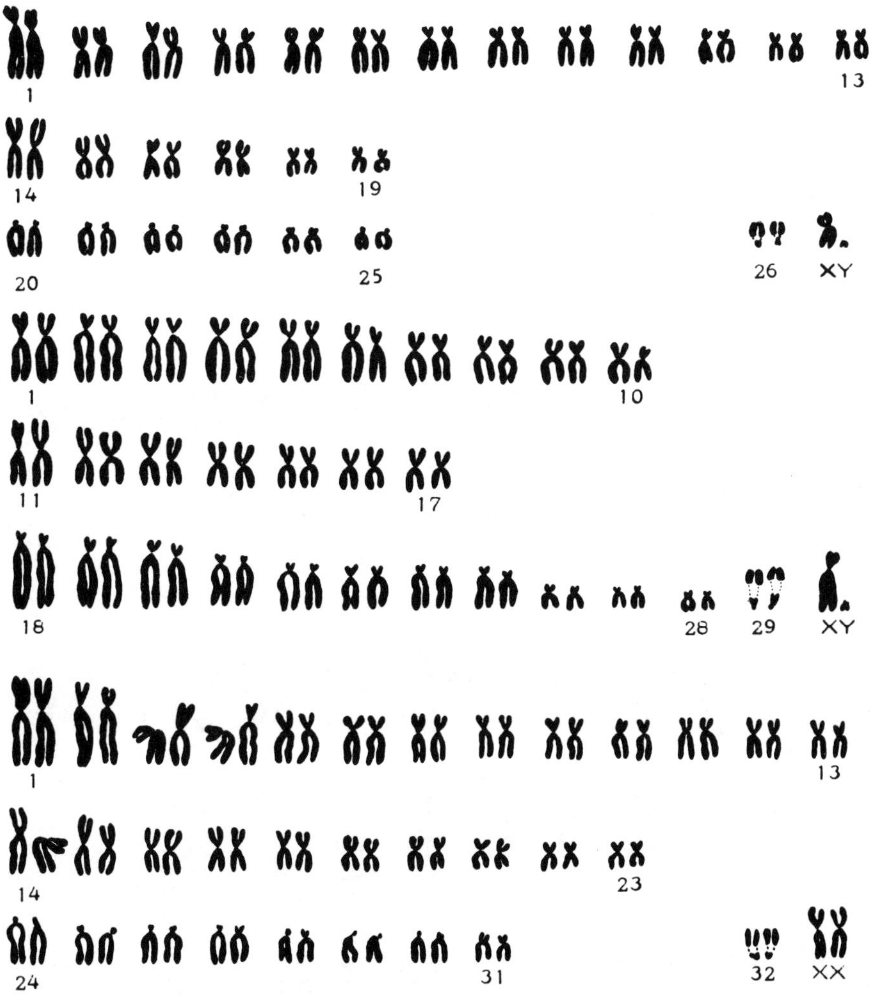

Fig.4. The chromosomes of three different species of the genus *Cercopithecus*.

sometimes be found in the same species. The origin of such a variation in the chromosome number among the species of a single genus is still mysterious and different hypotheses have been proposed for its explanation. Researches are now in progress to find the possible mechanism of such a variation. Chromosome morphology is not of help for taxonomic studies, metacentric, submetacentric and acrocentric chromosomes being present in various proportions.

No data are available for the genera Rhinopithecus and Simias; species which are now extremely rare, but which could furnish important karyological information.

Pygathrix nemaeus has 44 chromosomes as 44 is a common characteristic of Colobus polykomos, Presbytis (P. obscurus). They moreover present a large number of morphologically similar chromosomes.

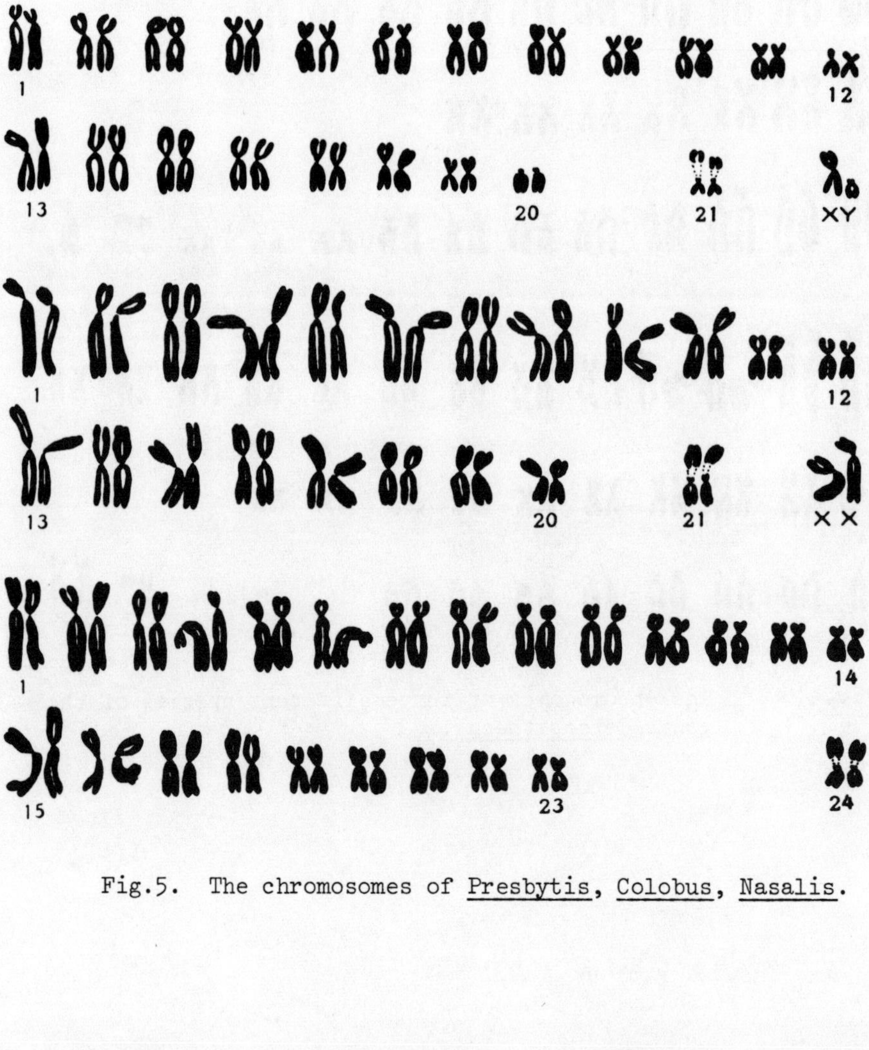

Fig.5. The chromosomes of Presbytis, Colobus, Nasalis.

The genus Nasalis has a diploid number of 2n=48 chromosomes, the morphology of these being similar to that of the chromosomes of the genera Colobus, Presbytis and Pygathrix.

The species of the genus Hylobates studied so far present diploid numbers of 2n=44, 50 and 52 chromosomes. Their chromosomes, however, appear to have many morphological characteristics in common with, or identical, to those of the species Colobus polykomos, Presbytis obscurus, Pygathrix nemaeus and Nasalis larvatus.

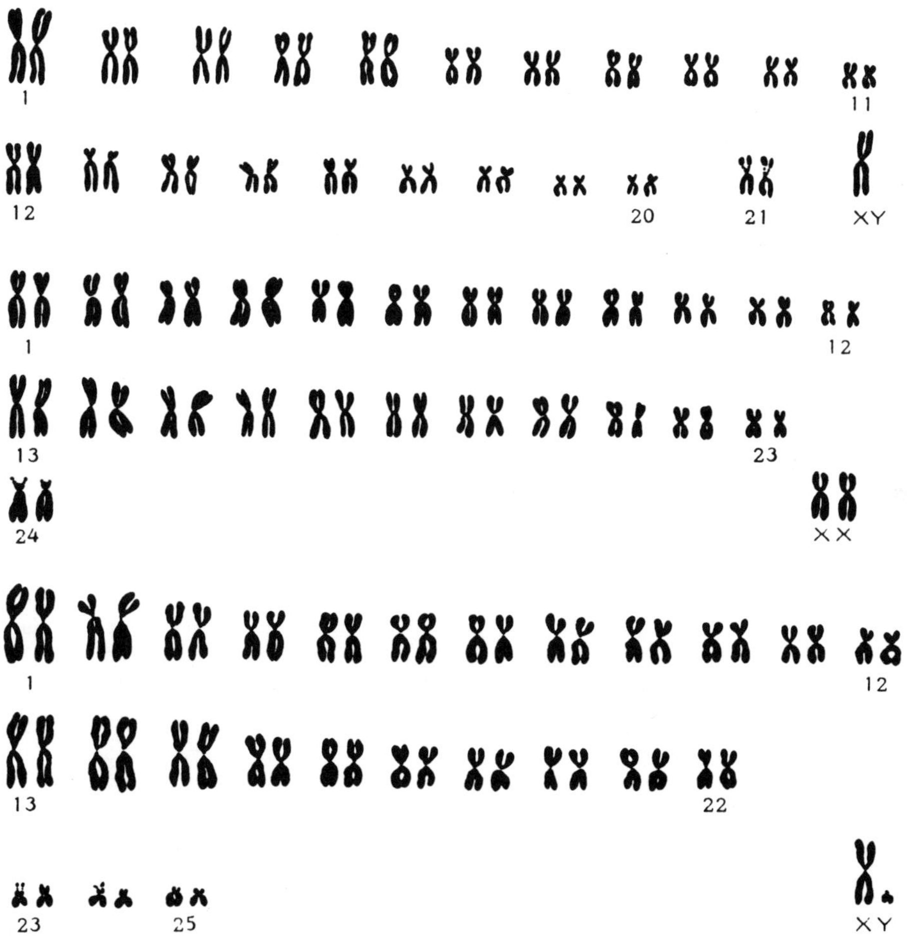

Fig.6. The chromosomes of Hylobates, Symphalangus and Nomascus.

These data and others which will be given later, disagree with the traditional taxonomic organization which allocates the Colobinae among the Cercopithecidae, and the Hylobatinae among the Hominoidae.

THE CHROMOSOMES OF THE ANTHROPOID APES AND THE ORIGIN OF THE HUMAN KARYOTYPE

The chromosome number and morphology of the true anthropoid apes (Orang, Gorilla and Chimpanzee) and Man are now known in detail.

The chromosome number of the three species of living anthropoids is 48. Man has 46 chromosomes.

The possible homologies of these structures between the Apes and Man can be tested in different ways. Strict morphological comparison can give an approximative idea of the possible gross variations (i.e. different number of metacentrics, submetacentrics or acrocentrics, different size etc).

The present approach developed by different authors (II, I3), consists in the direct comparison of the chromosomes after treatment which produce bands (quinacrine, trypsin, denaturating agents etc.).

In general it appears that chromosomes I, 3, II, I2, I4 and the X of the chimpanzee and gorilla possess banding patterns similar to the corresponding human chromosomes. Chromosomes 6, 8, IO and I3 of the chimpanzee also have a banding patterns, particularly in the long arms, closely resembling the human chromosome of the same number.

Morphologically the chimpanzee chromosome 2 appears similar to the human one, but the banding patterns are distinctly different. The same is for chromosomes 4, 5, I5, I6, I7, I8, I9 and for the Y. They seem to have no apparent human counterpart. It is possible that some structural change might have occurred in these chromosomes during the divergence of Man from Chimpanzee line.

The main difference, however, between the karyotype of Man and the Apes remain in the number. A centric fusion might account for a reduction of the chromosome number from 48 to 46 in an Ancestor of Man, which realized an important distinction between the karyotypes of the Anthropoid Apes and Man as we have hypothetized in early I962 (3).

The similarity of the banding patterns of two chromosomes of the G group in the chimpanzee to the human n.2 led some authors to support that this chromosome in Man originate from the centric fusion of two chromosomes of the G group type in the Ancestry of Man.

Can we speculate when and how could such a transformation did take place, thereby differentiating the human karyotype so sharply from that of the anthropoid apes?

Such mutations are not a rare event. Study of the karyotype of present-day human populations has uncovered, as we have noted before, cases of centric fusions between two acrocentric chromosomes

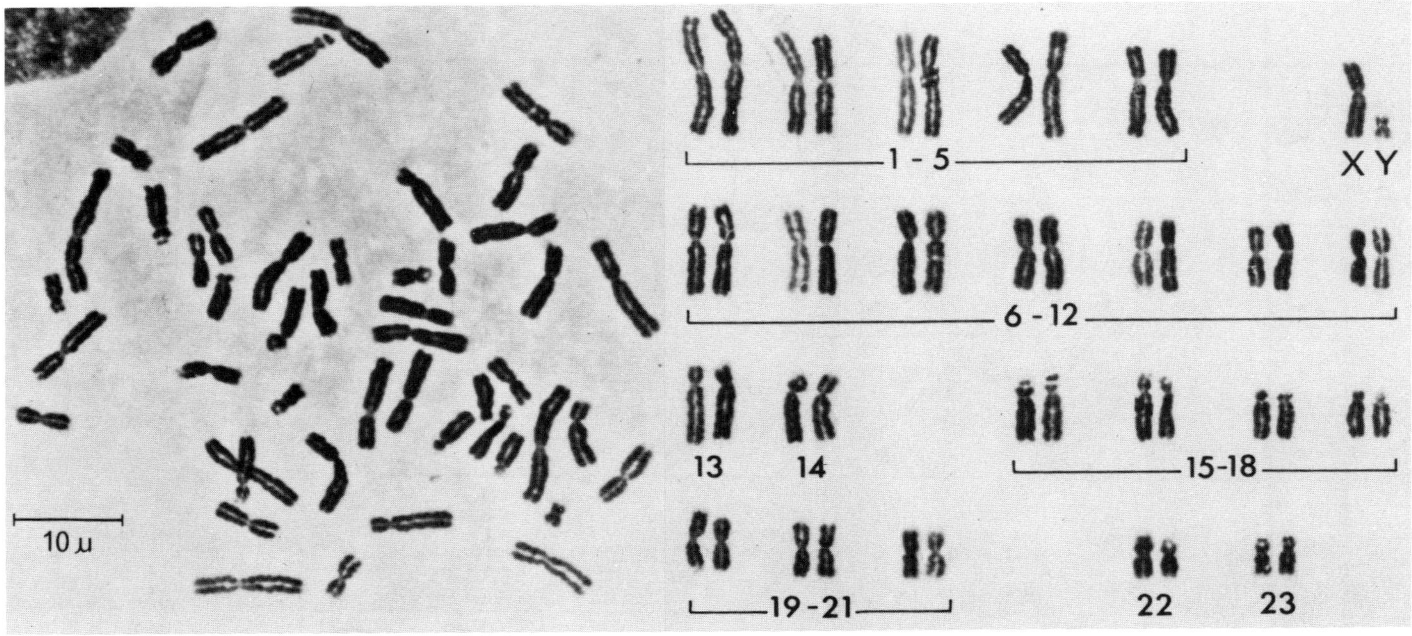

Fig.7. The chromosomes of a Chimpanzee and their possible karyotype.

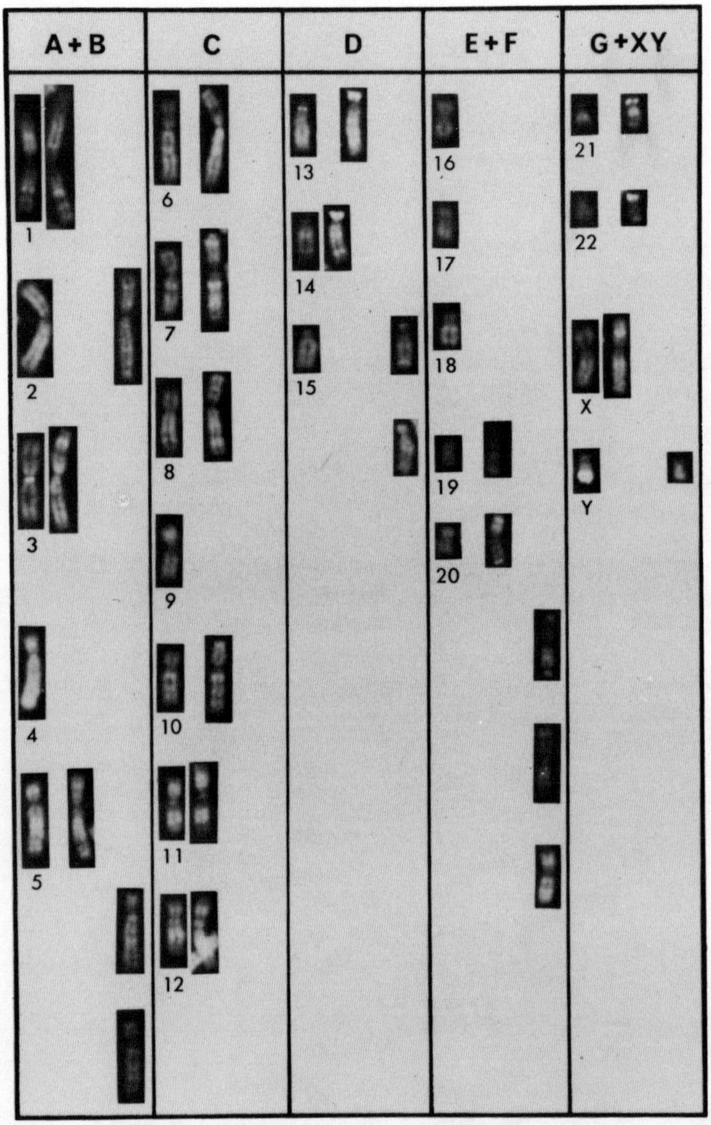

Fig.8. Direct comparison of human and chimpanzee chromosomes by fluorescent banding (Quinacrine treatment).

resulting in the formation of a karyotype of 45 chromosomes. Should two of these individuals mate offspring having 44 chromosomes would be produced, and would probably be perfectly normal. It is unlikely that this would happen in a contemporary population. Individuals with a karyotype of 45 chromosomes produced by centric fusion are fairly rare, and a mutation of this sort can easily be lost.

In order to understand how this could have occurred in an ancestral prehominid, we must consider the demographic dimensions of the populations at that time. They could not have exceeded twenty-odd individuals. They were family clans, very often ruled over a single male who enjoyed absolute dominion over the females of the group. Some of the males are likely to have frequently mated with their own daughters. Indeed it appears that the "leadership" of a group of chimpanzees lasts longer than ten years; the females reach sexual maturity at about eight-nine years.

A male with an analogous mutation, that is with a karyotype of 47, instead of 48 chromosomes had therefore two possibilities of establishing the new karyotype of 46 chromosomes. One possibility would have been that of mating with the various females of the group. Of the resulting offspring half would have had 47 chromosomes so that in the successive generation the reduction could have been completed. The other possibility would have been that of mating with his daughters by a sort of backcross. In this case the reduction of the karyotype would have been even more rapid and could have taken only two generations, therefore thirty years at most, considering the shorter time necessary for sexual maturity in our Apes-like ancestors.

Obviously the chances of establishing such a mutation, are conditioned not only by the existence of small reproductive communities but also, as already suggested, by the intrinsic power of survival of this mutation.

It is clear that there is little that can be said on this point, yet the massing of a large number of genes within the same chromosome may have represented some selective advantage at a certain moment in the evolution of the Hominids.

When did this event occur? As yet it is impossible to tell, but as knowledge of the karyotypes of the anthropoid apes increases, a means of attempting a calculation based on the frequencies of chromosomal mutations in natural populations and on the length of generations may be discovered.

AN ATTEMPT TO REVISE THE CLASSIFICATION OF THE OLD WORLD MONKEYS AND TO INTERPRET THEIR PHYLOGENESIS ON THE BASIS OF KARYOLOGICAL DATA.

The first data brought to light by many workers on the number and morphology of the chromosomes of the catarrhine monkeys, which are outlined, have led to an attempt to revise the taxonomic groupings, especially at the supergeneric level. Undoubtedly it is at this level that data of this type are most useful.

The genera *Macaca*, *Papio*, *Theropithecus*, and *Cercocebus* , as mentioned before, must be separated from the species belonging to the genus *Cercopithecus* and placed in a different subfamily to which the name Papinae could be assigned, leaving the name Cercopithecinae to the species belonging to the genus *Cercopithecus* only.

The genus *Symphalangus* must be separated from the various species of *Hylobates* and a clear *Nomascus* genus must be created for *Hylobates concolor*. These three genera should then be removed from the superfamily Hominoidea: they must be included in the superfamily Cercopithecoidea and constitute a family in themselves.

At the moment a detailed discussion of the group of Colobidea is not possible, yet they appear to be a fairly homogeneous group.

The superfamily Hominoidea should be restricted to the anthropoid apes (*Pongo*, *Gorilla*, *Pan*) and to man. Among these, man is sharply distinguished by the number of chromosomes and must be classified in the subfamily Hominidae; while the true apes will constitute the subfamily Ponginae.

In the subfamily Ponginae the karyotype of the orang-utan can be clearly distinguished from that of the gorilla and the chimpanzee. This difference could be taken into consideration at a supergeneric level, to discriminate the orangutan from the African apes. The anthropoid apes whose karyotype is most similar to the human one is the chimpanzee.

As regard the phylogenetic interpretation it is important to state again that the morphological variations in the chromosomes of the different species are the result of different mechanism, the most common of which are inversion and translocation. From a phyletic point of view however the variation in the chromosome number is of greater interest. The mechanisms which lead to these numerical variations, as we have shown, are centric fusions, centric fission and possibly polysomy.

Which of these mechanisms is responsible for the numerical variations in this group of species? Which was the original chromosome number of the ancestor common to all the Old World primates, accepting that we have had a single common ancestor? The chromosome number in the somatic cells in the living species of Old World primates varies from 42 to 72, but the greater part of this variability (from 48 to 72) are found in the diverse species of the genus *Cercopithecus*.

Extremely discordant opinions exist concerning the mechanisms which may have led to such a wide variation among the different species of a single genus, which moreover are very homogeneous from an anatomical-physiological standpoint. It is therefore impossible to formulate a hypothesis as to when this group was separated from the other species of Old World primates, or to which group of Cercopithecoidea they are most closely related.

In the other species of Old World primates the number of chromosomes varies from 42 to 50. The original number of chromosomes in the possible ancestor of the Old World Primates must probably be sought within the limits of these eight pairs. The fact that the groups of taxonomically diverse and phylogenetically ancient species such as Presbytis, Colobus, Pygathrix, and many species of Hylobates have 44 chromosomes suggests that this was the number of chromosomes of the species of primate that, in the middle Eocene about fifty thousand million years ago, was the original ancestor of the Old World Primates.

A centric fusion between two pairs of acrocentric chromosomes or a double translocation could have been responsible for the reduction from 44 to 42. For an interpretation of the chromosome number 48 in Nasalis, 50 in Symphalangus and 52 Hylobates (Nomascus) concolor recourse must be made to mechanisms of centric fission or of polysomy.

Probably the number and morphology of the chromosomes of the Anthropoid apes are not directly connected with the primate forms mentioned before.

The general morphology of the chromosomes is very different, and if there was a common ancestor he must be sought in the very remote past.

From the karyological point of view, therefore, we can distinguish three distinct lines of evolution in the group of catarrhine monkeys as schematized in figure 9: one concerning the group of species of the genus Cercopithecus; another for the various genera Macaca, Papio, Theropithecus, Cercocebus, Pygathrix, Presbytis, Colobus, Symphalangus, Hylobates and Nasalis; and yet another for the three anthropoid apes and man. The connections between these three distinct lines of evolution is ill-defined and controversial. Could the marker chromosome serve as a key for the correlation of the three lines?

The physiological homology of the achromatic portion of this chromosome in different species is supposed on the basis that this is the nucleolus organizing region (I5, 7).

As has already been shown (4), a chromosome pair with an achromatic region is present in all the species of Cercopithecinae, Papinae, Colobinae and almost all the Hylobatinae, being instead absolutely absent in the Ponginae and in Man.

The fact that this achromatic region plays the role of nucleolus organizer, and therefore must be a constant feature of the cell, stimulates the search for its possible location in the species which apparently do not share such a region and therefore for the establishment of possible homologies among the chromosomes involved.

The nucleolus organizing regions seem to be situated in the achromatic appendages, bearing satellites, in the short arms of acrocentric chromosomes I3, I5 and 2I (I6, 20). Direct comparison of these chromosomes to the nucleolus-organizing chromosome pair of Papinae is hazardous. However, when these two chromosomes are joined together in man or in the Anthropoid apes, their likeness with the marked chromosomes of a Macaque or a Baboon, both in size and shape is amazing (Fig. 10).

Apart from the physiological interest of this region, the perspective of such an homology in different species provides a means to face the question of the connection between the three different chromosomal lines which have been distinguished among the Old World Primates. This is, of course, primarily a matter of speculation. However, it could serve as a working hypothesis to develop further research.

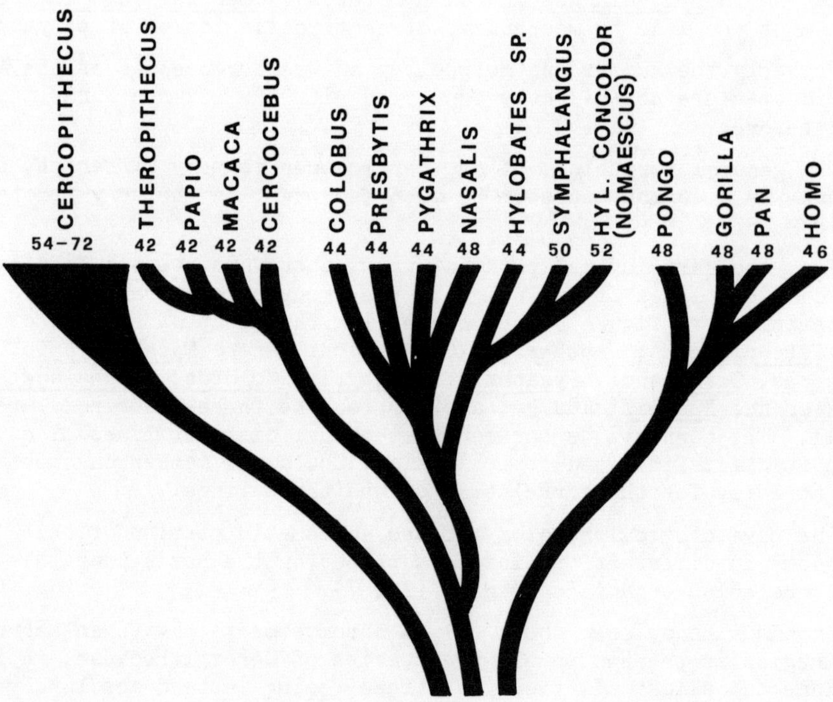

Fig.9. Possible phylogenetic relationship among the diverse genera of the Old World Primates

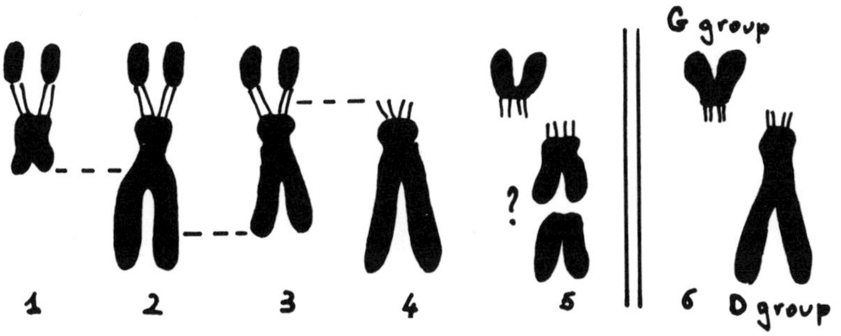

Fig.IO. Hypothetical relations among the so-called marked chromosomes in the Old World Primates. (I.Cercopithecus, 2.Papinae, 3.Hylobates, 4.Symphalangus, 5.Nomascus, 6.Hominoidea).

ADDENDUM

A substantial revision of our concept of chromosomal mutation is now in progress resulting from a reinterpretation of the ultrastructural organization of the chromosome and of the relation of the chromosomes in the nucleous of the Eukariotic cells.

According to these informations the metaphase chromosome results from the condensation of a single chromatin filament of about 200 Ä which is composed of the double helix of DNA compactly organized within a proteinaceous matrix. The filament does not only form a single chromosome but the whole haploid cellular set of chromosomes (genome) which is ring shaped in a dual existence, one of maternal and one of paternal origin.

These filaments are not free in the nucleoplasm, but they are attached at several points associated with the anuli of the nuclear envelope. Not all the attachment points with the nuclear envelope would have the same significance. Most of them would be relatively weak and would detach as soon as reduplication is over. On the contrary, a few others would be stronger and would last till the disappearance of the nuclear envelope. In the metaphasic chromosomal structure these latter attachment points could be assigned to centromeric regions of each chromosome.

If this interpretation result is verified, a change in the concept of chromosome mutation will result. The change of chromosomal morphology would not be due to mutation mechanisms, but simply to changes in the significance of one of the filament's attachment points to the nuclear envelope (from normal to telomeric or centromeric), bringing to a reinterpretation of mechanisms as centric fusion, tandem fusion, centric fusion and simple translocation.

The significance of this conceptual change in cytotaxonomic studies is evident. (Chiarelli, 1973, J. Human Evol., 2: 337; 1974, Boll. Zool., 40: 165.)

References

I. Boer L.E.M. de, I972, Genen en Phanen I5:I9-22

2. Chiarelli, B., I96I, Atti Ass.Gen.Ital. 6:2I3-220.

3. Chiarelli, B., I962, Caryologia, I5:99-I2I

4. Chiarelli, B., I966, Folia Primatologica 4:74-80

5. Chiarelli, B., I966, Amer.J.Phys.Anth. 24:I55-I69

6. Chiarelli, B. (ed.) , I968, Taxonomy and phylogeny of Old World Primates with references to the origin of Man. Rosenberg and Sellier, Torino.

7. Chiarelli, B. (ed.), I97I, Comparative genetics in Monkeys, Apes and Man. Acad. Press, London.

8. Chiarelli A.B., I972, Taxonomic Atlas of Living Primates. Acad. Press, London.

9. Chiarelli B., I973 , Evolution of the Primates. Acad. Press, London.

IO. Chiarelli, B. and E.Capanna (eds), I973, Cytotaxonomy and Vertebrate Evolution. Acad. Press, London.

II. Chiarelli, B. and C.C.Lin, I972, Genen en Phanen I5:2-3

I2. Chu, E.H.Y. and M.A.Bender, I962, Ann.N.Y.Acad.Science IO2: 253-266.

I3. De Grouchy, J.; C.Turleau; M.Roubin and M.Klein, I972, Ann. Génét. I5:79-84.

I4. Egozcue, J., I969, Primates. In K.Benirschke (ed.). Comparative Mammalian Cytogenetics. 357-389, Springer-Verlag, New York.

I5. Huang, C.C.; H.Habbitt and J.L.Ambrus, I969, Folia Primatologica , II:28-34

I6. Levan, A. and T.C.Hsu, I959, Hereditas 45:665.

I7. Manfredi Romanini, M.G., I972, J.Human Evol. I:23-40.

I8. Manfredi Romanini, M.G.; L.E.M. de Boer; B.Chiarelli and S.Tinozzi-Massari, I972, J.Human Evol. I:473-476.

I9. Rumpler, Y. and R.Albignac, I969, C.R.Soc.Biol. I63:I989

20. Slizynski, B.M., I964, In: Mammalian Cytogenetics and Related Problems in Radiobiology. (Pevan, Chagas, Frota-Pessoa, Caldas, eds.) I7I-I86. Pergamon Press, Oxford.

2I. Ying, K.L. and H.Butler, I97I. Canad.J.Genet.Cytol. I3:793-800.

IMMUNOGENETICS OF PRIMATES

Jacques Ruffie
Centre National de la Recherche Scientifique
Centre d'Hémotypologie
CHU Purpan
Toulose (France)

For a long time, the zoological definition of species was essentially morphological. Morphological characteristics are those which are immediately self-evident, which enable us to "recognize" an individual or a group, without the help of complex laboratoring examinations: they are the easiest to analyse.

When those characteristics reveal themselves to be immutable from one generation to another, they may be said to possess a control, which, at least to a great extent, is of genetical origin.

Those characteristics have made possible a definition of "species", "sub-species", biotypes and the establishment of a "link" between different groups. On them rests the Linnean classification. With regard to primates, morphological characteristics were, for a long time, the only ones studied. Morphological characteristics are used in the comparison of today's living forms with the fossil forms of species extinct long ago, in the integration of the latter in the systematics and in the shedding of some light on the trend of evolution. It is those morphological characteristics which make it possible to date, at least approximately, the main stages which have stood out as landmarks in the long history which ends in man. Paleontology, as a whole, is based on the analysis of fossil remains. Taxonomy is based on one fundamental law: the law of the constant transmission of specific racial or individual traits. In a word, it is the constancy of the "hereditary information" which assures the perenniality of groups.

Those traits which can be of the greatest use to the biologists are those which enable him to analyse more accurately this information. Now, morphological traits, as a whole, are poor material for the geneticist. Each trait possesses a certain amount of "hereditable" element but it is almost always difficult, if not impossible, to find out, with certainty, to what that hereditary element corresponds. Morphological traits are, in general, plurifactorial and conditionned by complex genetical patterns, most of which are not yet known and, without doubt, will not be known for a long time to come. Monofactorial mutations with morphological interpretation are, in truth, known in human or veterinary medicine (such as peromelies, zygodactylies, etc...) but they are rare occurences: more often than not they assume a monstrous character and hardly admit of a mathematical analysis - which is the only method which could enlighten us in the hereditary pattern of these mutations.

Besides, morphological traits are the result of a long ontogeny, a fact which keeps them remote from genetical information. Accordingly, they are responsive to environmental factors which have all the time they need to exert their influence. These data explain the disappointing results of biometry in the field of genetics, in spite of recourse to an increasing number of measurements.

Even if the human geneticist had succeeded in isolating a small number of simple, hence utilizable morphological traits, he would still come across insurmountable difficulties. First, in our species all experimental intercrossing is, by definition, impossible. The Mendelian method of cross-breeding between first generation hybrids (that is, between brother and sister) which yields valuable information, is unrealizable. The human geneticist does not bring about interbreeding: he is content with observing those cases already existing. At this point, a fresh difficulty, resulting from the span of human generations, arises: so, the compiling of utilizable observations takes a considerable time. These obstacles were to be cleared at the beginning of the century when Landsteiner discovered the first erythrocytic factors. Thanks to simple immunological methods, it became possible to identify, in the human blood, substances which have a monofactorial genetic control and which show a polymorphism and are found in all individuals. This material lends itself quite well to treatment by the mathematical patterns of population genetics which up to that point had been inapplicable to man.

That method was to prove very efficacious in tackling the study of human genetics.

Since the initial description by Landsteiner of the ABO system, discoveries have been increasing. This is particularly true since the time that Landsteiner and A.S. Wiener discovered the Rhesus system and thus proved that red corpuscles formed a real "mosaic of antigens" in which most of the elements were still passed over because they did not correspond to any antibody "naturally" present in human serum. It is only in the event of "immunity-conflict" (allo-immunization), begun in a subject who by accident receives a factor which he does not carry on the red corpuscles (pregnancy, transfusion), that antibody may appear (mother, receiver). This antibody will enable the detection of the antigen which is the cause of that immunization in all carriers. The systematic search for such reagents in certain persons (pregnant women, polytransfused patients) has, during the last 25 years, led to the discovery of a very large number of factors, and the number is still on the increase as is shown by the curve below:

A VISUAL REPRESENTATION TO SHOW THE NUMBER OF "PUBLISHED" ANTIGENS YEAR BY YEAR

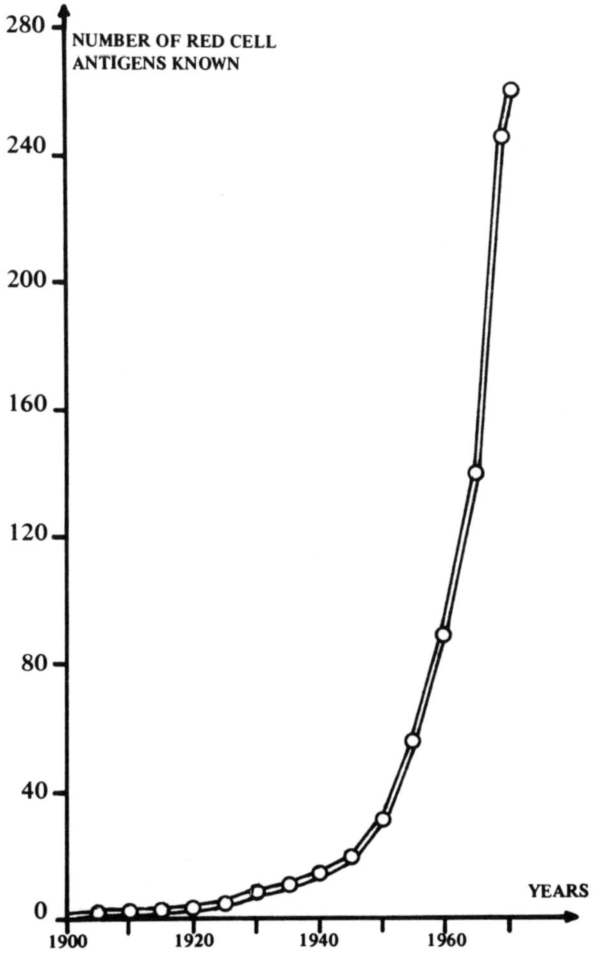

Figure 1

These factors are present in many other tissues besides the blood, but it is in the blood that they are the most easily disclosed, for the blood is a liquid tissue easy to sample, to handle and to preserve. It enables research to be conducted on masses of people, which is indispensable when the genetics of populations are studied.

Thus was a new science born; immunogenetics, which means the study of immunological factors with hereditary conditioning. The scientific study of human genetics has been made possible by the way of immunogenetics. However, as biochemistry and, in particular, molecular biology progressed, it appeared that blood contained many other factors with hereditary control but deprived of antigenic activity. The first discoveries in that field dealt with the formation of chains of globins which go to the making of haemoglobin molecules. Thanks to the Ingram method (1, 2), human haemoglobin has revealed an important structural polymorphism with a genetical conditioning, the full range of which is just beginning to be seen. Thus the immunogenetic concept was soon to make room for the larger hemotypology concept which includes the immune systems as well as others.

By the use of diverse methods, immunizing, physico-chemical or mixed, it is possible today to expose a great number of factors, cellular as well as plasmatic, belonging to autonomous genetic systems.

Taking into consideration only the family of principal systems, it is possible to adopt this classification:

1° - Immunological systems.

a) Cellular

1) Erythrocytic blood groups (ABO, MNSs, Rhesus, etc.)

2) Histo-compatibility groups (Leuko-platelets HLA)

b) Plasmatic

Immunoglobulin groups (Gm, Inv)

2° - Non-immunological systems.

a) Cellular

- hemoglobinic types

- enzymotypes: phosphatase acids: PGM_1, PGM_2 (Phosphoglucomutase), ADA (Adenosine deaminase), AK (Adenosine kinase), LDH (Locus A and B), G6PD (Glucose-6-Phosphate dehydrogenase), 6 Phosphogluconate dehydrogenase (6PGD).

b) Plasmatic

1) Non enzymatic proteins: haptoglobin, transferrins, group component (Gc system), Pi protein

2) Enzymotypes: alkaline phosphatase, cholinesterase (A.C.A.H.) (Locus E_1 and E_2).

It is due to the discovery of these systems that the number of genetic markers, not known, is very high. Each of them is the subject matter of research which is carried out at three levels:

1) Exposing the structure of the factors: whether only

one immunological definition is available (definition of a specificity by the corresponding antibody) or one may proceed with the "dissection" of the molecule and thus ascertain the contents and the order of the sequences of the peptides, the sequences of the amino-acids as well as the position of diverse non-peptidic groupings which are attached to them,

2) The study of the polymorphism of these factors: which corresponds -for a given genetic system- to the search for all the alleles which can be found in the same locus. Knowledge of the polymorphism often is dependant on the shrewdness of the techniques used. In quite a few cases the polymorphism increases pari passu with technological progress,

3) The definition of the genetic pattern governing each system studied (and the search for eventual links between the systems). This type of problem is resolved by the methods of population genetics; the only ones applicable to human populations for all the reasons set forth above.

From this can be gauged the importance of the discovery of blood factors in the development of human genetics. The techniques which had led to these discoveries were applied very early to the study of non-human primates. The first investigations were limited to erythrocytic groups.

As early as 1925, Landsteiner and Miller (3) pointed out that the red blood cells of the Chimpanzee belonged either to group O or to group A, those of the Orang-Utan to group A or B or AB. From that time many researches have been carried out, the most important of them by A.S. Wiener, J. Moor-Jankowski and E. Gordon. Today the distribution of the fundamental blood groups is known in almost all species of monkeys, so much so that to the traditional "sero-anthropology" is now grafted a "sero-primatology". In the sequel most of the systems, either immune or not in man, have been seen again, either identical or more or less modified, in the other primates.

Today it is possible to draw up hemotypological classification of primates. In quite a few cases this classification cuts across and defines more accurately the traditional classifications founded on morphology.

Moreover, these marker genes are expressed by blood groups very specific and easy to identify. In consequence it is possible, during the typing of extant species, to determine the approximate time of differentiation of each of these factors or of each of these systems, hence to trace a certain number of evolutive stages. Not infrequently, as will be seen at the end of this exposition, the immunogenetic evolution follows rather punctually the cytogenetic evolution. Matson went further: thanks to the progress of organic chemistry which makes possible a real molecular dissection, the evolution of certain peptide chain can be followed, the ancestral archetype from which are derived the peptides present in species now extant, can be defined.

The most remarkable example is given by the comparative analysis of globin chains to which we have refered earlier and to which we shall come back later. These research works are far too recent to enable one to know, in detail, the molecular evolution of primates. But results so far obtained prove, by their richness and precision, that much can be expected from this new "micromorphology".

In the following pages, we shall study the immunological blood group types with genetic conditioning at present known in primates. Then we shall discuss the conclusions which may already be drawn from these new acquisitions and examine in what direction research can be hopefully pursued.

IMMUNOLOGICAL SYSTEMS

They are divided into cellular antigens and seric antigens.

Cellular antigens: it is usual now to distinguish:

1) The erythrocytic factors which have at first been detected on red cells (where it is always easier to look for them) but which exist as well on other cells of the organism.

2) The histocompatibility factors present in leukocytes and in many tissues but absent in erythrocytes. Let us examine them seriatim.

The Erythrocytic Antigens

According to their distribution in primates, they are divided into two groups:

a) Blood factors of human type: thus called because they were first described in man, but several of which were subsequently encountered in many species of monkeys. In the latter case, they may be identical to human factors or show a few differences of structure or of mode of transmission. These systems, common to man and to other species, spread more or less in the systematics according to the date of their appearance. The ABO and MN systems and, to a lesser degree, Rhesus and Xg(a) fall within this category.

b) A second group of factors is represented by antigens peculiar to one species of primates: there are essentially the "simian types" described by A.S. Wiener and Moor-Jankowski, and their reagents are obtained by alloimmunization (Chimpanzee→chimpanzee) or by "cross-immunization" (Man⟶Chimpanzee) or further by hetero-immunization (Chimpanzee⟶rabbit). In this category we can include those factors which we have called "para-antigens" which produce paratypes and which are, in general, common to closely related species. They are shown up by natural antibodies present in human serum (4).

Blood factors of human type. These factors which fundamentally correspond to antigens of ABO, MN, Rhesus and Xg(a) systems, are shown up by human reagents after eventual elimination of non-specific hetero-agglutinins by absorption, dilution or elution. They have been the subject of special studies by J. Moor-Jankowski and A.S. Wiener (5, 6).

1) The ABO system: in the prosimii and, in particular, in the lemuridae: only one factor fairly close to human B factor (and capable of absorbing at least in part human anti-B serum) is present on the red blood cells. Wiener has suggested that it should be named B-like. This B-like factor, as A.S. Wiener and we ourselves have noted, is found again on the red blood cells of New World monkeys (Ceboidea).

In the ceboidea: depending on the species, it is possible to find factors A, B and H always in the secretions but never on the red blood cells. On the other hand, the B-like factor is almost certain to be always present on the red blood cells. It is advisable to retain the following notions:

a) the genetic pattern which controls the presence of ABH factors in ceboidea secretions is assuredly identical to the human pattern (series of three alleles: A, B, O, the first two being co-dominant between themselves and dominant to the third). However, the presence of B-like factor on the red blood cells (which seems to be invariable and does not accordingly reveal any polymorphism), certainly results from and independant genetic system. Almost always there are found in the serum, an agglutinin or agglutinins which do not correspond to the antigen or to the antigens present in saliva. This rule retains a rather general character but it is less strict than in man.

Moreover, the presence of B-like factor on the red blood cell does not prevent the appearance of anti-B antibody in the serum in the A or O animals. Besides, it must be noted that although the genetic pattern appears to be the same as in man, rather few species possess, simultaneously, the three mutations A, B, O. Most species appear to have only one or two. The table below (prepared after Wiener and coll.) sums up the blood groups of the Ceboidea which have, to date, been studied (Table 1).

In the cercopithecoidea, as in the two previously mentioned groups (Lemuroidea, Ceboidea), the B-like factor no longer exists on red blood cells. On the other hand, A, B and H factors are seen — according to the species — rather regularly in saliva and secretions. Again in this case, it is rare that the same species possesses the three mutations. The table below sums up the results obtained in the three species studied to date (Table 3).

Table 1. ABO Blood Groups in the Ceboidea (from J. Moor-Jankowski and A. S. Wiener, 1969)

SPECIES	BLOOD GROUPS ENCOUNTERED	ANTIGENS PRESENT IN RED CELLS	ANTIGENS PRESENT IN SALIVA
Spider monkeys Various species	"O", A, B	B-like	Sec A-B-H
Cinnamon capuchins Cebus albifrons	"O", B	B-like	Sec B-H
Squirrel monkeys Saimiri sciurea	"O", A	B-like	Sec A-H
Marmosets Various species	A	B-like	Sec A, H-like

Sec : secretor of A-B-H substances

If not otherwise indicated the above information is based on tests on at least 10 animals of each species

Table 2. ABO Blood Groups in the Cercopithecoidea (from J. Moor-Jankowski and A. S. Wiener, 1969)

SPECIES	BLOOD GROUPS ENCOUNTERED	ANTIGENS PRESENT IN RED CELLS	ANTIGENS PRESENT IN SALIVA
Baboons Papio cynocephalus and anubis	A, B, AB	Absent	Sec A-B-H
Papio ursinus	A, B, AB	Absent	Sec A-B-H
Drills Mandrillus leucophaeus	A	Absent	Sec A-H
Geladas Theropithecus gelada	O	Absent	Sec H
Celebes black apes Cynopithecus niger	O? , A, B	Absent	Sec A-B-H
Patas monkeys Erythrocebus patas	A	Absent	Sec A-H
Vervet monkeys Cercopithecus pygerythrus	A, B	Absent	Sec A-B-H
Rhesus monkeys Macaca mulatta	B	Absent	Sec B-H
Crab-eating macaques Macaca irus	O, A, B, AB	Absent	Sec A-B-H
Pig-tailed macaques Macaca nemestrina	O, B	Absent	Sec B-H
Stump-tailed macaques Macaca speciosa	B	Absent	Sec B-H

Sec : secretor of A-B-H substances

If not otherwise indicated the above information is based on tests on at least 10 animals of each species.

In the Pongidea: from this stage, the blood factors are present not only in the saliva but also on the red blood cells (except in the Gorilla in which cellular antigen remains very low compared to salivary antigen). A.S. Wiener has proved that Orang-Utans of the non-secretor type even may exist: with the antigen present on the cells but not in the saliva, those subjects are very rare whereas non-secretors attain a frequency varying between 20 and 30% in the human species.

The results of the tests of hominoidea groups is summed up in the following table.

Table 3. ABO Blood Groups in the Hominoidea
(from J. Moor-Jankowski and A. S. Wiener, 1969)

SPECIES	BLOOD GROUPS ENCOUNTERED	ANTIGENS PRESENT IN RED CELLS	ANTIGENS PRESENT IN SALIVA
Man	O, A, B, AB	H, A, B	Sec and nS A-B-H
Chimpanzees Pan troglodytes	O, A	H, A	Sec A-H
Orangutans Pongo pygmaeus	A, B, AB	A, B	Sec and nS A-B-H
Gibbons Hylobates lar	A, B, AB	H, A, B	Sec A-B-H
Siamangs Symphalangus syndactylus	B	H, B	Sec H-B
Lowland gorilla Gorilla gorilla	B-like	$\pm$	Sec B-H

Sec : secretor of A-B-H substances
nS : non-secretor of A-B-H substances
If not otherwise indicated the above information is based on tests on at least 10 animals of each species

The Chimpanzee which from the immunological stand point is the closest to man (as proved by cytogenetics) has only two groups: A and O. Group B has never been encountered. It is likely that this last group does not exist in this animal. However, the Chimpanzee, as well as man, possesses sub-groups of A, A_1 and A_2, which are fundamentally different, in respect of the amount of H substance (greater in A_2 than in A_1) and of the amount of A substance (greater in A_1 than in A_2). These two mutations are not exactly comparable to human mutations; in fact, the Chimpanzee A_1 type has constantly more H substance than A_1 human type, while the Chimpanzee A_2 type has less H substance than human A_2 type. On the other hand, it seems that there is less of A_1 factor in Chimpanzee A_1 group than in human A_1, while chimpanzee A_2 group has some A_1 substance in opposition to human A_2 which is entirely devoid of it.

The Chimpanzee A_1 and A_2 mutations are closer between themselves than human A_1 and A_2 mutations. Wiener is of the opinion that the Chimpanzee A_1 type, looked at from its immunological constitution, might correspond to the intermediate A type (A int.) encountered in man, particularly in negroid races (7). Ten percent of type A Chimpanzees belong to A_2 group. If the charge variations of H factor and A_1 factor in the four human and chimpanzee A_1 and A_2 types were compared, the following sequence would be obtained:

A_1 homo $> <$ A_1 chimp $> <$ A_2 chimp $> <$ A_2 homo

$>$ increasing load in respect of factor H
$<$ decreasing load in respect of factor A

Figure 2

The other anthropoid apes do not appear to present a polymorphism for A group, save, however, for the Gibbon which carries the A_2 gene, but at a very low frequency (A.S. Wiener). In conclusion, it must be stressed that the ABO system appeared very early in the evolution since factors almost identical to human antigens as well as corresponding antibodies are already seen in the Ceboidea. The immunological structure is comparable and the genetic pattern the same.

Evolutionarily the study of the ABO system makes it possible to retain three "levels":

1) a lower level, with presence on the red blood cells of B-like factors, connected immunologically to B-factor, but genetically independant. It is present in the prosimii and the ceboidea but absent in the others.

2) A mean level, characterised by the appearance in the saliva of A, B, H factors of human type and by the presence of anti-A and anti-B antibodies in the serum of subjects not carrying the corresponding antigen. This relation, though quite proper does not possess the strictly constant character which it is known to have in man. This level is proper to the Ceboidea and the Cercopithecoidea. The Ceboidea, then, simultaneously present characteristics proper to the first and the second group.

3) An upper level, proper to the Hominoidea (Pongidae + Hominidae) in which ABH antigens also appear on the red blood cells. Most of the representatives of that group have the ability to synthetize simultaneously, according to the structure genes present on their chromosomes, hydrosoluble factors (present in their secretions) and liposoluble factors (present on their red blood cells).

We have already referred to the way as A.S. Wiener showed that certain Orang-Utans have lost the ability to secrete the blood factors. This fairly uncommon type corresponds to the human non-secretor type. Further, in man, rare subjects are encountered who secrete in abundance A substance in their saliva but who do not carry it on their red blood cells. Their serum regularly contains anti-B agglutinin but never anti-A agglutinin. There it is a case of "reversive mutation" which restores an ancestral situation: that of the Cercopithecoidea. It is the reason why that type has been called Am (m = monkey) by A.S. Wiener (8). It is possible to represent the evolution of the ABO system by the following diagram (Figure 3).

This diagram shows how very remote the appearance of the ABO system is, seeing that the three mutations which condition it meet, according to the same genetic pattern, as far back as the Ceboidea stage. It is accordingly sound to consider that the three mutations were already present in the common ancestor of monkeys of the Old and of the New World, before those two branches split. The ABO system must have appeared, in transitional forms, between the prosimian stage, which does not possess it, and the Ceboidea stage, in which it is already fully materialized.

However, the preceeding tables which show the distribution of ABO groups in the main species of primates extant today, indicate that those who, like man, possess the three mutations are very rare. Most of the primates have only two (Chimpanzee: A and O; Gibbon: A and B, etc.).

It is accordingly possible to assume that, in some cases, one or two mutations got lost, without doubt, when the species were formed (Figure 5).

The preceeding is a powerful argument is a support of the hypothesis of B. Chiarelli (9) who sees, in the robertsonian type fusion which occurred between two acrocentric chromosomes, the first "accident" which led to hominization. That sort of modification must, initially, have been carried by one individual only. This theory has

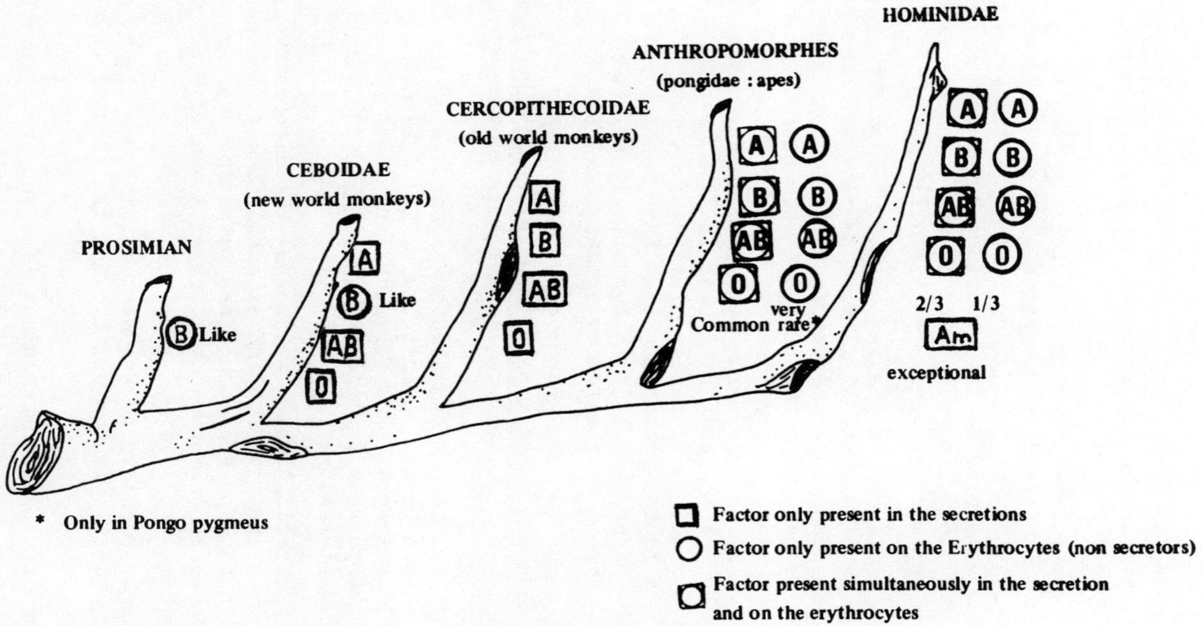

Figure 3. Evolution of ABO Groups in the Secretions and on the Erythrocytes of Primates

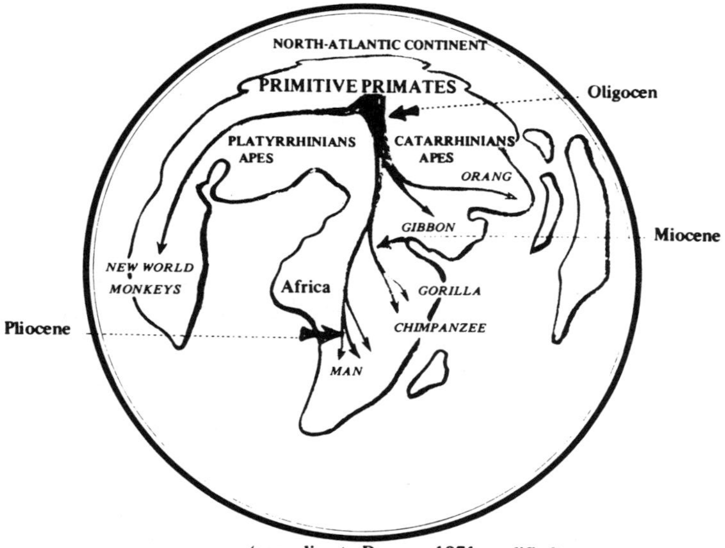

(according to Duroux, 1971 modified)

Green arrow : from this point ABO genes are differenciated. The common trunk splits in two branches : platyrrhinians and catarrhinians. (+ apes) (At the oligocen)

Yellow arrow : indicates the moment of the fixation of ABH on red cells. (at the miocene)

Figure 4. Geographical Evolution of Primates

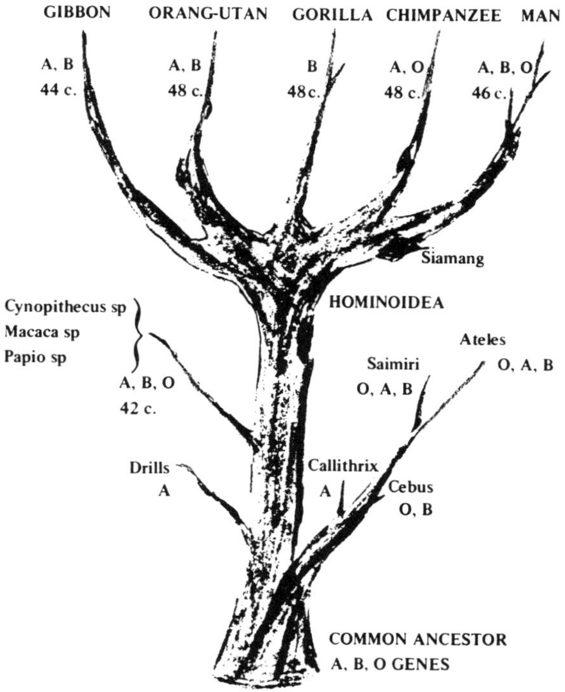

This table proves that there does not exist within groups a strict relation between the systematics of species and their ABO blood group system. This argues in favour of a random distribution of ABO genes in each collateral branch issued from the common trunk. (44, 48 means the chromosome number c stays for « chromosome »).

Figure 5. Distribution of ABO Factors in Different Species of Primates been considered elsewhere (10) which involves very strong genetic drift.

The fact that only one subject of hence an isolated pair, gave birth to a new karyotype (hence, to a new species) easily explains the frequent "loss" of a blood factor. The fact which emerge from the distribution of the antigens of the erythrocytic ABO system in the various species of primates, lead one to think that the diagram of B. Chiarelli may be applicable not only at the time of hominization but also at the appearance of the main groups of primates at present extant.

2) The MN system: from the comparative genetic standpoint, the MN system presents characteristics rather different from the ABO system. In fact the ABO system must have appeared (at least in its hydrosoluble form) at a very early stage in evolution. With the exception of B-like factor present in the prosimii and the Ceboidea, all the other A, B, H factors are first encountered at the Ceboidea stage; they are very closely related, if not identical, to the human factors. Further they conform to a same genetic pattern, as has been shown. Indeed, a few modifications may have manifested themselves in the course of the evolution (complexification of the molecule, appearance of a polymorphism of A factor which, from a certain stage, splits itself into A_1 and A_2, crossing of factors over to the red blood cells); but these variations are always of weak amplitude and do not implicate the whole of the immunogenetic system the pattern of which is already definitely realised in the New World monkeys.

In regard to the MN system, everything is different. The first antigens which appear (and which are related to M factor) are far from being identical to the human factor, at the most, they have a few common antigenic determinances. This is what is observed, initially in the Ceboidea, whose red blood cells are agglutinated by a small number of anti-M sera. All the rest of the molecule is probably made up of receptors very different from those present on the human antigen. Their resemblance increases progressively as there is a move up in the scale of Primates: the number of conflicting reactions, observed by means of a series of serum-tests, then diminishes. The N factor shows up much later than the M factor since it is only encountered in anthropoid apes (it would thus be peculiar to the hominoidea). However, the genetic pattern classically accepted for the human type (two independant genes able, on one hand, to produce M factor and then again N factor) is only encountered in man and oddly in the Gibbon (which, here again, expresses its antigenic consanguinity with man).

All this leads one to believe that the MN system -though of relatively ancient appearance- assumed its present form, at a more recent date than the ABO system, probably in the root stock common to hominoidea: before that date only a few motives peculiar to M factor are encountered on the red blood cells. Let us now examine the behaviour of these factors in the various groups of primates studied to date. Here again, most of the research work is by J. Moor-Jankowski and A.S. Wiener (5, 6, 7).

M FACTOR

The first traces of M factor, as we have previously mentioned, appear in the Ceboidea: in fact a few rare human anti-M sera can agglutinate ceboid erythrocytes; the others produce no reaction. On the contrary, in the Cercopithecoidea, all the species produce a positive reaction to a greater number of serum-tests. Thus the number of anti-M receptors, identical to those of human M factor is higher than in the preceding group. In M positive anthropoid apes, the antigen is still closer to the human factor as it is spotted by almost all the anti-M reagents. Further in anthropoid apes there exists a genetic polymorphism so far unknown. In fact, all the Chimpanzees and almost all the Gorillas carry the M factor, it is found again only in about 50% of the Orang-Utans and 25% of the Gibbons. The following table inspired by A.S. Wiener sums up this evolution:

Table 4. M Factors in Primates

SPECIES	REACTIONS TO ANTI-M SERA					
	n° 1	n° 2	n° 3	n° 4	n° 5	n° 6
Man						
Types M or MN	+	+	+	+	+	+
Type N	−	−	−	−	−	−
Chimpanzees	+	+	+	+	+	−
Cercopithecoidea						
Baboons	+	+	+	−	−	−
Rhesus monkeys	+	+	+	−	−	−
Green monkeys	+	+	−	−	−	−
Ceboidea						
White Ateles	+	−	−	−	−	−
Black Ateles	±	−	−	−	−	−
Capuchin monkeys	−	−	−	−	−	−
Lemurians	−	−	−	−	−	−

(according to Wiener)

This table allows of two remarks:

1) The number of antigenic receptors identical to those present on human M factor increases in the course of evolution, as evolution draws closer to man.

2) Anti-M antibodies do not all have the same zones of specificity: some will be able to detect a type of antigenic receptor (for example a type common to Cercopithecoidea and to Hominoidea): they will produce agglutinations with erythrocytes of numerous groups; on the other hand, others will have only specificities peculiar to anthropoid apes for instance and will prove incapable of detecting factors present on other groups.

This antigenic specificity of erythrocytes for the M factor is revealed by repeated absorption tests. If, for example, anti-M sera are absorbed by Gibbon erythrocytes, two fractions can be isolated: one fraction active against human erythrocytes only, the second fraction active at one and the same time, against human erythrocytes and against Gibbon erythrocytes. The first one must recognize a motive peculiar to human M factors and the second motives common to man and to hylobates (and perhaps to anthropoid apes as a whole).

N FACTOR

As has been shown, the appearance of N factor is more recent than the N, since no motive is found again either in the Ceboidea or in the Cercopithecoidea. On the other hand, a factor fairly close to human N factor is found in Gibbons (80%) and in almost 50% of the Chimpanzees. It seems that all the Gorillas carry N whereas this factor is not known on Orang-Utans erythrocytes. Most of the identifications have been carried out by a vegetable reagent, obtained from Vicia graminea (N^v) which has the advantage of not containing any heterospecific antibody.

THE PHENOTYPES

The Gibbon can -just like man- correspond to three types:

$(M)^{Gi}$ $(N)^{Gi}$ $(MN)^{Gi}$

(The index Gi implies that the reference is to M or N factor of the Gibbon type, which factors are not then absolutely identical to the human type).

In the Chimpanzee, it has been shown that all animals carried an M^{ch} factor fairly close to the human factor: 40% of them carry a N^{ch} factor. There are accordingly two phenotypes:

$(MN)^{ch}$: positive for the anti-M and the anti-N^v

$(M)^{ch}$: positive for anti-M only, negative for the anti-N^v.

In Orang-Utans, only anti-M antibodies produce a reaction sometimes positive, but anti-N^V never. There are accordingly two phenotypes:

$$(M)^{or} \quad \text{and} \quad (m)^{or}$$

Lastly, all the Gorillas are N or only a few M, which corresponds to two phenotypes:

$$(MN)^{gor} \quad \text{and} \quad (N)^{gor}$$

Figure 6 sums up these results.

THE GENETIC PATTERN

It has long been agreed that in man the MN system corresponds to two co-dominant alleles: M and N, which implies only one locus. It is possible to encounter three genotypes corresponding to three phenotypes (Table 5).

In apes, only the Gibbon presents these three phenotypes and could correspond to the human genetic pattern. It must be admitted that the Gorilla and the Chimpanzee present a different genetic pattern since, in these two, there might exist a mutation capable of producing simultaneously a synthesis of M factor and that of N factor. This type of gene might have appeared fairly early during evolution, subsequently a gradual "seclusion" might have taken place, which might have isolated the M factor by placing it under the strict dependence of a M gene and the N factor by placing it under the control of a solely N gene. However, that seclusion might not be absolute, even in man.

In fact it is now known that all M phenotypes contain also some amount of N factor. M human red blood cells can lower the strength of anti-N antibody. Furthermore, when injected into the rabbit, the M erythrocytes induce, by the side of the anti-M, the more or less constant appearance of a small amount of anti-N. The reciprocal is not true and N human factor would be completely devoid of M. This explains the occasional encounter of a "natural" anti-M in certain N subjects, whereas the appearance of an anti-N in M subjects is absolutely the exception. Lastly, we known of M_c mutation in man which appears to control simultaneously the synthesis of at least a part of M and N factors on the red blood cell. They might correspond to what is observed in the Chimpanzee or in the Gorilla and would take their place in the group of "reversive mutations", re-establishing an ancestral state in the human species.

However, from purely statistical consideration, A.S. Wiener and J. Moor-Jankowski have suggested a diagram, which postulates the existence of two pairs of independent genes, each carried by a pair of different chromosomes. The first would consist of M (assuring the synthesis of M factor) and a mute m gene. The second would consist of N (assuring the synthesis of N) and a mute n. M and N are

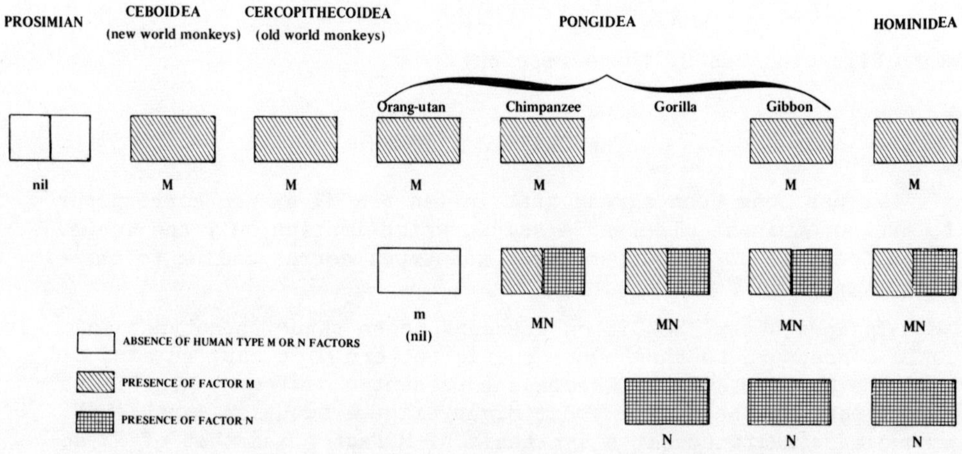

Figure 6. Activity of Genes M and N in Different Species of Primates (Classical Theory)

Table 5. Classical Genetic Pattern of MN System (Man and Gibbon)

GENOTYPE	PHENOTYPE
M M	M
M N	M N
N N	N

dominant to m and n (Table 6).

It is then possible to accept the following distribution in the groups of primates:
Ceboidea: only n chromosome would exist in couple N/n. In couple M/m, M or m might be encountered.
Cercopithecoidea: only n exists also in couple N/n. On the other hand, all would carry M, which implies perhaps the absence (or the loss) of m chromosome. Acording to the species, one or two chromosomes might be present.

The Chimpanzee would have only M, n, or N (loss of m), hence constant presence of M in all animals. The Orang-Utan: constant presence of n, never N. Presence of M or of m: loss of N. All Gorillas have N (loss of n) and a few M (presence of M or of m). Lastly the Gibbon and man have M or m and N (loss of n). If n has persisted, it would be possible to encounter M-N-subjects (equivalent to O group of ABO system, which is not the case). Table 7, which follows, sums up this distribution

The theory of Wiener agrees rather well with the experimental facts described earlier (in particular: constant presence of a certain amount of N factor in humans of M type). However, it would imply a partial epistasis of M on N: in fact, all men carry a double dose of N gene, but its expressivity would depend on the presence of the M factor which, in homozygous state, would be capable of inhibiting partially the synthesis of N.

In other respects, in the field of evolution, this diagram is more satisfactory than the classical theory. It implies, in effect, the formation (from the Ceboidea stage) of only one genetic pattern which will no longer vary. Only the mutations existing within the different groups will change, perhaps through loss of certain types of chromosomes (that is of certain genes) at the time of the speciation (Ex: M) or of the appearance of new mutations (essentially N which, at locus N/n, can at a certain moment take the place of n). As for the ABO system, this strong genetic drift agrees quite well with the chromosomic theory of hominization suggested by B. Chiarelli and which could, with some verisimilitude, be generally extended to the appearance of numerous species.

However, it is useful to stress a notion already referred to at the beginning of this chapter: that of the complexification of M antigen which semms to acquire new determinances as we move up the ladder of primates. Thus in man and in the higher species a series of "paleo-sequences" would be found, archaic antigens common to a great number of primates, and of "neo-sequences" which appeared later on and are peculiar to each zoological group having evolved according to its own lineage. The former are symboled by the letter M which corresponds to the presence of certain receptors, the latter are indicated by the coefficients (M^{ch}, M^{gor}, M^{gib}, etc.) which mark the particular specificities of each of these factors.

Table 6. Genetic Pattern of MN System

Classical Schema: M or N

Phenotype	M	MN	N
Genotype	M M	M N	N N

Wiener Theory: M or m; N or n

Phenotype	M	MN	N	m n
Genotypes (theoretical)	M M n n	M m N N	m m N N	
	M	MN	N	
	M m n n	M m N n	m m N n	
	m M n n	m M n N	m m n N	m m n n

1) The classical theory admits the existence of one locus only and of two allelomorphous genes M or N, each one responsible for the synthesis of the factor of the same name. Three genotypes correspond to three phenotypes.

2) The Wiener theory assumes the existence of two pairs of alleles (each with one active and one silent gene) carried by two pairs of independant chromosomes. Ten genotypes corresponding to four possible phenotypes (the last of which has not been described in man).

Table 7. Distribution of Chromosomes M/m and N/n in the Main Groups of Primates

	Genes	Blood groups
Ceboïdea	M or m and n n	M or –
Cercopithecoïdea	M M and n n	M or –
Hominoïdea Pan	M and N or n	M or MN
Pongo	M or m and n	M or –
Gorilla	M or m and N	MN or N
Hylobates	M or M and N	M, N or MN
Homo	M or m and N	M, N or MN

(according to the Genetic theory of A. S. Wiener and J. Moor-Jankowski)

3) The Rhesus system: the position of the Rhesus system in primates is rather paradoxical. The first anti-Rh antibody recognizing Rh_o factor or standard rhesus was discovered by Landsteiner and Wiener (12); it was obtained by the immunization of rabbits and guinea-pigs with the red cells of Macaca mulatta monkeys (Macacus rhesus). The red cells of closely related species (in particular of Papio) are able to stimulate identical antibodies. That first reagent was able to agglutinate the red cells of 85% of white-race New Yorkers, but could not agglutinate the remaining 15%. It is from these observations that Landsteiner and Wiener were led to postulate the existence of a new antigen (shown up by the "anti-Rhesus antibodies") present in 85% of the men studied and absent in 15%. For some time it was believed that there was identity between the antigen carried by the Macaca mulatta cells (and giving rise to the appearance of the initial anti-Rhesus in the rabbit or the guinea-pig) and the Rhesus factor present in Rh positive man. In truth, it soon became apparent that there was no absolute identity: in fact, whereas anti-Rh antibodies encountered in man are active on human Rh^+ red cells and on the red cells of most of the big anthropoids, they remain always inactive on the red cells of macaques and of baboons. On the contrary, the immune serum of anti-Rh rabbit is always active on human Rh^+ red cells and on those of Macaca and of Papio, but seems to be inactive on the red cells of most of the big anthropoid apes.

The above shows that there is no absolute identity between the two factors but that they can produce crossed reactions. Murray and Clark (13) have supposed that the Rh factor present in certain cercopithecoidea correspond to a precursor of Rh antigen of the hominoidea (big apes + man) and have suggested that this hypothetical factor be named L.W. factor, in honour of Landsteiner and of Wiener.

Such an antigen would be present on human Rh^+ red cells and on Rh^- red cells. In fact by heating the red cells of Rh^- subjects, Murray and Clark extracted a substance from them which, when injected into a guinea-pig, is capable of causing the appearance of an anti-Rh antibody identical to the initial reagent of Landsteiner and Wiener (or at least having common motives) (Murray-Clark effect). This antibody, as well as that obtained by Landsteiner and Wiener by immunizing rabbits with Macaca red cells, can agglutinate human Rh^+ cells and coat human Rh^- cells from which it can be removed by elution. Thus, there would exist an antigen common to Rh^+ and Rh^- red cells, present in almost all subjects of the human species (1% on the average would be devoid of it). In other respects, it seems that foetus and new-born of the Rh^-group have their red cells agglutinated by that anti-Rh antibody: everything happens as if, in the development process, there occurred a real "maturation" concealing the antigen which cannot then be coated by the antibody.

These facts show that the antigen, thus shown up, is not a L.W. precursor as was supposed by Levine, but really a part of the Rhesus antigen which could be symbolised by Rh^m (m for Macaca) and which is common to some cercopithecoidea and to Rh^+ or Rh^- man. In man, however, from post-natal life this antigen stays more or less concealed. Wiener considers that there exist between Rh^m and human Rh the same relations as between Tj (a) and P (only Tja+ subjects can be P+: hence the conclusion that motives proper to P cannot graft themselves only on Tj(a) antibody (15).).

However, this does not explain why anti-Rh^m antibody (initial Landsteiner and Wiener anti-Rhesus) does not agglutinate either the Chimpanzee red cells (which carry certain Rhesus motives also present in human Rh antigen) or human Rh- red cells. In the first case, it must be admitted that some of the sites, common to anthropoidea are absent from the Rhesus factor of the Macaque (and that the primitive anti-Rh antibodies do not then possess a corresponding specificity). On the contrary, human anti-Rh antibodies, produced by an allo-immunization, recognize certain sites proper to man and to the Pongidae, which, in man are, without doubt, more antigenic than the rest of the molecule. These facts show all the complexity of the development of the Rhesus system in primates -a complexity which only biochemistry will be able to enlighten some day.

Distribution of Rh Factors in Primates

If we make use of the antibodies of human origin and belonging to the five principal specificities at present known (anti-rh', anti-hr', anti-Rh_o, anti-rh" , anti-hr"), the following remarks can be made:

1) no cercopithecoidea red cell is agglutinated by these reagents. However, certain groups -as has been seen- carry a "standard Rhesus factor", Rh^m consisting of antigenic motives which is found again also on human Rh antigen (but which, without doubt, represents a small fraction incapable of normally assuring an allo-immunization process).

2) All the Gibbon red cells (Hylobatinae) react to anti-hr' human antibodies but react negatively to all other antibodies (including anti-Rh_o). The Gibbon accordingly carry on their red cells an hr' factor almost identical to human hr' factor (since the agglutinations seen are of the same titre as those obtained in hr' man) as these animals never have the Rh_o factor, their immunological constitution bring them very near to human rh type.

3) The red cells of Orang-Utan, of Chimpanzees and of Gorillas react constantly to anti-hr', often to anti-Rh_o but never to other antibodies usually encountered in man.

However, whereas the hr' factor seems to be constant in species of Pongidae and almost identical to the human factor, at least quantitative differences appear as far as Rh_o is concerned. In fact the force of agglutination produced by this factor with respect to specific antibodies varies a lot from one species to another. Moreover, the Gorilla (and possibly the Chimpanzee) shows a commencement of polymorphism Rh_o+/Rh_o-. Let us consider, in succession the different species of Pongidae:

- the Orang-Utan constantly presents hr' and Rh_o but the latter antigen produces very weak reactions, which indicates that it is quite different from the human factor. The Orang-Utan global Rh antigen calls to mind human Rh_o- antigen. For this reason it can be named $Rh_o{}^{-or}$ (Wiener suggests that italics should be used).

- In the Chimpanzee we constantly find hr', identical to the human type, as well as Rh_o. However, Masouredis and coll. (14) have shown that if a lot of Chimpanzee blood were treated with eluates derived from human O Rh+ red cells, previously coated with anti-Rh_o, positive subjects (for the most part) and a few negative subjects are obtained. In fact, J. Moor-Jankowski and A.S. Wiener have shown that that difference proceeded from the presence or the absence of one factor c^c belonging to the system of C.E.F. simian group, which is referred to later.

To sum up:

1° the chimpanzee Rh_o, in opposition to human Rh_o is never agglutinated by rabbit immune anti-Rh_o (rabbit immunized by the red cells of the macaque or of the baboon anti-Rh^m). It seems then that the Chimpanzee does not possess the antigenic motives common to the macaque and to man.

2° It is possible, by the absorption of human anti-Rh_o by Chimpanzee red cells to obtain fraction of anti-Rh_o antibody active solely on human Rh+ red cells, inactive on Chimpanzee red cells. This is proof of the heterogeneity of Rh_o antigen which, assuredly, is made up of many autonomous factors. Some are common to the Macaque and to man, others to man and to the Chimpanzee, others are peculiar to the Chimpanzee alone or to man alone. A.S. Wiener (17, 16) has shown that Rh_o human antigen was in fact composed of a series of autonomous factors: Rh^A, Rh^B, Rh^C, Rh^D, some of which could be lacking in rare subjects called Rh partial. It is likely that that "mosaic" structure spreads beyond strictly human factors and overlays all the sequences of the Rh system, from the Macaque to man. Maybe we are in presence of "step like" evolution.

Only an extensive study of Rh partial specificities in the Cercopithecoidea, the Pongidae and their comparison with human Rh could provide information on the meaning of that evolution. Unfortunatly partial anti-Rh antibodies are very rare.

In the Gorilla as in the Ponginae referred to above and in the Gibbon, all animals carry the hr' factor on their red cells with the same titre as in human hr' positive. Most animals also carry Rh_o factor and produce as strong reactions as those observed on human Rh+ red cells. A minority of Gorillas react much more weakly. If a human anti-Rh_o serum were absorbed with this "weak" red cells category a specific reagent would be produced which does not agglutinate strong Rh_o red cells of the Gorilla. Thus, this antibody can

Table 8

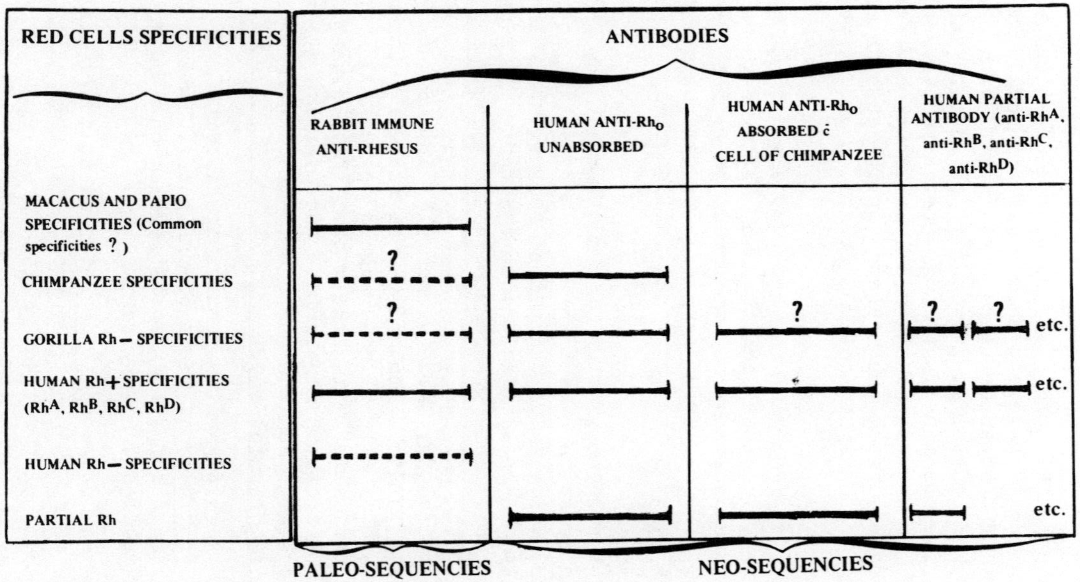

Connection of different anti-Rh antibodies and their specificity. The evolution of the structure of each factor shows that there exist motives common to several groups, undoubtedly inherited from a common ancestor (paleo-sequencies) and motives proper to each lineage, if not to each species (neo-sequencies).

—— CLUMPED CELLS - - - COATED CELLS

divide Gorillas into Rh_o+ and Rh_o-. Moreover, the study of Rh_o+ Gorilla red cells, coated by an anti-Rh_o serum proves that these cells contain about as many Rh_o antigenic receptors as human cells.

PARANTIGENS AND PARATYPES

We (18, 19) have thus named a number of specificities: X, U, W, Y, Z which are present on erythrocytes of certain primates and are spotted by the "natural antibodies" in the human serum.

Table 8b

FACTOR X	Found in all macaca, baboons, gibbons and chimpanzees studied
FACTOR Y	Proper to chimpanzees
FACTOR U	Found in the baboons
FACTOR W	Proper to gibbons W_1 : Strong agglutination W_2 : Weak agglutination W_o : no agglutination
FACTOR Z	Found in squirrel monkey and perhaps in some other species of new world monkeys. It seems to be completly independant of preceeding factors.

Legend : Parantigens found by natural antibodies occuring on human blood (J. Ruffié and Y. Marty, 1971)

1) Some are common to several phylogenically related species as for instance X factor which appears to exist in all species of Old World monkeys we have been able to study. It is the only factor seen in Macaca and corresponds to those parantigens previously defined by us.

2) Others which appear peculiar to one species or to only one group of closely related species, are constantly found in all individuals of that species -at any rate in the sample studied by us. They are antigens in the strict meaning of the term: Y factor in the

Chimpanzee, U factor in the baboon, Z factor in the squirrel monkey.

3) Last, others which are peculiar to only one species show a certain genic polymorphism as, for instance, W factor which may or may not exist in the baboon. Where it exist, it has two degrees with different agglutinating force. Accordingly three phenotypes can be encountered: W_1 which means strong agglutination, W_2, weak agglutination and W-, no agglutination.

The corresponding antibodies encountered in man, without doubt, are the result of a hetero-immunization process. In fact, it is likely that X, U, W, Y, and Z antigens are fairly widely distributed in nature.

Subjects studied to-date present 7 possible combinations. This phenomenon must depend on an antigenic polymorphism found in man but which is different from already known group systems.

Thus the nature and the intensity of the reactions observed by means of standard human anti-Rh and the type of Rh+/Rh- polymorphism shows that, as far as the erythrocytic system is concerned, the Gorilla is the anthropoid ape closest to man.

To sum up, if the evolution of the Rh system in primates is considered, the following facts can be noted:

a) the Rhesus system appears for the first time in some Cercopithecoidea (Papio, Macaca) in the form of a factor having, without doubt, a small number of antigenic motives in common with man.

b) It develops gradually in the Hominoidea: as it complifies itself it gives:

- the hr' factor, present in all anthropoid apes, from the Gorilla to the Gibbon. This factor is very close to the human factor (if not identical to it) since human anti-hr' antibodies give titers comparable to those obtained by human hr' erythrocytes whatever the species of anthropoid apes used;

- one Rh_o factor which appears only in the Ponginae. But then distinct differences, according to the species, are observed. It seems that there exists a fraction common to all the species (monkeys and man) and a fraction proper to each species. This factor is very weak in the Orang-Utan which shows it in undoubtedly a constant way. It is stronger in the Chimpanzee. But in the Chimpanzee, it is shown to consist of various fractions, one of which has been identified as c^c belonging to C.E.F. simian system;

- lastly, in the Gorilla, it is almost identical to the human factor and also presents a beginning of polymorphism (Rh_o+/Rh_o-);

- it must be noted that none of these animals possesses factors of the rh"/hr" series which seems to be peculiar to man.

3° It is only in the human species that the five principal factors are encountered: rh', hr', Rh_o, rh", hr" connected in 8 main antigenic types, and which present an important polymorphism.

Evolution of the Rhesus System

In the evolution of the Rh system is considered in the primates as a whole, a double process is observed:

a) evolution of the antigenic structure of main Rh (Rh_o) factor which, alongside motives common to several groups (without doubt inherited from the common ancestor and which may be called: paleo-factors), reveals motives proper to each lineage (if not to each species: they are the neo-factors).

b) Gradual complexification of genetic pattern with tendency towards acquisition of an increasingly wider polymorphism. These data are given on the following table:

Table 9. Distribution of Rh Antigens in Primates

	anti-Rh^m	anti-Hr'	anti-Rh_o	Antigens
Cercopithecoïdea Papio sp. Macaca sp.	+	−	−	Rh^m
Hylobatinea Gibbons	−	+	−	rh^{Gi}
Ponginea				
Orang	−	+	+	$\overline{Rh_o}^{Or}$
Chimp	−	+	++	$\overline{Rh_o}^{Chimp}$
Gorilla	−	+	+++ or −	$\overline{Rh_o}^{Gor}$ or rh^{Gor}
Hominoïdea Man	+	+ or −	+++ or −	Rh_1 Rh_2 Rh_o rh

There exists a gradual "complexification" of the genetic pattern, with addition of new fonctions at each evolutive stage and tendency towards a continually wider polymorphism.

This shows that the Boettcher theory cannot be retained (20, 21). This author had expressed the opinion that the three types defined above (Macaque, Ponginae, and man) correspond to three successive stages and that the passage from the one to the other was first by a mere duplication process, then by a triplication. The Rh genetic pattern could then be defined as follows:

a) one first pair of chromosomes exists, which carries genes responsible for the synthesis of the L.W. precursor. It possesses a "mute" mutation which, in a homozygous state, prevents all synthesis of the Rh factors and brings about the appearance of the serological phenotype called by Cepellini: Rh^{null} (---/---). The red cells of these subjects are entirely devoid of the Rh factor.

b) A second pair of chromosomes would exist, distinct from the first, responsible for the synthesis of proper Rh factors. It would consist of:

- one only locus in the Macaca and in the Papio occupied by the d mute gene. Those animals would carry only L.W. on their red cells,
- two loci in the anthropoids the first always occupied by one c (hr') synthetising gene, the second by a d mute gene in the hylobatinae, by a D in the Orang-Utan and the Chimpanzee, by a D or d in Gorilla,
- three loci in man, the first occupied by the D/d couple, the second by the C/c and the third by the E/e. This pattern is represented in Table 10.

Boettcher tries to explain the presence, in some men (in particular of the black race) of cD-/cD- subjects by a deletion bearing on the locus E/e or of -D-/-D- subjects by a deletion involving the whole zone carrying the two loci C/e and E/e. Moreover, certain rare subjects would have only one locus with d and would be devoid of any Rh factor. They could either have the L.W. precursor (---/---) or be without it (Rh^{null}). (In this case they have an anti-L.W. (anti-Rh^{m}) in their serum). A few "reversive" mutations (cD-) or (Rh^{null}) which re-establish the ancestral state, would then be encountered.

Such a theory may appear attractive; unfortunatly it does not correspond to the facts. In effect:

a) Rh_o(D) factor encountered in the different species of non-human primates is not questionable. Evolution does not affect the number of loci which remains the same: it affects the very structure of the locus which is accordingly modified. This modification affects essentially the Rh_o factor (and not rh' present in all Ponginae and Hylobatinae under a form identical, or almost identical, to the human).

b) This modification of Rh_o implies the acquisition of new antigenic specificities and sometimes by the loss of earlier specificities. There is accordingly a true "remodelling" of the molecule which shows, alongside constant structures found in all groups (paleo-factors) new structures (neo-factors) proper to certain species. This explains the crossed and non reciprocal reactions sometimes observed.

These views are illustrated by the serological behaviour of the Rh^{null} subjects. In fact, as has been previously indicated, if there exist subjects devoid of any human type Rh antigen (---/---) and also of L.W. (Rh^{m}) factor, there are also found rare subjects devoid of so-called L.W. (Rh^{m}) factor but carrying human Rh antigens (for

Table 10. Evolution of the Rh System in Primates

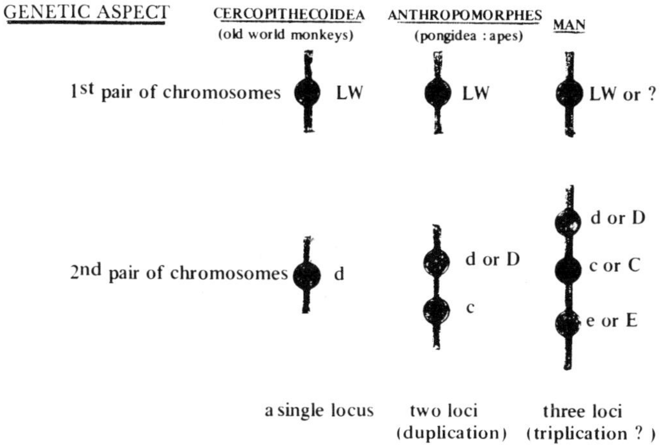

SEROLOGICAL ASPECT

CERCOPITHECOIDEA (OLD WORLD MONKEYS) : A SINGLE ANTIGEN : L. W.
No Rh FACTOR

ANTHROPOMORPHES (PONGIDEA : APES) : L. W. + ONE OR TWO FACTORS Rh
(c or c + D).

CHROMOSOME c d GIBBON
cD - ORAN-UTANG
- CHIMPANZEE
- GORILLA

MAN : L. W. + TWO TO FIVE FACTORS Rh : D, C, c, E, e
- 8 POSSIBLE MAIN TYPES OF CHROMOSOMES : CDe, cDE, cDe, Cde, CDE, cdE, cde, CdE:
- RARE NEGRO SUBJECTS HAVE c̄D CHROMOSOMES... OF ANCESTRAL TYPES
- RARE SUBJECTS HAVE NORMAL CHROMOSOMES BUT ARE LACKING LW FACTOR (A MUTATION ASSURING A TOTAL TRANSFORMATION OF THE PRECURSER).

(according to Bœttcher theory)

Table 11. Different Kinds of Antibodies Permitting to Show Up Simian Factors

1) Allo-immunization antibodies : immunization of an individual, of a species by the red cells of an animal of the same species (A. S. Wiener and coll).

Ex : A^c, B^c, C^c etc. in chimpanzees

2) Cross-immunization antibodies (or hetero-immunization) : immunization of a species by the red cells of another species (A. S. Wiener and coll).

Ex : V_1, H^c, K^c etc. in chimpanzees.

3) Natural hetero-antibodies : utilization of natural antibodies present in human serum to show up certain erythrocytic factors in monkeys (apes and monkeys). (J. Ruffié and Y. Marty).

Ex : X, W, U, Y, Z.

example Rh_1/rh) (22). The red cells of these subjects are agglutinated by anti-Rh of human origin but not by the antibody of rabbit immunized by the Macaque. Their serum contains an anti-Rh^m (anti-L.W.) which agglutinates almost all the human Rh+ red cells and capable of being absorbed by human Rh- red cells. This establishes that the antigen common to the Macaque and to man does not form a "precursor" of Rh_o factor as some have believed but presents an old antigenic grouping common to many species of primates (paleo-antigens) on which, in the course of evolution, are grafted new motives which produce certain species (neo-antigens) as has been suggested before. The only reservation which must be made is relevant to the hereditary mode of these factors. P. Tippet (23) has in effect shown that the primitive antigen (Rh^m or L.W.) and the factors of the human Rh system were carried by independent chromosome (23).

Xg(a) System

Xg(a) factor was discovered by Race and Sanger on the Gibbon red cells but not in other anthropoid apes (24). Twelve Hylobates lar lar and one Hylobate lar pileatus have been studied. But on these animals:

- 3 males were positive and 4 negative
- 4 females were positive and 2 negative

although these figures are not sufficient to enable definite conclusions to the drawn, that distribution suggests that, in man, the locus which conditions the Xg(a) system is carried on the Gibbon X chromosome.

Simian blood groups. The existence of blood groups proper to non-human primates was postulated, as far back as 1962, by A. S. Wiener and demonstrated in 1964 by A.S. Wiener, J. Moor-Jankowski and coll. (25). Subsequently, these authors have carried out much research work on the simian factors. The reagents were obtained by two methods:

1) allo-immunization: the immunization of subjects belonging to the same species (Ex. Injection into the Chimpanzee of cells from another Chimpanzee).

2) Hetero-immunization (or cross-immunization): immunization of an animal by the red cells of an animal belonging to another species (Ex. Injection of human red cells into the Chimpanzee).

We have suggested a third study consisting of the utilization of "natural" antibodies present in man to recognize certain simian factors. "Parantigens" causing paratypes are shown up. The three categories of antibodies thus obtained are illustrated in Table 11.

Let us examine the first two categories, and then the third, whose phylogenetic significance is not identical:

a) antigens produced by allo-immunization or by cross-immunization.

1) In the Chimpanzee

A.S. Wiener and J. Moor-Jankowski (26) have brought into evidence some ten factors named as follows:

Table 12. Some of the Blood Factors of Chimpanzees (Pan satyrus) and Their Frequencies

BLOOD FACTORS	SOURCE OF REAGENT		POSITIVE	NEGATIVE	TOTALS
A	Human anti-A serum	N	182	32	215
		%	85.0	15.0	
V_1	Cross-immune serum*	N	56	116	172
		%	32.6	67.4	
N^V (V)	Vicia graminea lectin	N	58	63	121
		%	47.9	52.1	
A^c	Isoimmune serum	N	72	86	158
		%	45.6	54.4	
B^c	Isoimmune serum	N	104	54	158
		%	65.8	34.2	
C^c	Isoimmune serum	N	86	72	158
		%	54.4	45.6	
E^c	Isoimmune serum	N	27	71	98
		%	27.6	72.4	
F^c	Isoimmune serum	N	52	46	98
		%	53.1	46.9	
G^c	Isoimmune serum	N	73	25	98
		%	74.5	25.5	
H^c	Cross-immune serum	N	75	23	98
		%	76.5	23.5	
I^c	Isoimmune serum	N	32	9	41
		%	78.5	21.5	

* Serum prepared in chimpanzees injected with human red cells.

(according to A. S. Wiener and J. Moor-Jankowski)

Some of these factors are linked to two main systems:

- V.A.B. system which includes four factors: V, V_1, A^c and B^c. A^c and B^c were obtained by allo-immunization: V_1 by cross-immunization (man-chimpanzee). Moreover, the V factor is identical to N^v factor of MN system and like it, is recognized by Vicia graminea extracts. V_1 and V have the same relations between them as A_1 and A antigens in the ABO system. It accordingly seems that the V A^c B^c system is more or less related to the MNSs system. By the use of three reagents (anti-V_1, anti-A^c, anti-B^c) Wiener and coll. obtain eight phenotypes corresponding to the following genotypes (27, 28):

Table 13. Simian-Type V-A-B Blood Group System of Chimpanzees (Pan satyrus) (according to A. S. Wiener and J. Moor-Jankowski)

PHENOTYPES DESIGNATION	NUMBER	%	REACTION WITH SERA ANTI-V_1	ANTI-A^c	ANTI-B^c	CORRESPONDING GENOTYPES[1]
v.O	4	3.0	−	−	−	vv
v.A	21	15.8	−	+	−	v^Av^A and v^Av
v.B	36	27.1	−	−	+	v^Bv^B and v^Bv
v.AB	25	18.8	−	+	+	v^Av^B
V.O	16	12.0	+	−	−	V^oV^o and V^ov
V.A	11	8.3	+	+	−	V^AV^A, V^Av^A, V^Av, V^AV^o and V^ov^A
V.B	9	14.3	+	−	+	V^BV^B, V^Bv^B, V^Bv, V^BV^o and V^ov^B
V. AB	1	0.8	+	+	+	V^AV^B, V^Av^B and v^AV^B
Totals	123	100.0				

1 For simplicity, the theoretically possible genes v^{AB} and V^{AB} are omitted from this table

- C.E.F. system: also described by the same authors consists of three factors: C^c E^c F^c. Of the eight possible phenotypes, only 5 have been encountered to-date: cef. CeF, CEF, and Cef. Table 14 gives the probable correspondence of phenotypes and of genotypes.

The serological behaviour is the same as that of the human Rhesus system to which it is probably related. We have already indicated how human anti-Rh_o absorbed by red cells of certain chimpanzees lost their anti-Rh_o specificity but preserved an antibody capable of agglutinating again the red cells of some Chimpanzees. This antibody recognizes a c^{-c} factor, antithetical to C^c; it accordingly belongs to CEF system which argues in favour of a connection between the Rh system and the CEF system (29).

Just as for the preceding system, it can be agreed:
a) that both human Rh and simian CEF share some paleo-factors

Table 14. Simian-Type C-E-F Blood Group System of Chimpanzees (Pan satyrus)

PHENOTYPES DESIGNATION	NUMBER	%	REACTION WITH SERA ANTI-C^c	ANTI-E^c	ANTI-F^c	CORRESPONDING GENOTYPES
cef	38	39.6	−	−	−	c/c
CEf	29	30.2	+	+	−	CE/CE, CE/C and CE/c
CeF	8	8.3	+	−	+	CF/CF, CF/C and CF/c
CEF	20	20.8	+	+	+	CE/CF, CEF/C, CEF/c, CEF/CEF, CEF/CE and CEF/CF
Cef	1	1.1	+	−	−	CC and Cc
Totals	96	100.0				

(according to A.S. Wiener and J. Moor-Jankowski)

which they inherited from a common ancestor and which have not been modified since;

b) that there exist neo-factors peculiar to men or to the chimpanzee, which appeared with the differentiation of the species.

There is no doubt that that diversification, shown in the table below, is connected with certain chromosomic changes which are characteristic of that speciation.

It is not possible to say yet if the other factors discovered to-date (H^c, L^c, I^c) group themselves in autonomous systems.

2) In the Gibbon

Thanks to allo-immunization, it has been possible to define three factors in the Gibbon: A^g, B^g and C^g. The first two are species antigens. In fact:

- A^g seems proper to Hylobates lar lar (all the representatives of that sub-species have it);
- B^g is proper to Hylobates lar pileatus where it seems there also to be general;
- C^g is encountered in some Hylobates lar lar (and, contrary to the two preceding factors, possesses a genetic polymorphism).

Table 15 sums up these results (30, 31).

The other anthropoid monkeys have not, to-date, been the subject of very advanced studies.

3) In the Baboon (Papio cynocephalus)

Two factors belonging to the same have been described: A^p and B^p. Probably three other factors exist.

4) In the Rhesus monkeys (Macaca mulatta)

Five different specific reagents have been produced; they recognize the factors called respectively, A, B, C, D, E. The first four are well identified. They have been named A^{rh}, B^{rh}, C^{rh} and D^{rh}. It is possible that these factors which present a certain polymorphism may be also present in the Macaca cynomologus and in the Macaca speciosa. Further, A^{rh}, B^{rh}, and C^{rh} seem to be constant in the Baboon (Papio cynocephalus) whereas D^{rh} is always absent in this animal. These factors accordingly do not present a polymorphism in that species.

Lastly in some Celebes black apes, A.S. Wiener and J. Moor-Jankowski have shown up by hetero-immunization of the rabbit, an A^{ha} factor which can be present or absent (32). (In 11 subjects studied, 4 were positive and 7 negative).

It is probable that such systems exist in all species, but it will be long before they are identified.

b) Parantigens and paratypes.

We have thus named a series of factors recognized by antibodies

Table 15. Simian Blood Groups of Gibbons

NAME	ANTIBODY	TYPE	SPECIES
Ag	Allo-immunization	Species specific (monomorphism)	Hylobates lar lar
Bg	Allo-immunization	Species specific (monomorphism)	Hylobates lar pibatus
Cg	Allo-immunization	Polymorphism	Hylobates lar lar

Table 16. Reactions of Four Isoimmune Rhesus Monkeys (Macaca mulatta) to Antisera[1]

TESTED AGAINST RED CELLS OF	SERA OF RHESUS MONKEYS			
	Rh n° 1 ANTI-A^{rh}	Rh n° 2 ANTI-B^{rh}	Rh n° 4 ANTI-C^{rh}	Rh n° 6 ANTI-D^{rh}
Rhesus monkeys (Macaca mulatta) - N. Y. U. laboratory animals				
Rh n°1	−	+	+	−
Rh n° 2	+	−	+	+
Rh n°3	+	+	+	+
Rh n° 4	+	−	−	+
Rh n° 6	−	+	+	−
Rh n° 8	+	+	+	+
Average titers (units[2])				
Saline agglutination	0	6	1	4
Antiglobulin	70	8	90	150
Ficinated red cell	0	3	80	70
Rhesus monkeys (Macaca mulatta)				
Dealer's stock animals				
Number positive	47	39	47	32
Number negative	0	0	0	15
Crab-eating macaques (Macaca cynomologous)				
Positive	8	2	12	10
Negative	4	1	0	2
Stump-tail macaques (Macaca speciosa)				
Positive	3	0	3	0
Negative	0	2	0	3
Baboons (Papio cynocephalus)				
Positive	20	3	20	0
Negative	0	0	0	20

1 Antiglobulin method except where indicated
2 The titer is the reciprocal of the highest dilution that produces agglutination.

(according to A. S. Wiener and J. Moor- Jankowski)

spontaneously present in human serum (4, 33, 34). These factors are either proper to one species (and may, in this case, present or not, a certain polymorphism: Ex. W in Gibbon) or common to related species (X for the old world monkeys and apes).

HISTO-COMPATIBILITY ANTIGENS

We shall make a brief reference to these systems. They are a recent discovery and their structure is today being studied. Every day new information is being gathered and we can only surmise that histo-compatibility factors will be of considerable interest in the study of human genetics.

Accordingly, only provisional schemas can be proposed: the best known system relates to HLA and to the evolution of primates. Very likely the antigens are all nucleate cells of the organism: all these details were recently set out at the Fifth International Histocompatibility Workshop Conference, Evian, 1972 (35).

Human HLA system consists of two series of allelomorphs (perhaps three) closely bound on the same chromosome. The first series includes: HL-A_1, HL-A_3, HL-A_{11}, HL-A_2, HL-A_9, HL-A_{10}, W-19, W-28, blank. The second series includes: HL-A_{12}, HL-A_{13}, HL-A_5, HL-A_7, HL-A_8, W-18, W-5, W-15, W-17, W-10, W-27, W-22, W-14, Da-24, Da-30, Da-31, blank.

The recombination rate between these two loci is slightly lower than 1%. The number of possible phenotypes should be much greater than 10,000 while the number of genotypes is quite below 20,000. "The HL-A system is the most polymorphic system ever to have been studied in man. It has almost specific configuration similar to finger prints" (J. Dausset) (36). The cause of that polymorphism are well as of its persistence is still unknown. It is possibly connected with the selective value of some heterozygotes which might be better equipped in the way of immune response.

However, research work so far undertaken to expose the relationship between certain diseases and certain HL-A types, has not yielded encouraging results.

Few non-human primates have, to-date, been studied. The most important studies by H. Balner and coll. (37) and by R.J. Metzgar (38) refer to the chimpanzee, the gorilla and the Rhesus monkey. It seems very likely that the chimpanzee and, probably, the gorilla possess a histo-compatibility antigen system very closely related to the human HL-A system, as they all present some common antigenic receptors and are probably controlled by the same bilocic or, possibly, trilocic model. However, while chimpanzee allo-immunization can produce can produce antibodies which identify a number of human factors, the contrary is not the case.

In Macaca mulatta, an Rh L-A system is found with a structure identical to that of apes or human HL-A. Some fifteen specificities have, to-date, been identified; they are controlled by two loci closely connected on the same chromosome. Of the known factors, some might be similar to those found in man and in the chimpanzee. At any rate, they appear to have analogous receptors. Rogentine and coll. (39) have proved that in these two cases the amino-acid composition is the same.

It is likely that the HL-A system, by its far reaching polymorphism, will shortly yield abundant information regarding the link between the higher species of primates and their evolution. Perhaps other closely connected systems related to other mammals will be discovered, as for example, the dog which presents a DL-A system.

Immunoglobulins

The distribution of allotypes of immunoglobulins in primates was recently studied providently valuable information in the field of evolution. Research work has, in particular, been limited to the Gm and Isf systems. This work is essentially due to L. Van Loghem (40), Ropartz (41), Blanc and Ruffie (43).

The distribution of factors. The Gm system.

The Gm system is carried by sub-classes of γG3 and of γG1 types of γG.

a) γG3:

- from the *Cercopithecoidea* stage (*Macaca mulatta*, *Papio sp.*) a first type of γG3 chains carry the four factors Gm 10, 11, 25, 24;
- in the *Hominoidea*, factor Gm(5), in addition, is seen. In the Orang-Utan it is added to the preceding factors. In the chimpanzee, Gm(6) appears in addition, whereas man appears to be the only one to carry Gm(14) on his γG3 chain;
- knowledge of the distribution of Gm factors on the other γG3 type is still inadequate. All that is known is that Gm(21) appears only in man.

b) γG1:

- in regard to the first type: it is the first Gm(1) which appears in the Macaque stage. Gm(17) is present in the baboon; it will be seen again in the Hominoidea. Gm(2) is proper to man;
- in regard to the second type: its distribution is as yet incompletely known. It is known that Gm(8) is present in all Hominoidea and that Gm(3) is proper to man.

Isf system. Isf_1 appears on γG1 from the chimpanzee stage. That distribution is illustrated on Figure 7.

Inv system. Inv allotypes have not been demonstrated either in apes or in monkeys although the amino-acid responsible for Inv(3)

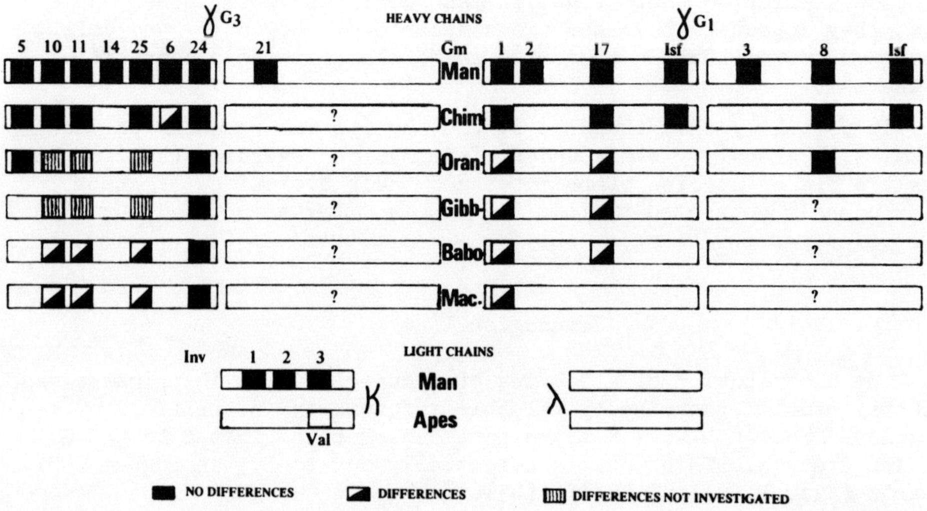

Figure 7. Evolution of Immunoglobulins Types (Gm) in Primates (γG_1 and γG_3 Chains)

Table 17. Difference of Structure between γ_1 Chains in Man Gm (1), in Chimpanzee, in Baboon, and in Man Gm (-1)

		Anti-Gm(1)	
		7	29
MAN Gm (1)	ASP - GLU - LEU - THR - LYS	+	+
CHIMPANZEE	ASP - GLU - LEU - THR - LYS	+	+
BABOON	GLU- GLU - LEU - THR - LYS	+	−
MAN Gm (-1)	GLU- GLU - MET- THR - LYS	−	−

allotype is present on the light K chains of chimpanzees.

Structure of Factors

It was at the orgin thought that the structure of Gm or Isf factors was the same in the different species of primates. In fact, it can change. Gm(17) revealed by anti-Gm(17) obtained by immunization of rabbit, produces negative reactions with the Macaque but positive reactions with the baboon, the chimpanzee and man; whereas anti-Gm(17) of human origin produces negative reactions with the Macaque, the Baboon, the Gibbon, the Orang-Utan, the Gorilla and positive with the Chimpanzee and man.

The antigen representing the Gm(17) would thus consist of two parts:

- one constant part, encountered from baboons to man $Gm(17^b)$ (but lacking in the Macaque)
- one special part, proper to the chimpanzee and to $Gm(17^h)$ man.

The factor present in the baboon could be symboled by $Gm(17^b)$. The human antigen would correspond to $Gm(17^b) + Gm(17^h)$. The same sort of difference has been observed by other authors and by ourselves in respect to Gm(1) when it is studied with animal antibody or human antibodies; it is known to-day that these differences are due to amino-acid substitutions. In fact, Baboon Gm(-1) becomes Gm(1) type through a substitution of leucine by methionine; baboon Gm(1) becomes chimpanzee or human Gm(1) by another substitution: aspartic acid for glutamic acid, as shown in Table 17.

The same phenomenon can be noticed on the factors carried by the chains of sub-classes $\gamma G3$: Gm(10), Gm(11), Gm(25). If these allotypes are studied by means of a rather numerous series of serum-tests, dissociation are observed between human type factors (Cercopithecoidea).

This shows that new structures appear on the Gm allotypes as we move up the primate ladder. Even the human factors considered today as equivocal are in truth composed of a mosaic of factors and might one day, in their turn, split into autonomous structures, as have already a number of Gm factors (mosaic of Gm(5)).

CONCLUSION

The comparative study of immunogenetics of primates is yet at its beginning, but it already reveals itself very useful. It confirms and defines accurately the main lines of the traditional classification:

1) Prosimii are a class apart: they form very ancient groups as no immunological factor, closely related to human antigens, is found in

Table 18. Reactivity's Difference of Anti-Gm Sera towards Primates' Sera

Anti-Gm (17)	Man* Gm(17)	Man* Gm(-17)	Chimpanzee (nineteen)	Baboon (sixty)	Macaca (two)
SNAGG CAS LAP 92	+ +	− −	+ +	− +	− −
Anti-Gm (1)	Man* Gm(1)	Man* Gm(-1)	Chimpanzee (nineteen)	Baboon (seventy five)	Macaca (two)
29 SNAGG 7 SNAGG	+ +	− −	+ +	− +	− +
Anti-Gm (10)	Man* Gm(10)	Man* Gm(-10)	Chimpanzee (nineteen)	Baboon (seventy five)	
2 1	+ +	− −	+ +	± −	
Anti-Gm (11)	Man* Gm(11)	Man* Gm(-11)	Chimpanzee (nineteen)	Baboon (seventy five)	
3 1	+ +	− −	+ +	+ −	
Anti-Gm (25)	Man* Gm(25)	Man* Gm(-25)	Chimpanzee (nineteen)	Baboon (seventy five)	
3 1	+ +	− −	+ +	± −	

* Europoïds - Mongoloïds - Negroïds

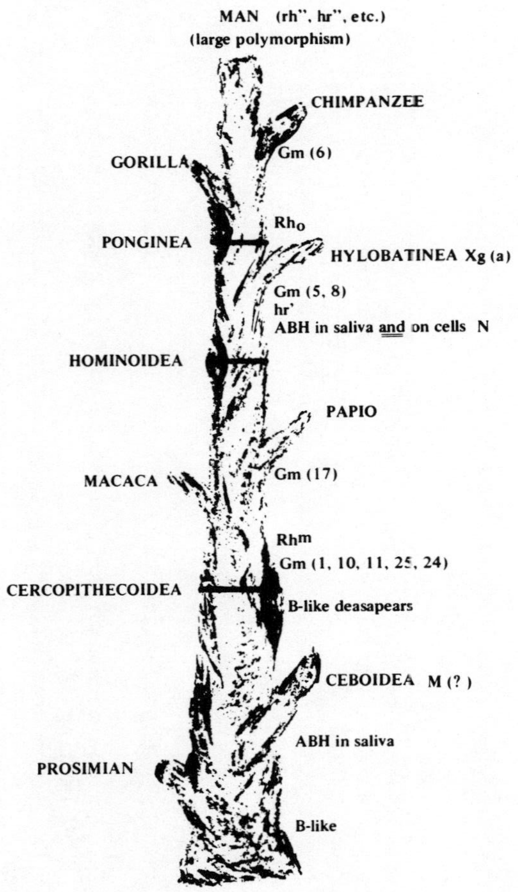

Figure 8. Scheme of Immunogenetic Evolution of Primates.

them, with the probable exception of a B-like present on the red cells.

2) The Ceboidea, confined to the New World, still have the B-like: but they present in their saliva the ABO genetic pattern which will be found again in man (with, however, the loss of certain genes in a few species).

3) The Cercopithecoidea no longer have this B-like. They retain the ABO system whose factors are in the secretions. Further, they carry several immunoglobulin allotypes presenting some fractions identical to what is found in man; the antigenicity of Gm factors increasing from the Macaques to the Baboon.

4) Anthropoid apes are, from the immunological standpoint, very close (if not identical) to the human group. It is in them that ABH factors cross over to the red cells: some present the first N factor and a Rh_o factor almost identical to that of man. Gm factors multiply and, at least in the chimpanzee, are almost identical to the human Gm factors.

But it is particularly in the field of the modalities of the evolution that comparative immunogenetics supplies to us the most important information.

1) The evolutive process has followed a common trunk which crosses a certain number of "immunological levels":

Prosimians	B-like
Ceboidea	ABH in saliva
Cercopithecoidea	Rh_m
	Gm(1, 10, 11, 25, 24)
Hominoidea	ABH in saliva and on cells
	rh'
	Gm(5, 8)
Ponginae	Rh_o
Man	rh", hr", Gm(2)

2) Each lineage was probably formed from only one individual or from a small number of them according to the robertsonian diagram by B. Chiarelli. This explains the loss, in several groups, of certain genes present in the common ancestor.

3) Each one of these lineages, after becoming autonomous, has continued to evolve on its own, giving rise to a diversifying evolution processus which we have studied elsewhere (44, 45).

This explains the appearance of simian blood fators which are either autonomous or related to systems already formed in the ancestral tree (Ex. VAB system of chimpanzees related to MNSs system: CEF system related to Rhesus system). The sequences of these simian systems, without doubt, form part of the same genetic whole.

The labelling of chromosomes by genic markers, by combining the efforts of cytogenetists and immunologists make it possible to throw light -in an altogether new way- on the modalities and the ways of evolution in primates.

References

1. Ingram, V.M. 1958. Biochim. Biophys. Acta, 28, 539-545. Biochim. Biophys. Acta, 36, 402-411

2. Ingram, V.M. 1959. Nature, 183, 1795-1798

3. Landsteiner, K. and Miller, C.P. 1925. J. Exp. Med., 43, 853-862

4. Ruffie, J. 1970. C.R. Acad. Sci., 271, 2437-2439

5. Moor-Jankowski, J. and Wiener, A.S. 1969. Primate in Med., 3, 64-77, Karger, Basel N/Y

6. Moor-Jankowski, J. and Wiener, A.S. 1971. Blood group of primates: their contribution to taxonomy and phylogenetics. Medical primatology 1970. Proc. 2nd Conf. Exp. Med. Surg. Primates N.Y.1969 Karger, Basel N/Y, Pp. 232-244

7. Wiener, A.S. and Gordon, E.B. 1960. Amer. J. of Phys. Anthrop., 18 (4), 301-311

8. Wiener, A.S. and Gordon, E.B. 1956. Brit. J. Haemat., 2, 305

9. Chiarelli, B. 1967. Caryological and hybridological date for the taxonomy and phylogeny of the old world primates. Proc. of a round table in Turin, June 1967."Taxonomy and phylogeny of old world primates with references to the origin of man". Edited by Rosenberg et Sellier

10. Ruffie, J. 1970. Revue Française d'Anthropologie, 10, cahier 4

I2. Landsteiner,K. and Wiener, A.S. I940. Proc. Soc.Exp.Biol.Med. 43, 223.

I3. Murray,J.; Clark, E.C. I952. Nature Lon. I69, 886.

I4. Masouredis,S.P.; Duprey,M.E.; Elliot,M. I967. J.Immunol.,28,8.

I5. Wiener, A.S. I968. Amer.Journal of Clin.Path. 49, I, I08.

I6. Wiener A.S., Unger,L.J. I959. J.A.M.A. I69,696.

I7. Wiener, A.S.; Lester,M.D. and Unger,L.J. I962. Transfusion, 2, 4,230.

I8. Ruffie,J.; Marty,Y. I97I. C.R.Acad.Sciences 273, I4, I244-I246.

I9. Ruffie,J.; Marty,Y. I972. Journ. of Med. Primatology, 3

20. Boettcher,B. I964. Vox Sanguinis 9, 64I.

2I. Boettcher,B. I965. Amer.Journ.of Hum.Genet. I7,4,308.

22. Levine,P.; Celano,M.J.; Wallace,J.; Sanger,R. I963. Nature London I98,596.

23. Tippett,P. Serological study of the Inheritance of Unusual Rh and other blood group phenotype. Ph.D. Thesis University of London I963.

24. Gavin,J.; Noades,J.; Tippett,P.; Sanger,R.; Race,R. I964. Nature 204, 4965, I322.

25. Moor-Jankowski,J.; Wiener,A.S.; Rogers,C.M. I964. Nature 202, 663.

26. Moor-Jankowski, J.; Wiener, A.S. I968. Blood groups of non-human primates: summary of currently available information. Primates in Medicine, S.Karger, Basel/N/Y. , 1, 49.

27. Wiener,A.S.; Moor-Jankowski,J. I965. Transfusion 5,64.

28. Moor-Jankowski,J. and Wiener,A.S. I965. Nature 4969, 369.

29 Wiener, A.S. Trans. of the N.Y.Acad.of Sc. Paris, II, 27,5,488.

30. Moor-Jankowski, J.; Wiener,A.S. and Gordon,E.B. I965. Transfusion 5,235

3I. Wiener,A.S.; Moor-Jankowski,J.; Gordon,E.B.; Daumy,O. and Davis, J.H. I966. Int.Arch.Allergy 30, 466.

32. Moor-Jankowski,J.; Wiener,A.S.; Gordon,E.B.; Guthrie,C.B. I965. Folia primat.

33. Ruffie,J.; Mme Dianorah-Schoenwetter. I97I. C.R.Acad.Sc. 272, I, I59-I62.

34. Ruffie,J.; Bouloux,C.; Ruffie,P. I969. C.R.Acad.Sc. 269, 25, 2624.

35. Ruffie,J. I972. The Hominid Evolution. Ve Conférence-Workshop Internationale sur l'Histocompatibilité, Evian 23-27 mai.

36. Dausset,J. I97I. Transplantation proceedings III, 3, I/39.

37. Balner,H. et al. I972. HL-A typing of chimpanzee lymphocytes Ve Conférence-Workshop Internationale sur l'Histocompatibilité Evian 23-27 Mai.

38. Dorf,M.E. and Metzgar,R.S. I970. The distribution of human HLA antigens in chimpanzees and gorilla. Med.Primatology - Karger, Basel I97I.

39. Rogentine,G.N.; Merritt,C.B.; Vool,L.A.; Ellis,E.B.; Darrow,C. I972. Transplantation proceedings IV, I, 55-62.

40. Van Loghem,E.; Shuster,J. and Fudenberg,H. I968. Vox Sanguinis I4, 8I-84.

4I. Rivat,L.; Rivat,C.; Ropartz,C.; Krupe,M.; Schmitt,J. and Dubouch,H. I968. Human Genet. 6, 2I5-224.

42. Ruffie,J. I97I. L'Anthropologie 75, I-2, 57.

43. Blanc,M.; Bouloux,C.; Ducos,J. I972. Différence de réactivité vis-à-vis de sérums de primates, d'anticorps anti-Gm I7 d'origine humaine et animale. Com.Soc.Franç. d'Immunologie, I3-I4 avril.

44. Ruffie,J. I966. The New Zealand Medical Journal 65, 4I2,844.

45. Ruffie,J. I966. Hemotypologie et évolution du groupe humain. Monographie du Centre d'Hémotypologie du C.N.R.S. -Hermann-

COMPARATIVE VIROLOGY IN PRIMATES

S.S. Kalter
Division of Microbiology and Infectious Diseases
Southwest Foundation for Research and Education
San Antonio, Texas, U.S.A.

Numerous species of nonhuman primates (Table I) are currently in use in laboratories throughout the world. These animals are obtained from various parts of Asia, Africa, South and Central America and number in the thousands each year. The diversity of the simian species used for research is only matched by the multitude of scientific disciplines involved. In general, the individual using an animal has little information about the animal's source or its past clinical history. Perhaps more unfortunate, most investigators fail to be concerned about this lack of information. They consider monkeys and apes to be "test-tubes", but are indifferent to the presence of any flora and fauna indigenous to the animal under study. It is only when some difficulty occurs, such as a threat to the health of the investigator or to the animal colony, that concern is expressed.

Within the last few years, several incidents have occurred resulting in deaths of both man and animal, necessitating a re-evaluation of the relationship between both parties. The 1934 description by Sabin and Wright (174) of a fatal human case of *Macaca* species-carried B virus (*Herpesvirus simiae*), as well as recognition of subsequent occurrences, did not do more than make the personnel involved in the events aware of them. The realization that chimpanzees transmit hepatitis to humans (82), followed by the explosive Marburg-Belgrade episode (140) did awaken the scientific community to the potential danger. Coupled to these events were the losses of large numbers of monkeys of various species as a result of infections in several major colonies in different parts of the world. In other instances, while deaths may not have occurred, large numbers of experimental animals were incapacitated, and this resulted in extensive loss of investigative time and in expenditures of funds not anticipated in initial planning. Included among these infections were such virus

Table 1. Genus and Species of Nonhuman Primates Used in the Virus Laboratory

Genus	Species	Common Name
Old world		
Gorilla	G. gorilla	Gorilla
Pan	P. troglodytes	Chimpanzee
	P. paniscus	Chimpanzee (pigmy)
Pongo	P. pygmaeus	Orangutan
Hylobates	H. lar	Gibbon
Papio	P. cynocephalus group: includes P. anubis, P. papio, P. ursinus (Chacma)	Baboon
	P. hamadryas	Baboon (hamadryas)
Theropithecus	T. gelada	Baboon (gelada)
Cercopithecus	C. aethiops	Grivet Aethiops group
	C. sabaeus	Green other groups-Mono,
	C. pygerythrus	Vervet Nicitans, etc.
	C. talapoin	Talapoin
Presbytis	P. entellus	Langur
	P. cristatus	Langur
Erythrocebus	E. patas patas	Patas
Macaca	M. mulatta	Rhesus
	M. fascicularis (preferred to M. irus, M. cynomolgus, and M. philippinensis)	Cynomolgus
	M. radiata	Bonnet
	M. nemestrina	Pigtailed macaque
	M. cyclopis	Formosan rock macaque
	M. speciosa	Stumptailed macaque
	M. fuscata	Japanese macaque
New world		
Saimiri	S. scieureus	Squirrel
Aotus	A. trivirgatus	Owl
Alouatta	A. belzebul	Howler
Ateles	A. paniscus	Spider
Cebus	C. capucinus	Capuchin
Lagothrix	L. lagothricha	Woolly
Pithecia	P. pithecia	Sakis
Callithrix		Marmoset
Cebuella	C. pygmaea	Marmoset (pigmy)
Saguinus	S. tamarin	Marmoset (tamarin)
Leontideus	L. rosalia	Marmoset (gloden lion tamarin)

diseases as monkeypox, Yaba, Yaba-like, simian hemorrhagic fever, and rubeola. Other viruses produced infection but not clinical disease as evidenced by seroconversion of humans and animals to positives.

At this time we will not consider non-viral infections, for example mycobacterium, salmonella, shigella, etc. The purpose of this chapter is to provide the interested student with information concerning how viral infections of nonhuman primates may influence an experimental program. We will outline the experiences (natural and experimental) of an animal from time of capture and attempt to define extraneous influences in terms of viral infections. Methods for minimizing extrinsic factors responsible for virus problems will be described as will suggestions for circumventing many of these events.

CAPTURE AND HOLDING IN EXPORTING COUNTRY

The majority of animals now in use obtained through commercial sources who have contacts in the country of origin. These contacts, for want of a better term, are referred to as "trappers". In most instances, this is a misnomer as the trapper obtains animals by employing natives who either collect the monkeys from other natives or actually capture animals by one or another method. The methods employed are similar throughout the world and consist of nets, snares, triggered-door traps, or killing a mother to obtain the young. Sites of capture are selected by information gathered from field surveys, information passed on from local citizens, and so on. This selection is often based upon the nuisance value of the animal to farmers. Accordingly, many animals that are described as coming from "the bush" without previous human contact have, in fact, had such contact. These animals, at least in Asia and Africa, live peripheral to farming areas. They derive their food by raiding crops, and they generally share the same water supply with the local human population. Very rarely are animals obtained that have had little or no contact with the human populations!

Supplementing routine trapping practices is the purchase of animals from natives who have kept them as pets or have trapped them to provide an additional source of family revenue. Generally, the individuals involved in these pracitces have limited education, extremely low incomes, and subsist under very primitive conditions. Sanitary conditions or any understanding of simple public health measures at this stage are simply nonexistent. Exchange of organisms in both directions (i.e., man-to-monkey or monkey-to-man) must be extreme, although it has not been studied. Nutrition is minimal, and large numbers of animals die. Surviving animals are transferred, either directly or through a series of transfers, in various types of boxes or cages to a holding facility in a major city. Time between capture and arrival at the holding facility is variable and is dependent upon distance and mode of transfer. In many instances, this period is of several days duration, with the animal living in boxes that are carried on foot, by boat, horse, wagon or truck.

Holding facilities very rarely provide improved hygienic conditions. The calibre of the individual handling the animals is generally of the same level as those involved in the capture, and the physical plant is maintained at only a superficially higher level of cleanliness. Adding to the health problems of the animals at the holding facility are two important considerations (a) use of the gang cages to hold large numbers of animals, and (b) contact of the animal with other species of animals (simian or otherwise) and/or with the same species from other geographic sites. Nutrition and watering are somewhat better but frequently are inadequate and dirty. The stress on these animals, especially those coming from broken troops or those that habitually live in small groups or as pairs, must be enormous. When sufficient animals are available to complete an order, they are individually caged in small wooden boxes for shipment via air; rarely are they shipped by boat. Cursory veterinary inspection may be provided prior to shipment.

SHIPMENT

An attempt to standardize shipping of animals has been made, but deficiencies still exist (2II). Inexpensive wooden cages made in the exporting country are used to house the animals. A small screened area permits air exchange, but few other conveniences are taken into consideration. Animals originating in India and Africa are generally transshipped through London where they are fed and watered. Animals from the Orient or South America are similarly transshipped. While credit should be given for good intentions, these transshipping facilities unfortunately give little recognition to the possible transmission or exchange of microorganisms. Animals of all species are brought together into a common holding area, although attempts to segregate them are made. None of these establishments have the capability to do more than provide food and water, remove dead animals, and see that the remainder are placed on the next carrier. Nothing in the way of recaging or sterilization is available. Common handlers are involved in all operations. In all fairness, it should be emphasized that these facilities provide what they can in the way of humane treatment.

IMPORTATION

Animals are usually received in the country of destination in a debilitated and stressed condition. A number of importers have recently attempted to improve conditions in their holding facilities. Clean cages, fresh water, and food are provided. The animals are allowed to settle down and recover some semblance of stability.

Table 2. Grouping of simian viruses, including biological characteristics and original host source

- Simian adenoviruses (DNA*, ether resistant, 70-80 nm, acid stable, nuclear inclusions, $HA^{\pm}$)
  - SV1, 11, 15, 17, 20, 23, 25, 30, 31, 32, 33, 34, 36, 37, 38---rhesus, cynomolgus
  - SA7, 17, 18---African green
  - V340, AA153---baboon
  - C-1, PAN 5, 6, 7---chimpanzee
  - Sq M-1---squirrel monkey
- Simian picornaviruses (RNA**, ether resistant, 18-38 nm, acid stable, cytoplasmic inclusions, $HA^{\pm}$, $MgCl_2$ stabilized)
  - Enteroviruses
    - SA2, 6, 16, 18, 19, 26, 35, 42, 43, 44, 45, 46, 47, 49---rhesus, cynomolgus
    - SA5---African green
    - A13---baboon
  - Unclassified picornaviruses
    - SV28---rhesus
    - SA4---African green
- Simian reoviruses (RNA, ether resistant, 70-77 nm, acid stable, cytoplasmic inclusions, HA+, $MgCl_2$ stabilized)
  - SV12 (Reo 1), SV59 (Reo 2)--rhesus
  - SA3 (Reo 1)---African green
- Simian papovaviruses (DNA, ether resistant, 40-57 nm, acid stable, nuclear inclusions, HA-)
  - SV40---rhesus
  - SA12(?)---African green
- Simian herpesviruses (DNA, ether resistant, 120-250 nm, acid labile, nuclear inclusions, HA-)
  - Type A
    - *H. simiae* (B virus)---rhesus
    - SA8---African green
    - *H. tamarinus* (platyrrhinae)---squirrel
    - SMV---spider
  - Type B
    - SA6---African green
    - *Herpes saimiri* (?)---squirrel
    - Liverpool vervet agent (?)---African green, patas
- Simian poxviruses (DNA, ether or chloroform sensitive, 200-325 nm, acid labile, cytoplasmic inclusions, $HA^{\pm}$
  - Vaccinia-Variola group
    - Monkey pox---cynomolgus
  - *Moluscum contagiosium* group (?)
    - Yaba---rhesus
    - Yaba-like disease (benign epidermal pox, 1121)---rhesus
- Simian myxoviruses (RNA, ether sensitive, 150-250 nm, acid labile, cytoplasmic, and nuclear inclusions, HA+)
  - SV5---rhesus (man?)
  - SV41---cynomolgus
  - SA10---African green
- Foamy virus*** (RNA (?), ether sensitive, 100-300 nm, acid labile, no inclusions, HA-)
  - Type 1---rhesus, African green
  - Type 2---African green
  - Type 3---African green
  - Type 4---squirrel
  - Type 5---galago
  - Type 6---chimpanzee
  - Type 7---chimpanzee
- Miscellaneous viruses
  - SA11---African green
  - SHF---rhesus
  - Marburg agent---African green

* DNA = deoxyribonucleic acid.
** RNA = ribonucleic acid.
*** Presently included with myxoviruses (myxo-like viruses).

Veterinary care is provided, but this assumes the form of superficial care of wounds and of obvious infections. Little capability is available to detect chronic disease or latent infections. Monitoring for tubercolosis may be done, usually in the form of a skin test, but no provisions are made for taking X-rays of animals. Attempts may be made to segregate animals but lack of adequate training in how microbes are transmitted substantially reduces the eddectiveness of these measures. Furthermore, animals of the same species are often pooled, regardless of their origin. Animals are kept at these facilities for varying periods of time, depending upon sales.

VIRUS INFECTIONS

It is obvious that the above means for the capture, holding and shipping of animals can only lead to the exchange of viruses or other organisms from man-to-monkey, monkey-to-man, or monkey-to-monkey. There is still need for additional study on the various species of simians used in the laboratory from time of capture through their experimental employment in order to evaluate this exchange of organisms more completely. It should be emphasized that only limited data are available concerning natural viral infections of nonhuman primates. Data collected to date, however, strongly demonstrate that this exchange of viruses occurs with regularity and with highly significant results in several instances. As a consequence, the following facts have been established: (a) a simian virus flora, specific for monkeys and apes, but capable of crossing species barriers, has been established; (b) viruses from other animals (primarily man) are capable of producing infection and/or disease (with death) in nonhuman primates; and (c) nonhuman primates carry viruses that produce infection and/or disease (with death) in man.

Simian Viruses

Since recognition by Rustigian et al. (172) of a virus infection of "normal" monkey tissues, a large number of indigenous agents have been described primarily through the efforts of Hull et al. (96,98) in the United States and Malherbe and his coworkers (132, 137) in South Africa. Viruses (for example measles) from other animal sources, primarily human, were also found in monkey tissue (170, 171). At times these findings have led to some confusion as the original source of the agent may not be immediately determined; however, studies by a number of different investigators clearly demonstrate that approximately 70 distinctly simian viruses do exist (78, 79, 59, 86, 94, 96, 98, 132, 137, 196). These have been shown to be separable into groups that are similar to, but are antigenically distinct from, those found in man and other animals (Table 2). Accordingly, these simian viruses are classified as counterparts of the following recognized animal virus families or groups: adenoviruses, herpesviruses, myxoviruses, picornaviruses, poxviruses, papovaviruses, reoviruses, and certain viruses that do not quite fit into any of these groups; for example, foamy viruses, Marburg virus, simian hemorrhagic fever virus, and others. Serologic studies also demonstrate that simians are often infected with viruses other than those considered to be of simian origin (9I, 92, 95, I07, I08, II2, II5).

In a few instances, virus (especially picornavirus) may be recovered from tissue, or other materials, primarily fecal samples, without antibody development in the host animal. Those viruses that are nonpathogenic are more of a nuisance than of serious concern; however, their mere presence may exclude the contaminated tissue or animal from experimental use. The presence of these agents in cultures used for vaccine production is of more than academic importance. Concern over the unknown pathogenic potential these agents may have in other host systems is another important consideration.

The simian viruses capable of producing recognizable disease in monkeys and apes are few in number; however, where disease has occurred, it has been devastating. Normally, such marked reactivities are associated with the crossing of an agent from one species to another. When this occurs, death is frequently the end result.

Herpesviruses

In their natural hosts, herpesviruses rarely produce overt disease. In the newborn, serious or even fatal disease may occur; however, the adult is latently infected and may act as a carrier of the virus with intermittent episodes of overt disease developing as a result of some additional unknown physiologic factor or factors. Of the simian herpesviruses, perhaps H. simiae has gained the greatest notoriety, because it transfers to man. This virus, endemic in macaques (primarily M. mulatta), produces little in the way of overt disease in its host species. Oral lesions may be found which, when cultured, generally yield virus, but other evidence of illness is minimal. The lesions heal without any sequelae except for production of antibody. Marked serologic cross reactions occur when antisera from such infected animals are tested with H. hominis antigens. Therefore, serologic differentiation is difficult, but it can be done when performed by experienced technicians. Burnet et al. (26) originally described a one-way serologic cross between H. simiae and H. hominis, but more recent and extensive studies have not substantiated this claim. In our laboratory, sera from numerous species of primates have indicated extensive serologic cross-reactions between the two viruses (II2, II5). Serologic testing of animals for antibody to the herpesviruses is of some value. Animals with antibody to herpesviruses may be assumed to have been previously infected and, therefore are potentially dangerous; however, that this antibody level reflects previous H. simiae infection is not clear. Current serologic findings suggest that the percentage of serologic positives is low, and this finding offers the investigator a choice of what to do with the animal. It may be assumed that a seronegative animal has not had previous contact with herpesviruses and, conversely, seropositive animals have had a previous exposure.

Infection of species other than macaques by H. simiae results in varied types of clinical response. It is to be emphasized that the full spectrum of reactions by all primates in response to this virus has not been ascertained. Thus, we must hold in abeyance any conclusions until experimental studies determine this point. It is though that infection of man with this virus results in 100% fatalities or is so devastating as to cause total decerebration of its human host.

In other primate species , H. simiae may be equally fatal. Infection of certain of the New World monkeys with this virus will also result in death. H. simiae, as well as H. hominis, will produce death in rabbits

H. hominis (herpes simplex, cold sore virus , etc.) has nor received the attention given to its simian counterpart primarily because it is a human virus and, like H. simiae, produces only limited disease in its natural host. Recently, the possible association of H. hominis (type 2) with cervical cancer has resulted in a renewed interest in this virus (105, 165) and attempts are currently underway to study its pathogeneticy for other primates (52, 110, 129, 155). The resistance of rhesus monkeys and the susceptibility of Cebus monkeys (C.capuchinus, C.apella) to cervical implantation of H.hominis type 2 has been described (52). Hunt and Melendez (101) were not able to induce lesions in Cebus albifrons with types 1 or 2 H.hominis. In our laboratory, the baboon (P.cynocephalus) was found to be resistant to this type 2 herpesvirus, but the marmoset (Callithrex sp.) as well as the capuchin, were found to be highly susceptible (110). In contrast to the capuchin which developed cervical lesions, the marmoset succumbed to this virus following a viremia, generalized localization of virus (even in the brain) and extensive destruction of the genito-urinary tract.

Interestingly, the resistance of the rhesus and susceptibility of the capuchin to herpes simplex was previously demonstrated by Zinsser (214) and McKinley and Douglass (142). In thei studies, the capuchin monkeys developed an encephalitis following intracerebral inoculation of virus. In contrast to our findings, as well as those of Hunt and Melendez (100), Deinhardt et al.(38) indicated that the marmoset was relatively resistant to herpesvirus. Hunt and Melendez (100) noted a fatal systemic disease in experimentally infected marmosets? Natural disease was not noted except in animals with experimentally induced lymphomas. According to Melendez et al. (145) , H.hominis causes a fatal systemic disease in the owl monkey (Aotus trivirgatus). The susceptibility of the gibbon (Hylobates lar) to herpes simplex was noted by several investigators (50, 186). The response of the gibbon to herpes simplex was variable: deaths in some instances, benign infection with reoccurences in others. For additional information concerning H.hominis infections of nonhuman primates, see reviews by Kalter (107, 108), Kalter and Heberling (112), Hull (95), Melendez et al. (146) and Hunt and Melendez (100).

H.tamarinus (herpes T, H.platyrrhinae) was first isolated from tamarins (Saguinus tamarin) by Holmes et al. (87) and by Melnick et al. (151),hence the name. Subsequent studies indicated, however, that this species, in which infection produces a systemic disease with numerous fatalities, was not the natural host. A number of investigators have now demonstrated that the hosts for this agent are squirrel monkeys (Saimiri sciureus), spider monkey (Ateles sp.) and Cebus monkey (Cebus albifrons) (148,88,118). Like the marmoset, the owl monkey is highly susceptible to this virus (99).

More recently, Melendez and his collaborators (142,143,145) have demonstrated the existence of a relatively large number of herpesviruses, primarily from New World monkeys. These viruses have been derived from such species as Saimiri sciureus, Aotus trivirgatus, Saguinus oedipus, Ateles geoffroyi. The majority of these herpesviruses are cytocidal, but at least two of them are oncogenic when introduced into other New World monkeys (marmosets, owl monkeys, and spider monkeys). The obvious concern over this finding needs no elaboration. This is the first demonstration, in primates of an oncogenic potential for a virus recovered from simians.

Poxviruses

This group of viruses is capable of producing infection and disease in a multitude of animal and avian hosts. As seen in Table 3, Andrewes and Pereira (4) have subdivided poxviruses into six subgroups, according to their immunologic, biologic, and histopathologic reactivities. Of the more than 25 poxviruses, at least four have been recovered from simian hosts -- monkeypox, yaba, yaba-like and molluscum contagiosum. Experimentally, a number of investigators have demonstrated the susceptibilities of large numbers of monkeys and apes to various other members of the poxvirus family -- variola, vaccinia, alastrum. Hahon (75) has reviewed the pertinent poxvirus literature and describes the reactions of various simian species to these viruses. Typical infection includes a local lesion resulting from the inoculation which may be followed by fever, generalized exanthem, encephalitis, orchitis, and occasionally death. It is quite possible that most of the poxviruses, given the appropriate opportunity, could cause serious overt infection. This generalization must be evaluated, however, in terms of some experimental data that demonstrate the resistance of certain of the simians to some poxviruses.

Table 3. The Animal Poxviruses*

Subgroup 1:	Viruses closely related antigenically, biologically, and morphologically to variola	
	Alastrim	Monkeypox
	Cowpox	Vaccinia
	Ectromelia	Variola
	Rabbitpox	
Subgroup 2:	Viruses related to orf	
	Paravaccinia (Milker's nodes)	
	Bovine papular dermatitis	
Subgroup 3:	Viruses affecting ungulates; some may belong to subgroups 1 and 2	
	Sheeppox	Swinepox
	Goatpox	Horsepox
	Lumpy skin disease	Camelpox
Subgroup 4:	Viruses of birds	
	Fowlpox	
	Canary and other bird poxes	
Subgroup 5:	Viruses producing tissue proliferation	
	Rabbit myxoma	Hare fibroma
	Rabbit fibroma	Squirrel fibroma
Subgroup 6:	Viruses that have not been classified	
	Molluscum contagiosum	
	Yaba monkey virus	
	Yaba-like monkey virus	

*Andrewes and Pereira[4]

The question of the existence of a poxvirus reservoir in one or another of the simian hosts is of more than passing interest. Such a possibility is intriguing, but little in the way of substantive data are presently available to support this hypothesis. Questions have been raised concerning the nature of the monkeypox virus. A number of investigators have suggested that this agent is nothing more than a vaccinia-variola virus passaging through simians. Nonetheless, extensive studies are currently underway in several laboratories to demonstrate such a relationship or lack of it.

Since its original description by von Magnus et al. (206), outbreaks of monkeypox have been described in a number of nonhuman primate colonies: however, Arita and Henderson (5) in a review of monkeypox epidemics emphasize that naturally occurring pox infections have been reported in simians since 1767. Variations in severity of monkeypox among different species of nonhuman primates have been recorded, and deaths are extremely rare. The clinical picture is similar to that occurring in man as a result of variola-vaccinia. Use of human vaccines prevents infection in monkeys and apes and probably accounts for the lack of human disease. Detection of serologic evidence of infection with monkeypox virus has been difficult. Recent clinical cases are detected rather readily by one or another of the usual serologic procedures -- complement fixation, hemagglutination inhibition, or serum neutralization. It is in survey work that one encounters problems. Thousands of sera obtained from a number of simian species have now bee assayed for antibody to poxviruses with equivocal results. Some HI positives have been obtained, but confirmation with SN procedures has usually failed to substantiate the original positive findings.

Yaba and yaba-like (benign epidermal monkeypox) are apparently distinct entities that are antigenically unrelated to the other poxviruses but are slightly related to each other. Both agents have the capability of producing benign tumor-like eruptions in a susceptible primate host including man. Both diseases have recently been described, yaba by Bearcroft and Jamieson in 1958 and yaba-like by Hall and McNulty (76) and by Casey et al.(29) in 1967. Both agents appear to demonstrate a species selectivity for Asian macaques, whereas New World species (at least those tested) are apparently resistant. Interestingly, Old World monkeys derived from Africa show an intermediate susceptibility -- the baboon, vervet, and patas monkeys born in the United States were successfully in inoculated with yaba virus by Ambrus and Strandstrom (2). It had been previously reported that African monkeys were resistant (1). While clinically similar to tumors produced by yaba virus, the yaba-like agent has been responsible for a number of extensive outbreaks in monkeys at various centers, again primarily in macaques.

Molluscum contagiosum was recently described by Douglas et al. (46) in chimpanzees. The clinical course was mild with development of proliferative skin (epithelial) lesions.

Arboviruses

No specific simian arbovirus has been described although some 200 plus viruses are recognized. Both man and monkey are highly susceptible to several of the viruses now considered arboviruses. Perhaps the most important of the arboviruses is yellow fever. Both Old and New World monkeys are highly susceptible to this virus, and devastating epidemics with high mortality rates have been described. It should be recognized, however, that monkeys and apes are extremely variable in their clinical response to yellow fever virus, with reactions ranging from mild viremia to rapid death (167, 168). The studies in Panama of Galindo and Srihongse (66) showed that spider and capuchin monkeys are highly resistant to yellow fever virus; the spider monkey developed a viremia, but with minimal symptoms, and the howler monkey, marmoset, and night monkey were highly susceptible.

African simians apparently do not respond to yellow fever virus as do monkeys from Central and South America. Clinical disease among several species (viremia, febrile reactions) has been reported , but deaths were apparently infrequent. Present information also suggests that even though the clinical course is mild, variation in intensity of the disease occurs.

Another arbovirus that has achieved a certain amount of notoriety because it is lethal for man and monkey alike is the Kyasanur Forest virus. It was first described in 1957 as an epizootic of man and monkey (Macaca radiata, Presbytis entellus), is tick-borne, and produces a nonsuppurative encephalitis (210). Under experimental conditions, other tick-borne, group B arboviruses have been introduced successfully into rhesus and/or spider monkeys; these include tick-borne spring-summer encephalitis virus, Japanese B, Kemerovo, and Langat virus. Generally, a viremia followed by encephalitis or the development of antibody were noted.

Additional arboviruses recovered from nonhuman primates include Manzanilla virus from howler monkeys in Trinidad (3) and eastern encephalitis from cynomolgus monkeys (128). Gerloff and Larson (67) reported that the rhesus monkey was susceptible to experimental infection with Colorado tick fever virus. Unpublished studies from our laboratory suggest that the baboon is resistant to certain arboviruses that are endemic to the United States (Keystone, Tahyna, Trivittatus, LaCrosse). According to Smithburn and Haddow (187), Semliki forest virus from Uganda was lethal for rhesus monkeys following intracerebral inoculation; apparently the red-tailed monkey is the reservoir for this virus. Smithburn et al.(188) indicated that both rhesus and African green monkeys were susceptible to Bunyamwera and Germiston viruses. These two species of monkeys were also susceptible to a new arbovirus (Lumbo) recovered from mosquitoes by Kokernot et al. (II9) in Lumbo, Mozambique. Cynomolgus monkeys were susceptible to several arboviruses, as reported by Verlinde (205), including: Mayaro and Mucambo (group A), Oriboca and Restan (group C), and Kwatta, an ungrouped South American virus. Rhesus monkeys were found to be susceptible to Chikungunya and Mayaro viruses, but not Onyong-nyong (I8). Bunyamwera virus was lethal for rhesus monkeys according to Boorman and Draper (2I). These investigators were also able to demonstrate the differences in susceptibility which existed between rhesus monkeys and the African green monkey (C. mona), which did not succumb to Bunyamwera virus. Type I dengue virus did not produce any evidence of overt disease in C. aethiops (I97), but the gibbon apparently is susceptible to all four strains of dengue virus (208).

Immunosuppression enhances susceptibility to the arboviruses. Spider monkeys developed a fatal encephalitis when treated with cyclosphosphamide and inoculated with Venezuelan equine encephalomyelitis (I56). Zlotnik et al. (2I5) demonstrated that cyclophosphamide enhanced the susceptibility of patas monkeys to louping ill virus. Similarly, rhesus monkeys treated with cyclophosphamide were more susceptible to western encephalitis.

For more complete information on arbovirus infectivity of simians, see the following reviews: Kalter and Heberling, Comparative virology of Primates, Bact.Rev., I97I; and Taylor, Catalogue of arthropodborne viruses of the world, U.S.Publ.Health Serv.Pub. I760, I967 (II2,I99).

Myxoviruses

This group of viruses has been subdivided to include orthomyxoviruses and the paramyxoviruses. For a time, we and others have considered the foamy viruses as myxoviruses (pseudomyxoviruses). This may not be correct, and until more information concerning their nature is available, we will simply indicate that there are seven types recovered from a variety of monkeys and apes. Their pathogenicity for animals (including man) is unknown. The simian viruses that have been isolated, namely SV5, SV4I, and SAIO, are considered as paramyxoviruses. SV5 is the most frequently isolated of these viruses, especially from kidney cells (49, 93, 98, 203, 2I3), but there is some question concerning its status as a true simian virus. The other two isolated, i.e. SV4I and SAIO, are only infrequently encountered.

SV5 is rarely, if ever, isolated from newly captured animals nor is there serologic evidence of its existence prior to capture of the animal (II2). Thus, the question has been raised concerning the origin of this virus -- human, canine, or other animal. Hull (95a) still feels, however, that SV5 is a simian virus.

Disease production in simians by SV5 has been reported in young baboons (I24) and vervets (77), but generally inapparent infection occurs. This virus has been infrequently found in humans (I77) and more frequently in dogs (32, I27).

Of perhaps greater importance to the monkey and ape population is the human myxovirus, measles (rubeola). The susceptibility of simians to this virus has been well documented. Early investigators attempting to isolate the etiologic agent measles utilized a number of nonhuman primate species in their investigations. In I940, Enders (5I) noted that measles research was made diffidult by the natural resistance of monkeys to the virus. Peebles *et al*. (I59) demonstrated that the vast majority of macaques examined had antibody to measles virus. It was also observed, however, that freshly imported animals were devoid of antibody to this agent. From the work of a number of investigators, it became apparent that animals coming out of India were generally infected as a result of human contact. Clinical disease was not seen in rhesus and cynomolgus monkeys; however, Potkay *et al*. (I63) did describe an outbreak of measles in a rhesus colony in which the clinical manifestations were similar to those described for man. Levy and Mirkovic (I25) reported that measles in New World marmosets (*C.jacchus*, *S.oedipus*, and *S.fascicollis*) was characterized by high susceptibility and numerous deaths.

The fact that viruses produce latent infections and persist within tissues for long periods of time is demonstrated by measles virus. It has been recovered from human cases of subacute sclerosing panencephalitis by inoculating cynomolgus monkeys (I60). Obviously, this is analogous to studies on slow viruses initiated by Gajdusek and his coworkers (I4,I5,58,65,68,69,I2I)-see below. Naturally occurring clinical disease that resembles subacute sclerosing panencephalitis has been reported in baboons (II6, 207).

Another human myxovirus, influenza (orthomyxovirus) has received mixed reports in the literature regarding infectivity for simians. A number of investigators have indicated that rhesus and cynomolgus monkeys could be infected with influenza virus without clinical disease; however, Saslaw and Carlisle (175) demonstrated that rhesus monkeys exposed to the Asian strain of influenza virus, developed clinical illness. Similarly, we have been able to produce clinical disease in baboons using a fresh isolate of influenza virus (A2/Hong Kong/68) (114). This same virus strain was shown to be infectious for gibbons by Johnsen and his collaborators (I03). The question of influenza virus infection of simians has not been helped by serologic studies because mixed results have been reported. Furthermore, there are a number of questions concerning the specificity of serologic reactions when using influenza virus antigens.

Other myxoviruses have been studied in nonhuman primates with variable results. Respiratory syncytial virus (chimpanzee coryza agent-CCA) was first recovered from the chimpanzee by Morris *et al*. (I53). It was soon shown to be a human agent by Coates and Chanock

(3I), who also demonstrated that marmosets and baboons were resistant to it. Rinderpest virus which is antigenically related to measles (rubeola) virus failed to infect various Cercopithecus monkeys and baboons (33). Findley and Mackenzie (57, 58) were able to produce a mild disease in rhesus monkeys with Newcastle disease virus.

Papovaviruses

Of the two simian papovaviruses, SV40 and SAI2, the former is by far the better known. SV40, which is frequently encountered in kidneys from apparently normal rhesus monkeys, is of concern because of its oncogenic capability after inoculation into newborn hamsters. Although no known tumor production in man or other primates can be attributed to this virus, the oncogenic potential is of obvious concern. It appears that SV40 is normally restricted to rhesus monkeys, although there is some evidence for its presence in cynomolgus monkeys. Experimentally, African green monkeys may become infected, and Ashkenazi and Melnick (7) were able to recover SV40 from this species for as long as six to eight months postinfection.

SAI2 was isolated from an African green monkey by Malherbe and Harwin (I33), but it evidently has not occurred with any frequency.

Picornaviruses

At least I4 enteroviruses (SV2, SV6, SVI6, SVI8, SVI9, SV26, SV35, SV42, SV44, SV45, SV46, SV47, SV49), primarily recovered from rhesus and cynomolgus monkeys, are now recognized. An additional two viruses, one (SA5) isolated from the African green monkey and the other (AI3) from the baboon, are also known. SV28 and SA4 have been isolated from rhesus and African green monkeys, respectively, and are generally included among the simian enteroviruses. To our knowledge, these latter two viruses have not been isolated from the intestinal tract and should not therefore be considered enteroviruses.

The simian enteroviruses are interesting because of their relatively frequent occurrence in simians but infrequent relationship to specific disease. It is also noteworthy that infection without antibody production has been recorded on a number of accasions for simian enteroviruses (80,86,95). For details regarding the presence of enterovirus antibody in primate sera, see Kalter and Heberling (II2).

Various species of monkeys and apes have been utilized in studies with human enteroviruses. Much of the early success in the isolation of poliovirus resulted from investigations utilizing monkeys and baboons (I22). "Natural" occurrence of poliomyelitis in chimpanzees was also noted in I935 (I54). The resistance of spider monkeys to poliovirus was reported by Mackay and Schroeder (I3I) and that of marmosets by Grossman and Kramer)7I). Trask and Paul (202) were able to recover poliovirus from the stools of rhesus monkeys following intracutaneous inoculation; whereas Sabin and Ward (I37) found that they could isolate poliovirus from stools of orally infected cynomolgus monkeys. Both the South American ringtail monkey (C.capucinus) and the African mustache monkey (C.cebus) were found to be satisfactory for the transmission of poliovirus (I52); however, it was shown, primarily through the efforts of Howe and Bodian (90), that the chimpanzee developed an infection with poliovirus similar to that seen in man. Melnick (I49) compared M.mulatta, M.cynomolgus, M.mordox, C.mona, C.sabaeus, C.cephus, C.capuchina, and P.satyrus for their susceptibility to poliovirus. All but M.mordox (insignificant number examined) were found to have virus in their stools. According to De Rodaniche (42), other South American monkeys (C.capucinus, A.palliata, and A.fusiceps) were useful in transmission studies of poliomyelitis. It will be recalled that Mackay and Schroeder (I3I) indicated that spider monkeys were resistant to poliomyelitis; however, Jungeblut and De Rodaniche (I06) successfully infected this species with poliovirus.

Overt and inapparent natural infections of apes with poliovirus has also been noted. Guilloud and Kline (72) described paralytic disease in gorillas, and Douglas et al. (45) recovered type 3 poliovirus from "normal" chimpanzee stools. The presence of an extensive enterovirus flora in chimpanzees has been reported (35, I89, I92).

Other human enterovirus studies in simians included infection without clinical disease of chimpanzees using I0 different types of coxsackieviruses (I50) and echovirus types 2, 3, 4, 6 (I02). The chimpanzee was also used for studies with coxsackie virus A7 (30, 34, 74, 89, I04) and echovirus type 9 (2I2). Experimental production of an endocarditis in cynomolgus monkeys with coxsackievirus B4 was suggested by Lou et al. (I30) and De Pasquale (4I). According to Sun and coworkers (I95), coxsackievirus B4 also produced a pancarditis in this same species. Theiler's encephalomyelitis virus (of mice) did not have any overt effect on rhesus monkeys (200, 20I). A new virus (encephalomyocarditis virus) capable of producing an interstitial myocarditis was recovered from a chimpanzee (8I), and a similar agent (meningoencephalomyelitis virus) was described in monkeys by Dick et al.(44).

Although no simian rhinoviruses have yet been described, there were some experimental studies with human isolates in several different simian species. Dick (43) described experimental infection of the chimpanzee with rhinovirus types I4 and 43. According to Martin and Heath (I39), the vervet may be used as a model for studies with human rhinoviruses. Pinto and Haff (I62) successfully infected gibbons with human strains of rhinovirus, but were not able to infect a patas monkey.

Reoviruses

Only limited comparative studies in simians have been reported with the simian reoviruses (SV12, SV59, SA3). Antibody to these viruses, especially type 3, are described in detail elsewhere (113). Masillamony and John (141) reported experimental infection in the bonnet monkey with type 3 reovirus.

Hepatitis

The linking of this virus to primates originates with the report by Soper and Smith (193) of a prevalence of hepatitis in humans receiving yellow fever vaccine stabilized with rhesus monkey serum. Epidemiologic association of hepatitis and simians was subsequently reported by Hillis (82) when he observed the occurrence of hepatitis in chimpanzee handlers. Bearcroft (9, 12) inoculated _E.patas_, _C. torquatus_, _C.aethiops_, _C.mona_, _C.erythrotis sclateri_, _C.erythrotis camerunensis_, _C.nictitans nictitans_, _C.nictitans erythrogaster_, _M. leucophagus_, _P.anubis_, _G.demidovii_, and _P.potto_ with infected human material and found that the patas monkey demonstrated evidence of infection. In similar studies, it was reported by Smetana (182, 185) that spontaneous hepatitis frequently occurs in chimpanzees and, to a lesser extent, in the patas monkey.

More recently, Deinhardt _et al_. (37) reported on the successful transmission of human hepatitis virus to the marmoset. There has been some question concerning the efficacy of these studies, and a number of laboratories are currently attempting to repeat these findings. A number of investigators (19, 36, 83, 126, 164, 180) have demonstrated the presence of both Australia antigen (20) and antibody in a variety of simian species.

Rubella

Because of its teratogenic effect on the human fetus, this virus has been experimentally studied to some extent in nonhuman primates. Habel (73) was able to transmit rubella virus to rhesus monkeys using patient's blood or throat washings. On the other hand, Sever _et al_. (178, 179) were only able to demonstrate an antibody rise in three of five rubella inoculated animals. African green monkeys were shown by several groups of investigators to be susceptible (27, 181). Uterine infection of rhesus monkeys (157, 158) and baboons (112) has been described. Draper and Laurence (47) found that the patas monkey was susceptible to both wild and vaccine strains of rubella virus. Delahunt and coworkers (39, 40) reported on the occurrence of cataracts and other embryopathies in rhesus monkeys after pregnant animals were experimentally infected. Our laboratory has provided extensive data on the presence of antibody to this virus in various species of simians (112).

Marburg

The singular occurrence of this virus in a population of African green monkeys shipped into Germany and Yugoslavia has been well documented (140). What is still not known, however, is the actual origin of this virus, and serologic studies by a number of investigators have not resolved the question. The resulting data are still disputed, and additional studies are necessary to define the source of the virus (112).

Simian Hemorrhagic Fever

Outbreaks of this new disease, which apparently has a propensity for Asian simians (macaques) have occurred several times during the past 10 years in various laboratories (Great Britain, Soviet Union, United States). Among susceptible species (*M.mulatta*; *M.fascicularis*, *M.nemestrina*, *M.assamensis*, *M.speciosa*) the mortality rate has been extremely high. Other primates, including man, have failed to develop apparent disease (198), and CF antibody to this agent has not been found in man, gorilla, chimpanzee, orangutan, gibbon, baboon, African green, rhesus, cynomolgus, patas, talapoin, stumptail, and marmoset. One lone rhesus monkey, which survived the original outbreak, was found to have CF antibody to this agent in its serum.

Rabies

This virus occurs more frequently than originally indicated among nonhuman primates. According to the Primate Zoonoses Surveillance, report no. 4, 1970, laboratory confirmed rabies has been reported for *Cebus* sp.,*M.fascicularis*, *Saimiri sciureus*, and *Leontideus* sp. Evidently, in Indonesia, the prevalence of rabies in *M.fascicularis* may be as high as 50%.

Lymphocytic Choriomeningitis

Although the original virus was recovered from a monkey (6), it is now recognized that this virus has a widespread host range and that its natural hosts are rodents. Very little has been done to study LCM in nonhuman primates, although well (112) have demonstrated the presence of antibody in sera from a number of species: chimpanzees, orangutans, African greens, rhesus, cynomolgus, marmosets and baboons. No antibody was detected in sera from the gorilla, gibbon, Japanese macaque, patas, talapoin, or squirrel.

Slow Viruses

This group of agents is rather ill-defined, heterogeneous collection of viruses having the common characteristic of producing infections with extremely long incubation periods. More importantly, the discovery of the virus etiology of kuru by Gajdusek and his collaborators (62) has opened up an entirely new concept in host-parasite (virus) relationships. As a result, nonhuman primates have played an important role in developing knowledge on these agents.

The original studies in chimpanzees (60, 61, 63, 64) were followed by a series of reports describing the pathology in chimpanzees (14), transmission from chimpanzee to chimpanzee but not to the gibbon (black, golden), macaque (rhesus, cynomolgus, Barbary ape, stumptail), African green, patas, spider (black, brown), squirrel, woolly, white-lipped marmoset, tree shrew, or slow loris (6I). The spider monkey was successfully inoculated with kuru following passage in the chimpanzee (65). Lesions in the chimpanzee and spider monkey has also been noted using electron microscopy (I2I). Gibbs et al. (68, 69) reported on the use of chimpanzees in studies on fatal spongiform encephalopathy (Creutzfeldt-Jakob disease). Beck et al. (15) described the neuropathology. For a comprehensive report on attempts to transmit subacute and chronic neurologic diseases to various species of primates and other animals, see Gibbs et al. (68). Rhesus monkeys were successfully inoculated with transmissible mink encephalopathy virus by Marsh et al. (138) and passaged to stumptail and squirrel monkeys (48).

CONCLUSIONS

We have attempted to describe, albeit briefly, virus studies in nonhuman primates. This was done to provide investigators with some comparative information about virus investigations that have been done with various monkey and ape species. No attempt was made to describe the various studies with tumor viruses, especially Rous sarcoma. These studies were avoided because it was felt that the relationship of viruses to primate tumors is just beginning to unfold, and hopefully a well-defined result will be forthcoming in the immediate future. For a review of the current status of this program see Kinard (117). We have also neglected to describe the various diseases and illnesses of simians attributed to viruses. These, too, have been documented in a number of texts and monographs (8, 16, 17, 22, 25, 28, 53, 56, 84, 85, II2, I20, I23, I6I, I66, I69, I76, I94, 204, 209).

An attempt was made, however, to correlate these infectious experiences with the history of the animal and to demonstrate the potential danger to animals from extrinsic sources. The use of nonhuman primates in experiments is also mentioned in order to provide an awareness of various types of research activities. It is quite evident that while much has been done, a great deal is still in need of study. Three or four major virus outbreaks have occurred demonstrating a potential for highly fatal disease in both man and simians. This suggests that future outbreaks may have similar outcomes. Of the large number of simian viruses recognized, the majority closely resemble their human counterparts and are generally not related to disease processes. On the other hand, several of these viruses have produced severe disease, and the first oncogenic primate virus was recently reported (144).

The potential usefulness of various simian species as models for studies of human disease has barely been exploited. Most of the studies recorded herein are based upon clinical observations, and the need for careful and detailed immunologic and histopathologic investigations is obvious.

It has not been possible to record all the known studies involving nonhuman primates in virology. Those that have been omitted from this overview were not included primarily because of space limitations. A more extensive review has been prepared and should be consulted for further details (109).

ACKNOWLEDGMENTS

This study was funded in part by U.S.P.H.S. grants AI05374, RR00278, RR00361, RR00451, RR05519, and contracts NIH 69-93 and NIH 71-2348 and WHO grant Z2/181/27. This laboratory serves as the World Health Organization's Regional Refernce Center for Simian Viruses.

References

1. Ambrus, J.L., Feltz, E.T., Grace, J.T. and Owens, G. 1963. Natl Cancer Inst. Mono., 10, 447-453

2. Ambrus, J.L. and Strandstrom, H.V. 1966. Nature, 211, 876

3. Anderson, C.R., Spence, L.P., Downs, W.G. and Aitken, T.H.G. 1960. Amer. J. Trop. Med. Hyg., 9, 78-80

4. Andrewes, C. and Pereira, H.G. 1967. Viruses of vertebrates, 2nd ed., Williams and Wilkins Co., Baltimore

5. Arita, I. and Henderson, D.A. 1968. Bull. Org. Mond. Sante., 39, 277-283

6. Armstrong, C. and Lillie, R.D. 1934. Publ. Hlth. Rep., 49, 1019-1027

7. Ashkenazi, A. and Melnick, J.L. 1962. Proc. Soc. Exp. Biol. Med., 111, 367-372

8. Balner, H. and Beveridge, W.I.B. (Eds.) 1970. Infections and immunosuppression in subhuman primates, Munksgaard, Copenhagen

9. Bearcroft, W.G.C. 1963. Nature, 197, 806-807

10. Bearcroft, W.G.C. 1964. J. Path. Bact., 88, 511-519

11. Bearcroft, W.G.C. 1969. Brit. J. Exp. Path., 50, 56-65

12. Bearcroft, W.G.C. 1969. Brit. J. Exp. Path., 50, 327-330

13. Bearcroft, W.G.C. and Jamieson, M.F. 1958. Nature, 182, 195-196

14. Beck, E., Daniel, P.M., Alpers, M., Gajdusek, D.C. and Gibbs, C. J., Jr. 1966. Lancet, 1056-1059

15. Beck, E., Daniel, P.M., Matthews, W.B., Stevens, D.L., Alpers, M.P., Asher, D.M., Gajdusek, D.C. and Gibbs, C.J., Jr. 1969. Brain, 92, 699-716

16. Beveridge, W.I.B. (Ed.) 1969. Using primates in medical research. Part I, Vol. 2, S. Karger, Basel

17. Beveridge, W.I.B. (Ed.) 1969. Using primates in medical research. Part II, Vol. 3, S. Karger, Basel

18. Binn, L.N., Harrison, V.R. and Randall, R. 1967. Amer. J. Trop. Med. Hyg., 16, 782-785

19. Blumberg, B.S. 1972. Vth Inter. Symp. on Comparative Leukemia Research, Padova/Venice, Karger, Basel

20. Blumberg, B.S., Alter, H.J. and Visnich, S.A. 1965. J. Amer. Med. Ass., 191, 541-546

21. Boorman, J.P.T. and Draper, C.C. 1968. Trans. Roy. Soc. Trop. Med. Hyg., 62, 269-277

22. Bourne, G.H. (Ed.) 1969. The chimpanzee, Vol. 1, Univ. Park Press, Baltimore, Md.

23. Bourne, G.H. (Ed.) 1970. The chimpanzee, Vol. 2, S. Karger, Basel

24. Bourne, G.H. (Ed.) 1970. The chimpanzee, Vol. 3, Univ. Park Press, Baltimore, Md.

25. Bourne, G.H. (Ed.) 1971. The chimpanzee, Vol. 4, Univ. Park Press, Baltimore, Md.

26. Burnet, F.M., Lush, D. and Jackson, A.V. 1939. Austr. J. Exp. Biol., 17, 41-51

27. Cabasso, V.J. and Stebbins, M.R. 1965. J. Lab. Clin. Med., 65, 612-616

28. Carpenter, C.R. (Ed.) 1969. Proceedings of second international congress of primatology, Vol. 1, S. Karger, Basel

29. Casey, H.W., Woodruff, J.M. and Butcher, W.I. 1967. Amer. J. Path., 51, 431-446

30. Chumakov, M.P., Voroshilova, M.K., Zhevandrova, V.I., Mironova, L.L., Itzelis, F.I. and Robinson, I.A. 1956. Probl. Virology, 1, 16-18

31. Coates, H.V. and Chanock, R.M. 1962. Amer. J. Hyg.,76, 302-312

32. Crandell, R.A., Brumlow, W.B. and Davison, V.E. 1968. Amer. J. Vet. Res., 29, 2141-2147

33. Curasson, G. 1942. Traite de pathologie exotique veterinaire et comparee, Vol. 1, 2nd ed., Vigot Freres, Paris, Pp. 12-169

34. Dalldorf, G. 1957. J. Exp. Med., 106, 69-76

35. Day, P.W., Soike, K., Levenson, R.H. and Van Riper, D.C. 1966. Lab. Anim. Care, 16, 497-504

36. Deinhardt, F. 1970. In Balner, H. and Beveridge, W.I.B. (Eds.). Infections and immunosuppression in subhuman primates, Munksgaard, Copenhagen, Pp. 55-63

37. Deinhardt, F., Holmes, A.W., Capps, R.B. and Popper, H. 1967. J. Exp. Med., 125, 673-688

38. Deinhardt, F., Holmes, A.W., Devine, J. and Deinhardt, J. 1967. Lab. Anim. Care, 17, 48-70

39. Delahunt, C.S. 1966. Lancet, 1, 825

40. Delahunt, C.S. and Rieser, N. 1967. Amer. J. Obstet. Gynec., 99, 580-588

41. DePasquale, N.P., Burch, G.E., Sun, S.C., Hale, A.R. and Mogabgab, W.J. 1966. Amer. Heart J., 71, 678-683

42. De Rodaniche, E.C. 1952. Amer. J. Trop. Med. Hyg., 1, 205-209

43. Dick, E.C. 1968. Proc. Soc. Exp. Biol. Med., 127, 1079-1081

44. Dick, G.W.A., Smithburn, K.C. and Haddow, A.J. 1958. Brit. J. Exp. Path., 29, 547-558

45. Douglas, J.D., Soike, K. and Raynor, J. 1970. Lab. Anim. Care, 20, 265-268

46. Douglas, J.D., Tanner, K.N., Prine, J.R., Van Riper, D.C. and Derwelis, S.K. 1967. J. Amer. Vet. Med. Ass., 151, 901-904

47. Draper, C.C. and Laurence, G.D. 1969. J. Med. Microbiol., 2, 249-252

48. Eckroade, R.J., ZuRhein, G.M., Marsh, R.F. and Hanson, R.P. 1970. Science, 169, 1088-1090

49. Emery, J.B. and York, C.J. 1960. Virology, 11, 313-316

50. Emmons, R.W. and Lennette, W.H. 1970. Arch. Ges. Virusforsch, 31, 215-218

51. Enders, J.F. 1940. In Gordon, J.E. et al. (Eds.) Virus and rickettsial diseases, with special consideration of their public health significance. Symposium held at the Harvard School of Public Health, June 12-17, 1939, Harvard University Press, Cambridge, Mass., 237-267

52. Felsburg, P.J., Heberling, R.L. and Kalter, S.S. 1973. Arch. Ges. Virusforsch

53. Fiennes, R.N.T.-W. (Ed.) 1966. Symp. Zoological Soc. London, No. 17, Academic Press, London

54. Fiennes, R. 1967. Zoonoses of primates, Cornell Univ. Press, Ithaca, N.Y.

55. Fiennes, R.N.T.-W. (Ed.) 1972. Pathology of simian primates. Part I, General pathology, S. Karger, Basel

56. Fiennes, R.N.T.-W. (Ed.) 1972. Pathology of simian primates. Part II, Infectious and parasitic diseases, S. Karger, Basel

57. Findlay, G.M. and Mackenzie, R.D. 1937. Brit. J. Exp. Path., 18, 146-155

58. Findlay, G.M. and Mackenzie, R.D. 1937. Brit. J. Exp. Path., 18, 258-264

59. Fuentes-Marins, R.A., Rodriguez, A.R., Kalter, S.S., Hellman, A. and Crandell, R.A. 1963. J. Bact., 85, 1045-1050

60. Gajdusek, D.C. 1967. Kuru and experimental Kuru in chimpanzees, Acad. of Med. Sci. USSR

61. Gajdusek, D.C. 1967. Curr. Tropics Microbiol. Immunol., 40, 59-63

62. Gajdusek, D.C., Gibbs, C.J., Jr. and Alpers, M. 1966. Slow, latent and temperate virus infections, U.S. Publ. Hlth. Serv. Publ., 1378

63. Gajdusek, D.C., Gibbs, C.J., Jr. and Alpers, M. 1966. Nature, 209, 794-796

64. Gajdusek, D.C., Gibbs, C.J., Jr. and Alpers, M. 1967. Science, 155, 212-214

65. Gajdusek, D.C., Gibbs, C.J., Jr., Asher, D.M. and David, E. 1968. Science, 162, 693-694

66. Galindo, P. and Srihongse, S. 1967. Bull. Wld. Hlth. Org., 36, 151-161

67. Gerloff, R.K. and Larson, C.L. 1959. Amer. J. Path., 35, 1043-1054

68. Gibbs, C.J., Jr., Gajdusek, D.C. and Alpers, M.P. 1969. Int. Arch. Allergy, 36, 519-552

69. Gibbs, C.J., Jr., Gajdusek, D.C., Asher, D.M., Alpers, M.P., Beck, E., Daniel, P.M. and Matthews, W.B. 1968. Science, 161, 388-389

70. Goldsmith, E.I. and Moor-Jankowski, J. (Eds.) 1969. Ann. N.Y. Acad. Sci., 162, 1-704

71. Grossman, L.H. and Kramer, S.D. 1936. Proc. Soc. Exp. Biol. Med., 35, 345-347

72. Guilloud, N.B. and Kline, I.C. 1966. J. Amer. Phys. Therapy Ass., 46, 516-518

73. Habel, K. 1942. Pub. Hlth. Rep., 57, 1126-1139

74. Habel, K. and Loomis, L.N. 1957. Proc. Soc. Exp. Biol. Med., 95, 597-605

75. Hahon, N. 1961. Bact. Rev., 25, 459-476

76. Hall, A.S. and McNulty, W.P., Jr. 1967. J. Amer. Vet. Med. Ass., 151, 833-838

77. Heath, R.B., El Falaky, I., Stark, J.E., Herbst-Laier, R.H. and Larin, N.M. 1966. Brit. J. Exp. Path., 47, 93-98

78. Heberling, R.L. and Cheever, F.S. 1960. Ann. N.Y. Acad. Sci., 85, 942, 950

79. Heberling, R.L. and Cheever, F.S. 1964. Amer. J. Epidem., 81, 106-123

80. Heberling, R.L. and Cheever, F.S. 1966. Amer. J. Epidem., 83, 470-480

81. Heldwig, F.C. and Schmidt, E.C.H. 1945. Science, 102, 31-33

82. Hillis, W.D. 1961. Amer. J. Hyg., 73, 316-328

83. Hirschman, R.J., Shulman, R.N., Barker, L.F. and Smith, K.O. 1969. J. Amer. Med. Ass., 208, 1667-1670

84. Hofer, H.O. (Ed.) 1969. Proceedings of the second international congress of primatology, Atlanta, Ga. 1968, Vol. 2, Recent advances in primatology, S. Karger, Basel

85. Hofer, H.O. (Ed.) 1969. Proceedings of the second international congress of primatology, Atlanta, Ga. 1968, Vol. 3, Neurology, physiology and infectious diseases, S. Karger, Basel

86. Hoffert, W., Bates, M.E. and Cheever, F.S. 1958. Amer. J. Hyg., 68, 15-30

87. Holmes, A.W., Caldwell, R.G., Dedmon, R.E. and Deinhardt, F. 1964. J. Immunol., 92, 602-610

88. Holmes, A.W., Devine, J., Nowakowski, E. and Deinhardt, F. 1966. J. Immunol., 96, 668-671

89. Horstmann, D.M. and Manuelidis, E.E. 1958. J. Immunol. 81, 32-42

90. Howe, H.A. and Bodian, D. 1941. Bull. Johns Hopk. Hosp., 69, 149-181

91. Hsiung, G.D. 1968. Bact. Rev., 32, 185-205

92. Hsiung, G.D. 1969. Ann. N.Y. Acad. Sci., 162, 483-498

93. Hsiung, G.D. and Atoynatan, T. 1966. Amer. J. Epidem., 83, 38-47

94. Hsiung, G.D. and Melnick, J.L. 1958. Ann. N.Y. Acad. Sci., 70, 342-360

95. Hull, R.N. 1968. The simian viruses, Virology monographs, Vol. 2, Springer-Verlag, New York, Pp. 1-66

95a. Hull, R.N. Personal communication

96. Hull, R.N. and Minner, J.R. 1957. Ann. N.Y. Acad. Sci., 67, 413-423

97. Hull, R., Minner, J.R. and Mascoli, C.C. 1958. Amer. J. Hyg., 68, 31-44

98. Hull, R., Minner, J.R. and Smith, J.W. 1956. Amer. J. Hyg., 63, 204-215

99. Hunt, R.D. and Melendez, L.V. 1966. Path. Vet., 3, 1-26

100. Hunt, R.D. and Melendez, L.V. 1969. Lab. Anim. Care, 19, 221-234

101. Hunt, R.D. and Melendez, L.V. 1972. PAHO Sci. Publ., 234, 174-188

102. Itoh, H. and Melnick, J.L. 1957. J. Exp. Med., 196, 677-688

103. Johnsen, D.O., Wooding, W.L., Tanticharoenyos, P. and Karnjanaprakorn, C. 1971. J. Infect. Dis., 123, 365-370

104. Johnsson, T. and Lundmark, C. 1957. Lancet, 271, 1148-1149

105. Josey, W.E., Nahmias, A.J. and Naib, Z.M. 1968. Amer. J. Obstet. Gynec., 101, 718-729

106. Jungeblut, C.W. and De Rodaniche, E.C. 1954. Proc. Soc. Exp. Biol. Med., 86, 604-606

107. Kalter, S.S. 1969. Ann. N.Y. Acad. Sci., 162, 499-528

108. Kalter, S.S. 1971. Natl. Acad. Sci., Washington, D.C., Pp. 481-527

109. Kalter, S.S. In press, Virus research. Bourne, G. (Ed.), Primates in biomedical research, Academic Press, New York

110. Kalter, S.S., Felsburg, P.J., Heberling, R.L., Nahmias, A.J. and Brack, M. 1972. Proc. Soc. Exp. Biol. Med., 139, 964-968

111. Kalter, S.S. and Heberling, R.L. 1968. Natl. Cancer Inst. Mono., 29, 149-160

112. Kalter, S.S. and Heberling, R.L. 1971. Bact. Rev.,35, 310-364

113. Kalter, S.S. and Heberling, R.L. 1971. Amer. J. Epidem., 93, 403-412

114. Kalter, S.S., Heberling, R.L., Vice, T.E., Lief, F.S. and Rodriguez, A.R. 1969. Proc. Soc. Exp. Biol. Med., 132, 357-361

115. Kalter, S.S., Ratner, J., Kalter, G.V., Rodriguez, A.R. and Kim, C.S. 1967. Amer. J. Epidem., 86, 552-568

116. Kim, C.S., Kriewaldt, F.H., Hagino, N. and Kalter, S.S. 1970. J. Amer. Vet. Med. Ass., 157, 730-735

117. Kinard, R. 1970. Science, 169, 828-831

118. King, N.W., Hunt, R.D., Daniel, M.D. and Melendez, L.V. 1967. Lab. Anim. Care, 17, 413-423

119. Kokernot, R.H., McIntosh, B.M., Worth, C.B., DeMorais, T. and Weinbren, M.P. 1962. Amer. J. Trop. Med. Hyg., 11, 678-682

120. Kratochivil, C.H. (Ed.) 1968. Primates in medicine. I. 1st Holloman symposium on primate immunology and molecolar genetics, S. Karger, Basel

121. Lampert, P.W., Earle, K.M., Gibbs, C.J., Jr. and Gajdusek, D.C. 1969. J. Neuropath. Exp. Neurol., 28, 353-370

122. Landsteiner, K. and Popper, E. 1909. Z. Immunforsch., 2, 377-390

123. Lapin, B.A. and Yakoleva, L.A. 1963. Comparative pathology in monkeys, Thomas, Springfield, Ill.

124. Larin, N.M., Herbst-Laier, R.H., Copping, M.P. and Wenham, R.B. M. 1967. Nature, 213, 827-828

125. Levy, B.M. and Mirkovic, R.R. 1971. Lab. Anim. Sci., 21, 33-39

126. Lichter, E.A. 1969. Nature, 244, 810-811

127. Lief, F.S. Personal communication

128. Livesay, H.R. 1949. J. Infect. Dis., 84, 306-309

129. London, W.T., Jr., Nahmias, A.J., Fuccillo, D.A. and Sever, J. L. 1971. Obstet. Gynec., 37, 501-509

130. Lou, T., Wenner, H.A. and Kamitsuka, P.S. 1961. Arch. Ges. Virusforsch., 10, 451-464

131. Mackay, C.M. and Schroeder, C.R. 1935. Proc; Soc. Exp. Biol. Med., 33, 373-374

132. Malherbe, H. and Harwin, R. 1957. Brit. J. Exp. Path., 38, 539-541

133. Malherbe, H. and Harwin, R. 1963. S. Afr. Med. J., 37, 407-411

134. Malherbe, H., Harwin, R. and Ulrich, M. 1963. S. Afr. Med. J., 37, 407-411

135. Malherbe, H. and Strickland-Cholmley, M. 1969. Lancet, 2, 1300

136. Malherbe, H. and Strickland-Cholmley, M. 1969. Lancet, 2, 1427

137. Malherbe, H. and Strickland-Cholmley, M. 1970. Lancet, 1, 785

138. Marsh, R.F., Burger, D., Eckroade, R.J., ZuRhein, G.M. and Hanson, R.P. 1969. J. Infect. Dis., 120, 713-719

139. Martin, G.V. and Heath, R.B. 1969. Brit. J. Exp. Path., 50, 516-519

140. Martini, G.A. and Siegert, R. (Eds.) 1971. Marburg virus disease, Springer-Verlag, New York

141. Masillamony, R. and John, T.J. 1970. Amer. J. Epidem., 91, 446-452

142. McKinley, E.B. and Douglass, M. 1930. J. Infect. Dis., 47, 511-522

143. Melendez, L.V., Barahona, H.H., Daniel, M.D., Hunt, R.D., Fraser, C.E.O., Garcia, F.G., King, N.W. and Castellanos, H. 1972. PAHO Sci. Publ., 235, 145-151

144. Melendez, L.V., Daniel, M.D., Hunt, R.D. and Garcia, F.G. 1968. Lab. Anim. Care, 18, 374-381

145. Melendez, L.V., Espana, C., Hunt, R.D., Daniel, M.D. and Garcia, F.G. 1969. Lab. Anim. Care, 19, 38-45

146. Melendez, L.V. and Hunt, R.D. 1966. Current views on herpes-T infection in South American monkeys, Proc. 5th Panamerican Cong. Vet. Med. & Zootechnique, 1966, Caracas, Venezuela

147. Melendez, L.V., Hunt, R.D., Daniel, M.D. and Trum, B.F. 1970. New World monkeys, Herpes virus and cancer, Balner, H. and Beveridge, W.I.B. (Eds.), Infections and immunosuppression in subhuman primates, Munksgaard, Copenhagen, Pp. 111-117

148. Melendez, L.V., Hunt, R.D., Garcia, F.G. and Trum, B.F. 1966. A latent herpes-T infection in Saimiri sciureus (squirrel monkey), Fiennes, R.N.T.-W. (Ed.), Some recent developments in comparative medicine, Academic Press, London, Pp. 393-397

149. Melnick, J.L. 1956. J. Immunol., 53, 277-290

150. Melnick, J.L. and Kaplan, A.S. 1953. J. Exp. Med., 97, 367-400

151. Melnick, J.L., Midulla, M., Wimberly, I., Barrerra-Oro, J.G. and Levy, B.M. 1964. J. Immunol., 92, 596-601

152. Melnick, J.L. and Paul, J.R. 1943. J. Exp. Med., 78, 273-283

153. Morris, J.A., Blount, R.E., Jr. and Savage, R.E. 1956. Proc. Soc. Exp. Biol. Med., 92, 544-549

154. Muller, W. 1935. Maschr. Kinderheilk, 63, 134-137

155. Nahmias, A.J., London, W.T., Catalano, L.W., Fuccillo, D.A., Sever, J.L. and Graham, C. 1971. Science, 171, 297-298

156. Nathanson, N. and Cole, G.A. 1970. Clin. Exp. Immunol., 6, 161-166

157. Parkman, P.D., Phillips, P.E., Kirschstein, R.L. and Meyer, H. M., Jr. 1965. J. Immunol., 95, 743-752

158. Parkman, P.D., Phillips, P.E. and Meyer, H.M. 1965. Amer. J. Dis. Child., 110, 390-394

159. Peebles, T.C., McCarthy, K., Enders, J.F. and Holloway, A. 1957. J. Immunol., 78, 63-74

160. Pelc, S., Perier, J.-O. and Quersin-Thiry, L. 1958. Rev. Neurol., 98, 1-24

161. Perkins, F.T. and O'Donoghue, P.N. (Eds.) 1969. Lab. Animal Handbook, No. 4

162. Pinto, C.A. and Haff, R.F. 1969. Nature, 224, 1310-1311

163. Potkay, S., Ganaway, J.R., Rogers, N.G. and Kinard, R. 1966. Amer. J. Vet. Res., 27, 331-334

164. Prince, A.M. 1971. Hepatitis associated antigen; longterm persistence in chimpanzees, Goldsmith, E.I. and Moor-Jankowski, J. (Eds.), Medical Primatology, Proceedings of 2nd Conference on Experimental Medicine and Surgery in Primates, S. Karger, Basel, Pp. 731-739

165. Rawls, W.E., Tomkins, W.A.F., Figueroa, M. and Melnick, J.L. 1968. Science, 161, 1255-1256

166. Reynolds, H.H. (Ed.) 1969. Primates in medicine. 4. Chimpanzee: Central nervous system and behavior; A review, S. Karger, Basel

167. Rosen, L. 1958. Amer. J. Trop. Med. Hyg., 7, 406-410

168. Rosen, L. 1958. Amer. J. Trop. Med. Hyg., 12, 924-928

169. Ruch, T.C. 1959. Diseases of laboratory primates, W.B. Saunders Co., Philadelphia, Pa.

170. Ruckle, G. 1958. Arch. Ges. Virusforsch., 8, 139-166

171. Ruckle, G. 1958. Arch. Ges. Virusforsch., 8, 167-182

172. Rustigian, R., Johnston, P.B. and Reihart, H. 1955. Proc. Soc. Exp. Biol. Med., 88, 8-16

173. Sabin, A.N. and Ward, R. 1942. J. Bact., 43, 86-87

174. Sabin, A.B. and Wright, A.M. 1934. J. Exp. Med., 59, 115-136

175. Saslaw, S. and Carlisle, H.N. 1969. Ann. N.Y. Acad. Sci., 162, 568-586

176. Sauer, R.M. (Ed.) 1960. Ann. N.Y. Acad. Sci., 85, 735-992

177. Schultz, E.W. and Habel, K. 1959. J. Immunol., 82, 274-278

178. Sever, J.L., Meier, G.W., Windle, W.F., Schiff, G.M., Monif, G. R. and Fabiyi, A. 1966. J. Infect. Dis., 116, 21-26

179. Sever, J.L., Schiff, G.M. and Traub, R.G. 1962. J. Amer. Med. Ass., 182, 663-671

180. Shulman, R.N. and Barker, L.F. 1969. Science, 165, 304-306

181. Siqurdardottir, B., Givan, K.F., Rozee, K.R. and Rhodes, A.J. 1963. Can. Med. Ass. J., 88, 128-132

182. Smetana, H.F. 1965. Lab. Invest., 14, 1366-1374

183. Smetana, H.F. 1969. Amer. J. Path., 55, 65A-66A

184. Smetana, H.F. and Felsenfeld, A.D. 1969. Virchows Arch. A. Pathol. Anat.-Path., 348, 309-327

185. Smetana, H.F. and Felsenfeld, A.D. 1969. Gastroenterology, 56, 1222

186. Smith, P.C., Yuill, T.M., Buchanan, R.D., Stanton, J.S. and Chaicumpa, V. 1969. J. Infect. Dis., 120, 292-297

187. Smithburn, K.C. and Haddow, A.J. 1944. J. Immunol., 49, 141-157

188. Smithburn, K.C., Haddow, A.J. and Mahoffy, A.F. 1946. Amer. J. Trop. Med., 26, 189-208

189. Soike, K.F., Coulston, F., Day, P., Deibel, R. and Plager, H. 1967. Exp. Mol. Path., 7, 259-303

190. Soike, K.F., Coulston, F. and Douglas, J.D. 1969. Exp. Mol. Path., 11, 323-332

191. Soike, K.F., Douglas, J.D., Plager, H. and Coulston, F. 1969. Exp. Mol. Path., 11, 333-339

192. Soike, K.F., Krushak, D.H., Douglas, J.D. and Coulston, F. 1971. Exp. Mol. Path., 14, 373-385

193. Soper, F.L. and Smith, H.H. 1938. Amer. J. Trop. Med., 18, 111-134

194. Starck, D., Schneider, R. and Kuhn, H.-J. (Eds.) 1967. Progress in primatology. First congress of the international primatological society, G.F. Verlag, Stuttgart, Germany

195. Sun, S.C., Sohal, R.S., Burch, G.E., Chu, K.C. and Colcolough, H.L. 1967. Brit. J. Exp. Path., 48, 655-661

196. Sweet, B.H., Hatgi, J. and Polise, F. 1969. Bact. Proc., 69-160

197. Sweet, B.H. and Hilleman, M.R. 1960. Proc. Soc. Exp. Biol. Med., 105, 420-427

198. Tauraso, N.M., Myers, M.G., McCarthy, K. and Tribe, G.W. 1970. Simian hemorrhagic fever, Beveridge, W.I.B. (Ed.), Infections and immunosuppression in subhuman primates, Munksgaard, Copenhagen, Pp. 101-109

199. Taylor, R.M. 1967. Catalogue of arthropod-borne viruses of the world, U.S. Publ. Hlth. Serv. Publ. 1760

200. Theiler, M. and Gard, S. 1940. J. Exp. Med., 72, 49-67

201. Theiler, M. and Gard, S. 1940. J. Exp. Med.,72, 79-90

202. Trask, J.D. and Paul, J.R. 1942. Ann. Intern. Med., 17, 975-978

203. Tribe, G.W. 1966. Brit. J. Exp. Path., 47, 472-479

204. Valerio, D.A., Miller, R.L., Innes, J.R.M., Courtney, K.D., Pallotta, A.J. and Guttmacher, R.M. 1969. Macaca mulatta. Management of a laboratory breeding colony, Academic Press, N.Y.

205. Verlinde, J.D. 1968. Trop. Geogr. Med., 20, 385-390

206. Von Magnus, P., Anderson, E.K., Peterson, K.B. and Birch-Anderson, A. 1959. Acta Path. Microbiol. Scand., 46, 156-176

207. Voss, W.R., Benyesh-Melnick, M., Singer, D.B. and Nora, A.H. 1969. Lab. Prim. Newsl., 8, 10-11

208. Whitehead, R.H., Chaicumpa, V., Olson, L.C. and Russel, P.K. 1970. Amer. J. Trop. Med. Hyg., 19, 94-102

209. Whitney, R.A., Johnson, D.J. and Cole, W.C. 1967. The subhuman primates: A guide to the veterinarian, Med. Res. Laboratory, Research Laboratories, Edgewood Arsenal, Md.

210. Work, T.H. 1958. Prog. Med. Virol., 1, 248-277

211. World Health Organization 1971. Health aspects of the supply and

use of non-human primates for biomedical purposes, WHO Tech. Rep. Ser., No. 470

212. Yoshicka, I. and Horstmann, D.M. 1960. New Engl. J. Med., 262, 224-228

213. Yoshida, E.H., Yamamoto, H. and Shimojo, H. 1965. Jap. J. Med. Sci. Biol., 18, 151-156

214. Zinsser, H. 1929. J. Exp. Med., 49, 661-670

215. Zlotnik, I., Smith, C.E.G., Grant, D.P. and Peacock, S. 1970. Brit. J. Exp. Path., 51, 434-439.

COMPARATIVE PRIMATE LEARNING AND ITS CONTRIBUTIONS TO UNDERSTANDING DEVELOPMENT, PLAY, INTELLIGENCE, AND LANGUAGE

Duane M. Rumbaugh
Department of Psychology
Georgia State University and
Yerkes Regional Primate Center
Emory University, Atlanta, Georgia (U.S.A)

The purpose of this chapter is to summarize our current understanding of the comparative learning processes and capabilities of primates. The reader's attention is directed to recent reviews of comparative primate learning (19, 20) for additional information and formulations that are related directly to the present discussion. The studies herein discussed, and those reviewed in the foregoing references, are all embraced within the domain of classic, comparative psychology.

THE NEED FOR A VALID COMPARATIVE BEHAVIORAL PRIMATOLOGY AND PSYCHOLOGY

In years past, proper concern has been expressed as to the validity of the comparative psychological framework and of the methods used by psychologists in their conduct of studies purported to be of a comparative-psychological nature. In particular, Beach (1) expressed grave reservations as to whether or not a comparative psychology existed or could ever be built. He concluded that comparative psychological studies were so fraught with methodological problems that valid comparisons, necessary for comparative-psychological studies, were beyond reach. Beach concluded that what existed, in point of fact, was the field of animal behavior, which embraced the study of psychological processes, such as those of learning and perception, and the study of species-specific behavior and the correlates thereof, as espoused by the ethologists.

Breland and Breland (3) reached similar conclusions underscoring a fact, long overlooked by comparative psychologists, that behavior

( ) This paper is supported in part by NIH Grant RR-00165

is not subject to capricious modification, for the species-specific character of animal behavior, delimited by genetic and morphological factors, resulted in response tendencies being differentially potent (probable).

More recently, Lockard (I2) and also Hodos and Campbell (9) have leveled their artillery and fired point-blank at the framework of comparative psychology. Lockard, in essence, jointed the camp tented earlier by Beach and by Breland and Breland. Hodos and Campbell charged, and with considerable validity, that comparative psychologists had erred in their quest for a comparative psychology because of ignorance of evolutionary processes and naivete in assuming that the long-touted phyletic scale was a valid way of conceptualizing the evolution of increasingly complex forms of life.

The well-intended and thoughtful arguments of the foregoing theoreticians are not be lightly dismissed, but I must express concern that the above referenced critics have piped many of those who would seek the chalice of a valid comparative psychology into a state of confusion and depression. It is my thesis that a true comparative psychology not only can but does exist, notably within the field of behavioral primatology. It is true that the building of a comparative psychology has been encumbered by incompetent methodology; but it is one thing to say, and responsibly so, that the building of a comparative psychology is irrevocably dependent upon the development of defensible, equitable research methods and tools, and quite another thing to say, and irresponsibly so, that those methods do not exist and cannot be devised.

Behavioral scientists will persist in comparing the behavioral responses and repetoires of diverse animal forms to the end of better understanding the correlates and adaptive advantages of animal behavior. Whenever they do so, they are, intentionally or not, within the proper domain of comparative psychology. The proper question is, "On what bases will the behaviors of animals be compared?" Particularly within the domain of comparative primate behavior defensible methods for comparing learning and perceptual processes are emerging. These methods will be reviewed in the present chapter. Too, the relative importance of early environment, as reflected in experiments where primate forms are systematically deprived of certain experiences and information, are known to be of differential significance for various primate species (I8, 27).

In short, there is only one significant barrier to our establishment of a valid comparative psychology: our willingness to commit the energy and intelligence necessary for the refinement of methods that yield measurements of an order that warrant the formulation of conclusions regarding the emergence and adaptive functions of animals' behavior.

We should recognize that we have no choice but to pursue the development of the required methodology and rules of comparison, for a comparative psychology is the sine qua non for the eventual understanding of man and his behavior. Our attempts to cope with the adjustment of the human individual in the very unnatural milieu within which man now lives and within which his very survival now depends are destined to fail to the degree a) that we fail to understand the biological predicates of man's behavior and b) that we fail to discern the genetically-determined behavioral units which surely have at least subtle, though nonetheless real and reliable, influence upon the development of human personality and social organization. Only in part can we discern the facts of these matters by the study of man alone; only in part can we discern them through the study of animal behavior; but by studying both man and extant animal forms, and in particular his nearest living relatives, the nonhuman primates, we enhance the probability of achieving the evolutionary perspective of the course of events which have produced modern man.

The study of nonhuman primates over the past few decades has already taught us many important principles, not the least of which is that the species variable is extremely important in the determination of a wide variety of behaviors, including those of a social, cognitive, and perceptual nature. A variety of behaviors are reliably associated with the species variable patterns of social organization, copulation, mother-infant relationships, problem-solving skills, learning processes, the dimensions for transmission of proto-cultural behaviors, etc. This knowledge implies that given the species, Homo sapiens, we shall eventually be able both to understand and to predict his behavioral/cogntive maturation and his propensities for a variety of social interactions and organizations. Man's behavior is not totally divorced from the consequences of evolution that provided for its initial emergence.

Let the feint-of-heart abandon the study of comparative psychology. The problems are great enough for us, in addition, to shoulder with patience and forbearance their discouragement and depressions. Let us encourage and support those who recognize the profound necessity for achieving a valid comparative-psychological framework if we are to cope effectively with the forces that now deny us the quality of life that we need and which tretens our very lives along with generations of all life forms yet unborn.

SOCIAL BEHAVIORAL PRIMATOLOGY AND ADULT COMPETENCE

Consideration now will be given to the study of primate behavior as part of the broader field of animal behavior and comparative sociobiology, and the contributions made by early experience and learning to the development of competent individual primates as adults.

Animal Behavior and Comparative Sociobiology

The study of primate behavior, as part of the larger field of animal behavior, is exemplified best by the plethora of field studies conducted, primarily, since 1955. A small army of anthropologists, zoologists, psychologists, and conservationists, has gone to the field to study various non-human primate forms to discern, in the main, how they are adapted to the varied ecological niches which they occupy. As a rule, all of these studies have emphasized behavior, for behavior is fundamental to adaptation. Behavior transports the animal through space so as to obtain the elements and conditions necessary for life--food, water, shelter, protection from danger, etc.-- and is also critical, of course, for reproduction, without which extinction is assured. As many forms of behavior are facilitated by the presence of a group of genetically similar animals, there has been an understandable emphasis placed on the study of social behavior.

Each field study contributes, in its own right, to our understanding of the behavior manifested by individual primate species. Studies of a given species in various habitats help define the role of learning, tradition, and proto-cultural developments in the achievement of successful ecologic adaptation. Collectively, studies on a variety of species in relation to ecological factors provide the foundation for the development of a comparative primate social behavior. The task is not an easy one, but the varied behavioral responses/adaptations to both common and unique problems faced by all primate species, in relation to selected dimensions that describe the habitats adjusted to, yields trends and principles that serve to integrate present knowledge and guide future social research both with feral and captive groups. The reader particularly interested in recent developments of a comparative primate social behavior is directed to chapters by Bernstein and Carpenter (this volume).

Social Competence of the Nonhuman Primate as Influenced by Early Environment

Perpetuation of a species is contingent upon competent behavior. Over the past twenty years a rich literature has accumulated that attests to the fact that early experiences are critical to the development of competent social behaviors in maturing monkeys and apes (25, 15, for reviews). Primates deprived of social interaction as infants are likely to manifest behavioral deficiences and aberrations and, also, inadequate integration of responses necessary for copulation when juvenile and adults (13). Further, they are likely to be incompetent in their interpretation and response to the social signals and communications of other conspecifies. Seay (27) reports evidence that the effects of social deprivation during infancy are not constant for all primate species. Accordingly, a comparative psychology is emerging that focuses upon the definition of the conditions and experiences in relation to the species variable that are necessary for behavioral competence in the adult form. This psychology, in due course, surely will contribute richly to our understanding of the developing human child and his needs for experiences that will permit him to achieve social competence as an adult. To the degree that we fail to understand and fail to take into account the biological predicates of man's social behaviors, we are ill-equipped to deal competently with the individual and social maladies of our own species. As we systematically study and come to understand the biological predicates of the species-related social behaviors of non-human primates, we equip ourselves both with valuable techniques and insights for understanding man's social behaviors, for man, too, is a primate form.

Cognitive Competence of the Nonhuman Primate as Influenced by Early Environment

Very recently it has been discerned that early experiences also affect the cognitive skills of adult monkeys and apes. Impoverished rearing conditions preclude the development of full cognitive capability. As these cognitive deficiencies promise to be life-long in effect, it behooves us to determine the critical parameters of early experience that differentially facilitate/deter cognitive growth. As these conditions likely interact with the species variable, systematic-comparative inquiry is necessary.

Harry F. Harlow's distinguished reasearch career at the University of Wisconsin has included many pioneering inquiries into the determinants of the Rhesus monkey's (Macaca mulatta) cognitive and social skills. The monkey's simple and complex learning processes develop in a lawful and orderly manner with maturation in a manner analogous to the development of intelligence in the human child (32, for a review). (That is not to imply, however, that the dimensions of monkey and human intelligence are identical). Competence for mastery of complex tasks increases with the monkey's maturity to adulthood, but that competence diminishes in old age in a manner that bears similarities to the assault of senility on man's intellectural competence. At the Yerkes Regional Research Center we have recently attempted the assessment of the complex learning and transfer-of-training skills of several old chimpanzees in their thirty's and forty's. Their incompetence was so profound as to totally frustrate our attempts to even quantify what abilities remained! Their effort was impressive, but their mastery was not.

For many years Harlow was of the opinion that even though early social deprivation might, at the extreme, render monkeys forever incompetent in contexts of a social nature, in mating, and in motherhood, such deprivation did not deny the normal development of their complex-learning skills and intelligence. Recently, however, Harlow et al. (8) have reported that both social isolates and control animals, housed individually in cages but in visual and auditory contact with one another, were inferior in learning the oddity concept to monkeys raised with parents and siblings. The latter monkeys, with the benefit of parental and family interactions, were superior to the social isolates and control animals. The performances of the social isolates and control animals were essentially identical. The cognitive cost, then, of early rearing was experimentally sensed when an enriched rearing condition was included for comparison and when a more demanding learning task (oddity concept formation) was employed. The earlier assessments of cognitive competence had required the less demanding tasks of discrimination learning set (7) and delayed response, which requires of the monkey that he remember through the passage of time the one object of two or more under which he saw a food incentive placed.

In cooperation with Charles M. Rogers and Richard K. Davenport, I pursued the question regarding possible cognitive deficits incurred by circumscribed/restricted early rearing with chimpanzees as subjects. At the Yerkes Field Station there is housed a group of chimpanzees, half of which were feral-born and reared in a style that provided for social interaction and half of which were maintained for upwards of the first two years of life in a manner that was tantamount to profound restriction. The latter animals were denied all social experience during those years and were allowed either no or, at best, modest environmental stimulation. Over their first 14 years of life, these two groups of chimpanzees were studied, primarily by Davenport and Rogers (5 for a project report), to determine their cognitive and social competence. The long-term costs incurred by the early restriction to half of the animals was pronounced. In every way, except physically, they were inferior to their feral-born, group reared colleagues.

There was always the concern, however, that the inferior performances of the restricted-reared chimpanzees on cognitive tasks might be artifacts of non-cognitive deficits, such as attentiveness and motivation to persevere in the demands of the training tasks. To clarify the matter of interpretation of the reliably obtained differences, we decided to test both groups with a new technique that yields equitable assessment of cross-species complex-learning and transfer-of-learning skills, a line of inquiry in which the same concerns exist regarding attentiveness and motivation, along with other factors relating to differences in morphology and perceptual characteristics.

The technique in reference I have titled the Transfer Index (20). The procedure entails criterional mastery on each of a series of two-choice discrimination problems prior to the administration of the test trials, on which the initial cue values are reversed and from which come the prime data for quantification of cognitive skills. It is maintained that the criterional training ensures engagement--activation--of the animal's cognition to a predetermined degree, set

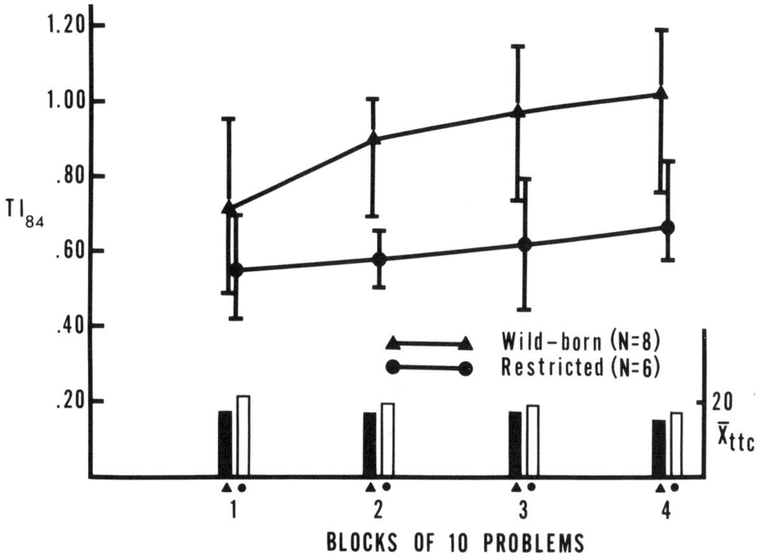

Figure 1. The Transfer Index of two groups of chimpanzees: wild-born and restricted-reared. The mean ($\overline{X}$) trials to criterion per block for the two groups are given in the lower portion of the figure (see right axis).

by the criterional training schedules, and that the test trials answer the question as to what the animal is able to do with what it has learned by way of transfering that training/learning to a similar, yet different, situation--one in which the initial cue values are reversed.

The experiment was revealing in every way. Fortuitously, on the average both the restricted-reared and feral-born groups achieved criterion on the problem series with nearly identical facility. Yet on the test trials, subsequent to the criterional training, the feral-born group was profoundly superior to the restricted-reared group. Figure 1 portrays the basic data. We can now confidently conclude in the affirmative: Early environmental experiences/rearing can profoundly retard the cognitive development of the chimpanzee. The implications for one of the chimpanzee's nearest living relatives--man--are obvious. One might conclude that primates are never too young to learn, and that if denied the opportunity to benefit from encounters from an enriched environment that includes social interactions, they will never be old enough to learn (efficiently).

The chimpanzees of the study just discussed were on the average 14 years of age. For the restricted-reared chimpanzees the years of life from two to 14 were not sufficient to offset the deficits incurred by the first two years! A widely held view, of probable validity, is that man's personal adjustment, along with his values, prejudices, and beliefs, are shaped and set by experience of infancy and childhood. Only in very recent years have we come to suspect that the pre-school years influence learning and achievement in school; only in very recent years have we came to suspect that educational methods of our schools might work at cross purposes with the development of creative thought processes and of full expression of the intellect. The results of our chimpanzee study and of those studies recently reported by Harlow with monkeys should compel us to be deeply concerned (if not frankly alarmed) about how we rear children from the day of birth.

We should insist on assigning the highest possible priority to the definition of those conditions which contribute maximally to the early cognitive development of our children. Whereas the genetic determination of our children's intelligence is set at the time of conception, their environment is not. Although intervention into the environment of the child *in utero* is problematic, it is by no means beyond our influence. The importance of prenatal care is now recognized, but it is not as generally recognized that the home/social environs of the child are of equal importance. True, in recent years we have seen rather haphazard, half-hearted, stat and stop efforts to launch environmental enrichment programs, such as HEAD START, but we have failed to mobilize an irrevocable commitment to doing the ground work (research) necessary for such programs to be caste optimally and to be implemented in a sustained way. I suspect that if the early years are important to the life-long cognitive competence of the Rhesus monkey and to the chimpanzee that they are absolutely critical to the intelligence of the human child. Can there be any question as to the contribution of these and related primate studies in terms of demonstrating our need for concern for our children's early experiences?

PLAY: THE FOUNTAINHEAD OF COMPETENCE AND CREATIVITY?

A recent trend in the literature of behavioral primatology is to view *play* activities of young nonhuman primates as a prime milieu for the efficient learning of both social and nonsocial skills (4).

Most of the social interactions of young monkeys and apes can be properly classified as forms of relatively unstructured play. In that play, I suspect, along with others, monkeys and apes learn all sorts of things about themselves and their friends. The challenge/stimulation afforded by the presence of other conspecifies elicits in due course most, if not all, of the behaviors which a given level of neurophysiological development can mediate, delimited by the morphology of the body.

It is problablv not due to chance alone that responses become serialized in certain chains so as to have the net effect of enhancing the competence of play. They are predisposed to certain kinds of actions, and not others, because of their morphology and because their neurophysiology is what it is. It is certainly easier for a monkey, for example, to be a successful player with other monkeys than it is for a man. We not only are not monkeys, a cherished fact but nonetheless a handicap in such play, but we lack the response proclivities essentially natural for monkeys because of their genetic endowment.

We lack the response repertoire of the monkey which is so powerful in eliciting the broad spectrum of responses from monkeys in a group, and we lack the ability to respond appropriately to the "let's play" signals of monkeys and apes so as to sustain their play behaviors. Monkeys need monkeys, apes need apes, and children need children first and foremost for their response limits to be exhausted and for the desired serializing of responses into organized chains to occur. This is not to say that in other ways infant primates, regardless of their species, have no need for mother and other adults, but it is to suggest that infants are better "tutors" of learning to play for other conspecifics than are adults. Why is this likely so?

Not only are adult forms, in general, physically unable to sustain the gamboling antics and rough-and-tumble play of infant monkeys and apes, but their response units are already organized into more-or-less competent chains. The competence of these chains, or patterns, are likely too overwhelming, too successful, for them to be long tolerated by the infant disposed to play at the moment. By contrast, age peers are near equals on the dimension of motor/response patterned competence/incompetence. As they interact they do not excessively challenge their collective competence other than for fleeting instants of time.

But even more is likely involved. To a major degree these infants have yet to learn the full meaning of the social signals--facial expressions, vocalizations, postures, odors--which they emit because of the fact that they are monkeys, or apes, or human. Yet those signals are emitted and quite naturally so, but because interpretation of their meaning is deficient, they do not, at least initially, serve to inhibit responsiveness and, consequently, the generation of still other signals as they do in the adult whose social education is presumed to be complete. The relative inability of infants to inhibit responding, or even to know that they should inhibit responding in the interests of safety should the signal generator be an adult male (!), serves to sustain the freewheeling antics that otherwise would be thwarted and suppressed.

Harlow's work has served to underscore the competence of the infant both as a tutorer for other monkeys learning play and, consequently, social skills. Total social isolation for the first year of the Rhesus monkey's life precludes its social competence as an adult. But to the degree that duration of social isolation is short of total devatation of the ability to learn social skills, the effects of the isolation are best counter-acted by none other than a normal infant monkey (28). The normal infant monkey is the best therapist, the best tutorer for the monkey not irreparably damaged by social restriction during early months of life. The challenge of the behaviors upon which it insists (because of its age), including clinging, are not so complex, not so overwhelming but that in due course the socially incompetent monkey responds in kind. And social competence emerges. Perhaps there is an interesting principle not to be passed over: No psychotherapist, no psychiatrist should be more integrated, better adjusted than those whom they wish to help therapeutically. While this principle is offered somewhat frivolously, it does not deny the fact that the competent therapist first defines "where the patient is" in terms of adjustment and works with him/her from that point. To do otherwise makes the disparity between the communications of the patient and therapist too great for much of anything desired to ensue.

It is frequently asserted in the primate literature that apes are more playful as adults than are monkeys and that the most playful adult of all primates is man. Why might this be so?

I suggest that the trend toward increased playfulness, reported in relation to evolutionary development as one progresses from the prosimians, to the monkeys, apes, and man, is the result of an interaction between a) number of available response units, b) the probability of those units becoming organized into fixed-action-patterns during the preadult years, and c) intelligence. Monkeys, I submit, are primarily responsive to signals of a social nature and relatively nonresponsive to novelty of the environment. By comparison, infant apes, though possibly as responsive to signals of a social nature, are more curious and more responsive to elements of the novel environment. Finally, I submit that the human child, though not unresponsive to stimuli of a potentially significant social type (the human smile and the human facial expressions, known for their high incentive value in the near-universal game of "Peek-a-boo!), is far and away, compared to monkey and ape, more responsive to the nonsocial aspects of its environment. Consequently, given reasonable opportunity to do so, the human child learns to be creative with the elements of its environment. This creativity is surely dependent upon intelligence provided for by the most advanced brain ever evolved.

As the probability of the response units of the child becoming serialized into fixed-action-patterns is less than that for ape and monkey, respectively, the units may be recombined, reorganized into any number of play behaviors. Play eventually evolves into games, with agreed-to-rules. Freed from species-predicted fixed-action-patterns that appear to be instinctive in quality, yet not necessarily independent of experience as provided for in social play, and equipped with a brain that can create new combinations for response-unit organization/serialization, the adult human remains playfully inclined until the assault of age denies the competence to perform the units reliably over time, if at all.

Psychologists have never been found short in explaining behavior, though admittedly they are at times found lacking in predicting behavior accurately. Explanations of behavior are of merit to the degree they allow us to preduct what will happen, given conditions and manipulations thereof. What are some of the implications of the foregoing speculations with regard to the generation of playfulness and creativity? One implication is that if in some manner we increase the number of response units that will remain free of species-characteristic fixed-action-patterns, we will enhance creativity and sustain through life the propensity to innovate activities, to devise games, etc. The following study, admittedly of a limited scope, yielded observations in support of this speculation.

In cooperation with Austin H. Riesen and Sue C. Wright, I (24) conducted a study to determine whether young great ape infants trained with a variety of materials on a series of tasks would be more creative in their unstructured play with new materials than would other infants trained on the same tasks but with the materials held constant for each task of the series. The training tasks entailed extraction of a food incentive from a container, carrying objects across the room, serially inserting small cannisters into larger ones constituting a graduated series of size, and placing appropriate plastic forms into form boards. In the first training condition a variety of materials were used in each task; in the second condition the materials used within a given task were held constant. A third group of apes had only unstructured and random encounter with the training materials during free-exercise sessions. At the end of nine months of daily training, the animals of the three conditions were given modified tests of all of the above and, in addition, were given opportunity for free-play with new materials in a new setting.

The observations were consistent with the predictions: Training with a variety of materials on each task enhanced the responsiveness and creativity of apes in a free-play situation. In particular, one orangutan in the multiple-materials training condition was both responsive and creative on test sessions. She incorporated into her play with new objects/materials well organized elements of response units accrued during the nine months of prior training. As the study entailed only six animals, one must be reserved about drawing firm conclusions, but the results did support the expectancy that with the acquisition of new response units (during training) the creativity of play is enhanced. Were it possible to sustain such study through the entire development of the apes, I would predict that as adults they would remain playful, creative, and inventive within the limits afforded by their brains, advanced but primitive compared to man's.

It is through play that social skills, patterns of varied responsiveness, and the principles of inter-relationships between social/physical elements is learned. Learning provided for by play is the origin of social adjustment and fountainhead of invention and creativity. Play should be the rule, not the exception, to both early environment and education.

THE EVOLUTION OF HUMAN INTELLIGENCE

The precursors of man are extinct. It is impossible to study directly the phyletic evolution of intelligence, though it is possible to infer within gross limits the intelligence of primitive man from estimates of his cranial capacity made possible by fossilized fragments of his brain encasements and the implements found in association with his abodes.

Comparative psychologists interested in the evolution of human intelligence can study only extant living forms behaviorally, and no contemporary primate can be said to have evolved directly from another contemporary form. This restriction, and it is a profound one, does not preclude our study of complex learning processes in their relation to brain development as evolved and differentially manifested in the array of primates which still survive. Elsewhere I have discussed this area of study in greater detail (20-21) than is warranted in this particular treatise. Briefly, contemporary primates can be arranged along the dimension of brain development, essentially identical to the series defined by Le Gros Clark (11) as the one that increasingly approximates man--prosimian, tarsier, New World monkeys, Old World monkeys, lesser apes, great apes, and, finally, man. Although none of these extant forms evolved from other of the series, they, nonetheless, likely approximate critical stages of the evolutionary chain of events that rendered modern man.

Studies which I have directed over the past 12 years, coupled with those of other investigators, support the conclusions that a) complex learning skills are enhanced by brain development, b) transfer-of-training skills are similarly enhanced, and c) of the nonhuman primates the great apes are superior in transfer-of-training skills, if not complex-learning skills *per se*.

In an earlier portion of this chapter reference was made to the Transfer Index (TI). This measurement was devised to allow equitable cross-species estimates of intelligence. To achieve such estimates it is necessary to make allowance for the varied characteristics which a species array of primates manifests, any one of which might differentially prejudice the equity/fairness of the test situation and produce *performance* rather than *ability* differences.

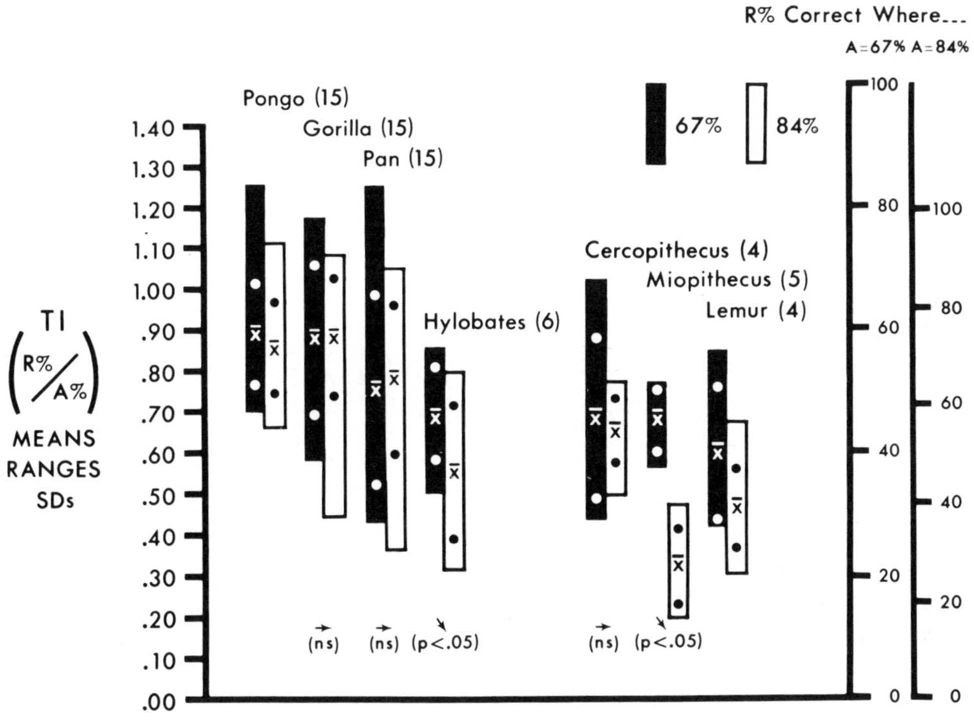

Figure 2. The TI is calculated by dividing the percentage of responses correct on the reversal trials (trial I deleted) of IO problems by the prereversal criterional lev --67 or 84. Two TI values for each subject at both the 67% and 84% training levels provide the basis for this figure.

The TI testing procedure entails criterional training on each of a series of two-choice visual discrimination problems prior to the test trials on which the measurements of prime significance are taken for the purpose of estimating cognitive capacity. Earlier in this chapter it was noted that the restricted-reared versus the feral-born chimpanzees were equivalent on criterional mastery of the TI problems but markedly different on the subsequent test trials, yielding a great and reliable difference in TI measurements. Empirical search has led to the conclusion that within the limits of 11 to 60 trials to reach the desired criterion, numbers of trials _per se_ are not reliably associated with performance on the test trials. Accordingly, it has been concluded that differences obtained between average numbers of trials for species to achieve criterion are of no major moment in accounting for differences on the critical test trials, when such are obtained.

To date, two pre-test (pre-reversal) criterional training schedules have been employed--the 67% and 84% (see 20 for details). Each TI measurement is based on 10 problems, where each problem has been mastered within the limits of 11 to 60 trials. Over the years we have succeeded in testing numbers of primates of various species in a program that entailed procurement of, first, two successive TI values at the 67% level and, then, two successive TI values at the 84% level. Figure 2 portrays the basic data obtained to date. These data (see 22 for additional discussion) are the most extensive and defensible cross-species data reported to date. They reveal that when TI measurements are based upon the 67% pre-reversal criterional training schedule great apes have a higher range limit than any other primate form tested. Moreso than for other species tested, the great apes manifest a positive relationship between pre-reversal mastery (determined by the criterional training schedule employed) and reversal performance. This conclusion is based on the fact that their average TI values were essentially unchanged as the pre-test criterional level was advanced from 67% to 84% and by the fact that the axes on the right of the figure reveal that their reversal performance in an absolute sense increased from 15-20% as the criterional level was advanced from 67% to 84%. The more they knew pre-test (pre-reversal) the better thay did on test (reversal). By contrast, the other primate forms either manifested a smaller increment in their reversal performance (_Cercopithecus aethiops_), no increase (_Hylobates lar_ and _Lemur catta_), or a tremendous decrement (_Miopithecus talapoin_). The great apes with their superior brains benefitted on test in direct accordance with the mastery level required pre-test; by contrast, the talapoins with their relatively primitive brains were compromised. The more the talapoins knew regarding the relatively cue values of the discriminada prior to test (84% rather than 67% task mastery), the worse they did on the reversal test trials (reversal only 30% correct rather than 45% correct)! The talapoins negatively transferred their learning, whereas the great apes positively transferred it.

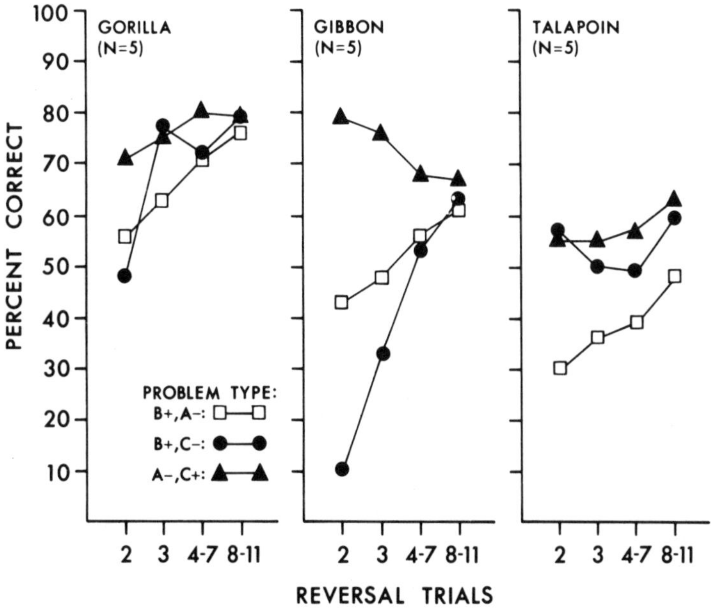

Figure 3. Evidence that primates of different brain development have different learning processes. In relation to the discussion of the text, the control condition is B+, A–, the second reversal condition is B+, C–, and the third reversal condition is A–, C+.

Now, one can imagine the dilemna that man would be in if his transfer-of-training skills were like the talapoins. If the more he knew the less adept he would prove to be in coping with his next cognitive challenge, to be creative and an effective problem-solver in the face of the increasingly complex problems of the modern day he would need always to know less and less to do well! That course of logic is not one which I would choose to defend, but it is offered to help the reader see how far the apes and man have come from the distant past when our ancestors' brains were approximate to that of the present-day talapoin.

QUALITATIVE DIFFERENCES IN LEARNING IN RELATION TO BRAIN DEVELOPMENT

In another study it was determined that qualitative differences in learning processes are probably associated with different levels of brain development (21). Groups of gorilla, gibbon, and talapoin were once again criterionally trained on each of a series of two-choice object discrimination problems. All groups achieved the criterion at comparable rates. Upon achievement of criterion for a given problem the cue values were reversed--the food-rewarded object became nonfood-rewarded. This first reversal trial served as a cue to these test-sophisticated subjects that on the second and following reversal trials some alteration in object choice would have to occur if additional reward was to be obtained. In the control reversal condition, both objects used in the pre-reversal criterional training remained but with their relative cue values reversed, as instituted on the first reversal trial of all problems. Performances of the groups on this type of reversal test revealed that the gorillas were superior to the gibbons and that the gibbons, in turn, were superior to the talapoins (Fig. 3). The better the brain, the better the reversal, even though criterional acquisition rates were similar. In the second type of reversal, the object correct during criterional training and incorrect on the first reversal trial was deleted and in its stead a new object was substituted and assigned the value of "incorrect"--non food--rewarded if chosen. If criterional acquisition had entailed the formation of a habit to the initially correct object and if efficient reversal is deterred by trials being required to break or extinguish that habit, then this condition should facilitate reversal performance. If on the other hand, criterional acquisition had been on the basis of some abstractive form of learning, this condition should neither facilitate nor deter reversal, for choice of the initially incorrect object would reliably net reward on reversal--a condition very familiar to all of these animals. Performances revealed that only for the gorillas was this reversal condition neither more nor less difficult than the control reversal condition. This reading was taken as the basis for inferring that they had, in fact, an abstractive learning process which neither of the other two groups had. The third, and final reversal condition was similar to the second except that a new object was substituted for the object that was initially incorrect during criterional training but correct on the first reversal trial. This condition was used to assess the degree to which a tendency not to respond to the initially unrewarded object deterred selection of it on reversal trials, this tendency is termed "inhibition" in learning theory. If inhibition did deter efficient reversal by making the subject disinclined to choose an object that was initially not rewarded, then deletion of that object and substitution of it with a new object should facilitate reversal. Only for the talapoins were the second and third reversal conditions easier (higher reversal performances) than the control condition. This reading was taken to mean that the talapoins achieved criterional acquisition pre-reversal by means of a habit formation process--the building up of a response tendency to one of a tendency not to respond to the other. Gibbons were different from both gorillas and talapoins. Their tendency to select the new stimulus, as used in the second and third reversal conditions, was so strong as to allow no inference as to the nature of the learning process whereby criterional learning occurred.

Again, it is noted that the method devised and employed allowed for the defensible drawing of comparisons between diverse species and inferrences regarding the nature of learning processes. The method allowed for a comparative-psychological investigation that was revealing as to differences in learning processes associated with brain development, a subject of considerable conjecture compromised by lack of defensible obtained comparative data through the literature of the past 50 years.

If brain development provides for abstractive learning processes to emerge, what might be done to enhance brain development (and/or refinement) in even the normal human child? How? When? Such inquiries are surely possible. If man has the genius to go and to return from the moon, in due course he should be able to reduce the frequency of mental retardation and to enhance the intelligence of his kind.

THE RELATIVE INTELLIGENCE OF THE GREAT APES

Let us return to the basic TI data presented in Figure 2 and consider the relative intelligence of the great apes. It has been widely claimed that the chimpanzee (*Pan*) is the most intelligent of the nonhuman primates, superior to the other great apes--gorilla (*Gorilla*) and orangutan (*Pongo*). If it is assumed that the ability to transfer training in a positive manner is germane to the processes of intelligence as we know them in man, then examination of TI data, which in essence measure transfer-of-training from pre-reversal criterional training to the reversal-test trials, for the great apes should provide at least a partial answer to the question of relative great ape intelligence (22).

Figure 4 portrays the trial-by-trial reversal performance for the 45 great apes and six lesser apes (Hylobates) when the pre-reversal criterion was 67%; Figure 5 portrays their reversal performances when the pre-reversal criterion was 87%.

The performances of the chimpanzees and gibbons, while not differing significantly from one another, are significantly below those of the gorillas and orangutans when TI measurements are obtained through use of the 67% criterional training schedule (Fig. 4). When TI measurements are obtained through use of the 87% training schedule (Fig. 5), no statistically reliable differences are obtained among the great ape genera, but all three of these genera are superior to the gibbons. On the average, then, chimpanzees are somewhat inferior to gorillas and orangutans. Another analysis for age-matched groups of the three great ape genera yielded the same findings and conclusions, and still another analysis revealed that the five best gorilla, chimpanzee, and orangutan learners were essentially identical in their reversal performances. Conversely, the lowest five chimpanzees were well below the five lowest gorillas and orang-utans.

General impressions regarding the intelligence of nonhuman primates can be deceiving! Systematically collected data from a formal test situation that ensures cognitive engagement of the subjects prior to the procurement of the data used for comparative assessment, as in TI testing procedures, reveals that the chimpanzee is not the intellectual giant of the nonhuman primates as generally has been asserted. The great apes are, however, collectively superior to the lesser apes, the gibbons.

INDIVIDUAL DIFFERENCES AMONG NONHUMAN PRIMATES

Humans differ one from another in various ways, including intelligence. The same is true for nonhuman primate genera. Within each genera of monkey and ape that we have studied, there are marked differences in performances. We have observed some squirrel monkeys (Saimiri) that exceeded the performances of the poorer great apes, and within each great ape genera there are those that do extraordinarily well and those that have no intellectual distinction whatsoever.

To date no investigator has capitalized upon the distribution of learning skills, as typically obtained within a sample of a nonhuman primate species, to discern neurophysiological correlates of those performance differences or to discover how a variety of psychological/cognitive attributes might cluster among subjects of varying "intelligence" levels. This line of inquiry with nonhuman primates should be pursued since we may attempt experimental interventions to modify their expressed achievement levels through methods disallowed with humans (and for good cause!).

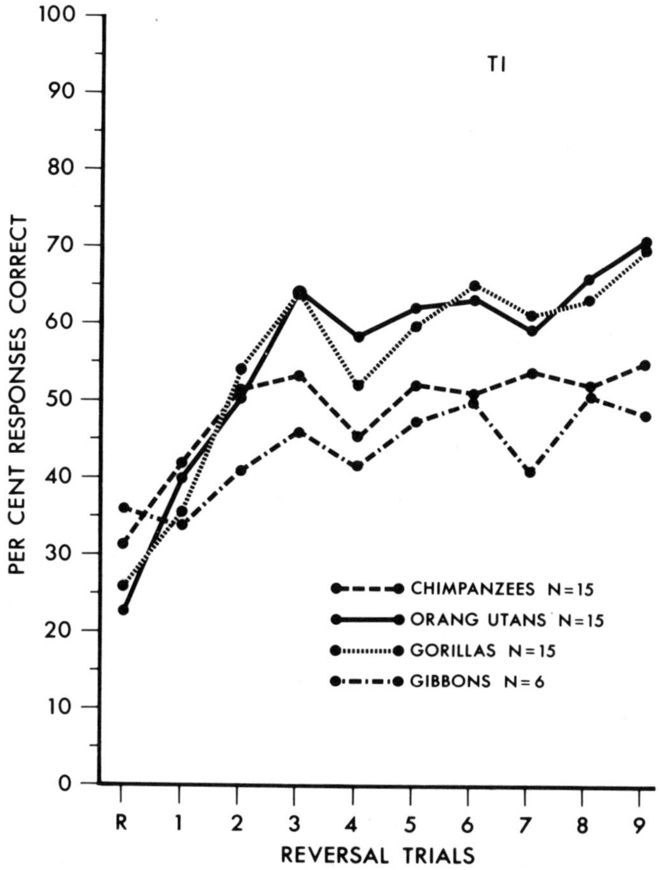

Figure 4. Reversal performance subsequent to criterional mastery to the 67% correct level.

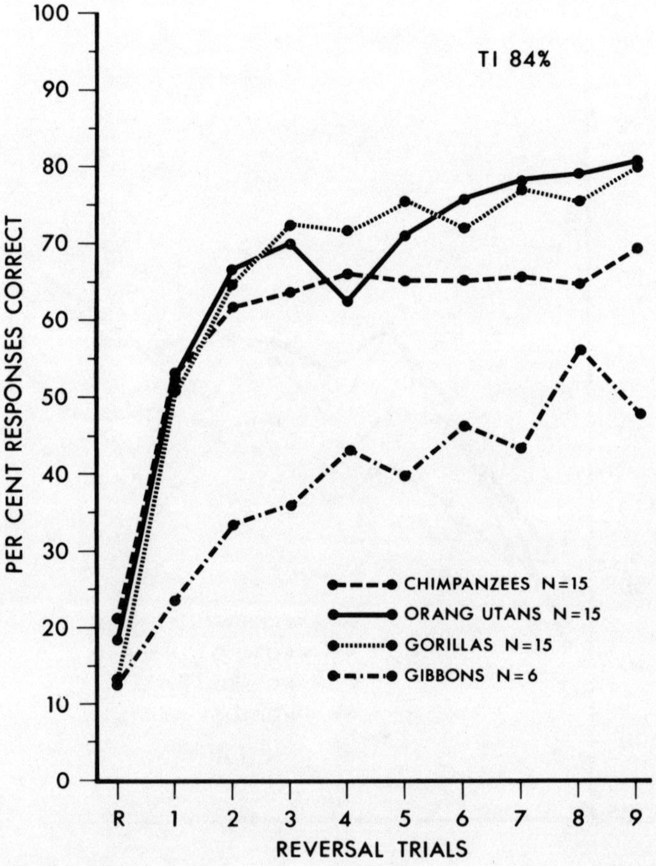

Figure 5. Reversal performance subsequent to criterional mastery to the 84% correct level.

READINESS TO ATTEND TO VISUAL FOREGROUND CUES: AN ECOLOGICAL ADAPTATION?

The vast majority of primates, including man, rely heavily upon visual information for adaptation to their environs. Totally blind nonhuman primates simply do not survive long in the field!

Young and Farrer (31) have made detailed studies of primate vision and have concluded that although the eyes of the great ape genera have some anatomical differences, their visual acuity is functionally equivalent and their color vision very similar. It would seem, then, that gorillas, chimpanzees, and orangutans would "see" the same field, the same things if the visual environment is held constant in a formal test situation. But this is not the case.

We have demonstrated that in a formal test situation orangutans are much more inclined to attend to the distractions of systematically-introduced irrelevant foreground cues than are, in turn, chimpanzees and gorillas (23). This rank ordering is of interest as it aligns perfectly with how arboreal (tree dwelling) these apes are in their natural ecological niches. Orang-utans, morphologically ill-equipped for quadrupedal locomotion on a flat surface, spend most of their time in the trees and descend to the ground only rarely. Chimpanzees are well equipped for locomotion both in the trees and on the ground, and move freely and confidently from one milieu to the other. Gorillas are cautious climbers and, except when young, seem to prefer the ground, which probably reflects their relative massime proportions.

The study which yielded the data for the above-stated finding came from a formal testing in which each animal learned each of a series of two-choice object-quality discrimination problems to a criterion prior to the administration of the test trials which were designed to tell us the degree to which they were inclined to attend to irrelevant foreground cues. Each group of great apes had eight animals selected so as to constitute three groups with equivalent TI ability.

On a given problem, when criterion was achieved the first of three test trials was given. On these test trials, sections of 1/2-inch wire mesh were interposed between the objects, which constituted the problem just mastered, and the glass which formed the fronts of the bins which contained the objects. On these test trials, then, the subjects had to not only look through the glass fronts of the bins, but also through the wire-mesh sections to view the objects of the problem. The more the animals dropped in performance accuracy on these test trials, the more they must have attended to the wire mesh, *per se*, for otherwise the problem was unchanged. (Between the second and third test trials on each problem, five training trials without the wire-mesh sections were given so as to support the initial discrimination). All animals in each of the groups were

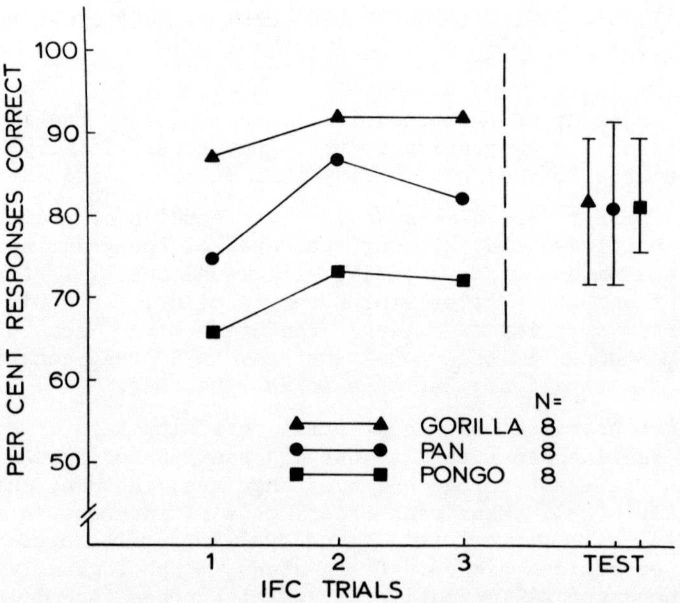

Figure 6. Performances for three genera of apes on the three test trials of the first 10 problems. On these trials irrelevant foreground cues, defined by the wire mesh, were present. Values presented in the right-hand portion of the figure were obtained from the 50 two-trial problems in which the irrelevant foreground cues were present only on the second trial.

sustained in this testing procedure until such time as they had no more than three errors on the test trials of 10 consecutive problems (three errors or less in 30 test trials).

On the average the three groups of apes reached the pre-test criterional performance at comparable rates. They differed markedly, however, on the test trials (Fig. 6). On the test trials of the first ten problems the gorillas' performances dropped 10%, the chimpanzees' dropped 20%, and the orang-utans' dropped 30%. Further, the gorillas achieved the inter-problem criterion (not more than three errors over the course of 10 problems) with the first 11 problems! By contrast, the chimpanzees required 14 problems and the orang-utans required 22 problems.

Subsequent to achievement of the inter-problem criterion, all animals were given 50 two-trial problems, the first trial without the mesh and the second trial with the mesh in place. On these problems the three great ape genera performed quite equivalently (right-hand portion of Fig. 6). The chimpanzees and orang-utans did learn to ignore the introduction of the irrelevant foreground cues, but it was more problematic for them than for the gorillas. Recall that these groups were equated for TI proficiency. Acordingly, we negate cognitive differences as the probable source for the observed differences. Too, in an identical experiment it was observed that the restricted-reared chimpanzees of the Yerkes Field Station were no different from their feral-born colleagues so far as distraction to the introduction of the foreground cues was concerned, though the latter group was significantly superior to the former in TI--additional evidence that cognitive capabilities are not the source for the obtained differences in readiness to attend to foreground cues.

What is the possible adaptative value to readiness to attend to foreground cues? Quite likely it serves to protect the eyes from damage that might be incurred by twigs and branches common to the arboreal environment. At any rate, some process of perception has made for a difference in attention that cannot be attributed to differences in visual anatomy, so far as we know, or to intelligence. Has evolution selected for behavioral propensities quite apart from anatomical/neurophysiological characteristics? The results of the foregoing study would seemingly answer that question in the affirmative.

LEARNING AND LANGUAGE

The task of mastering language, with all its ambiguities and abstractions, probably taxes learning capacity to a maximum. Indeed, many intelligence tests for humans are primarily measurements of language mastery.

Many people have been intrigued with the possibility that non-human primates, notably apes, might be able to develop language skills. Early attempts to teach chimpanzees language failed, because they were attempts to teach human speech. We now know that the vocal apparatuses of monkeys and apes are such as to preclude the production of the phonemes from which our spoken language is comprised. The King's English is not to be aped!

In recent years the work of the Gardners (6) and of Premack (16) has served to revive efforts to determine the language-relevant capacities of the chimpanzee and other apes. The Gardners' work, in particular, has given us reason to expect that apes might possess the intelligence and learning skills necessary for the acquisition of a language system. Only in man is language a species-characteristic skill. Whereas the majority of life forms have signal systems of varied complexity, no form other than man has developed a communication medium with the flexibility and efficiency of language, the processes whereby symbols of arbitrary meaning are recombined to communicate novel messages.

The Gardners demonstrated that their chimpanzee, Washoe, was able to learn a rather extensive vocabulary of hand signs and gestures, derived from American Sign Language for the deaf. Over time they also observed that Washoe spontaneously chained signs rather appropriately. Since none of Washoe's language training entailed rules of language structure or syntax, the chained, spontaneous expressions were possibly (probably?) due to chance. But research proceeds in small steps, and the Gardners took a giant step as they undertook and succeeded, to the degree they did, in their work with Washoe. They re-opened a research area of great interest and significance, one that had been put to rest by the majority who concluded that such work would be to no avail.

This past year several colleagues (Professors E. von Glasersfeld and P. Pisani from the University of Georgia; Professor H. Warner from Yerkes; Professor J. Brown from Georgia State University; Mr. T. Gill and Mr. C. Bell from Yerkes) and I have undertaken the development of a computer-controlled environment within which we hope to quantify language acquisition in infant apes. If the system becomes proven, it will also provide the technology necessary for replication and systematic inquiry into the parameters of language formation.

Our subjects are Lana, chimpanzee, and Biji, orangutan, now two years old. The language training environment consists of a large, specially designed test chamber, complete with food and liquid vendors, hi-fi speakers, movie screen, and toy/blanket dispensers. A console, replete with keys individualized by their colors and geometric signs, is also in the chamber. Each key symbolizes a word. In due course, depression of the keys in accordance with the rules of sentence structure, programmed into and monitored by the computer, will be necessary for the ape subject (worked with one at a time) to obtain the incentive and environmental alteration which it seeks.

Initially, as with language acquisition in children, the apes will learn different holophrases ( ), then the individual words in series which comprise them. As language develops (if it does), the apes will be able to obtain whatever they desire--choice of drinks, foods, playmates, movies, music, people, blankets, toys, even an outing! The learning of language is reinforcing for it provides control over the environment, so we maintain. To the degree that

( ) A holophrase is a single expression, such as "Cookie!", which is an abbreviation for a longer sentence, such as "Please, give me a cookie." Children typically use holophrases as they begin to talk.

our apes learn language, they will obtain for themselves the "Good Life", and simultaneously give us a better understanding of language-relevant processes.

All that happens will be stored for analysis by the computer. The system includes the capacity for two-way conversation between the experimenter and the apes. The computer will always intervene between the two as a mediator, to monitor and to record all that transpires.

We view the computer as "Mama" in the situation, but she is an improvement on the human mother, we trust, when it comes to the instruction of language skills. She will always be "on line" 24 hours a day, she will be infinitely patient and errorless in her tutoring, she will be replicable in her training methods, and she will save us endless hours of tedious data analysis.

Finally, it should be noted that the technology we are developing promises to allow for very defensible comparative studies of language acquisition which, in due course, might include all great ape genera and well as the human child. The technology for building a comparative psychology in within our reach.

LEARNING AND THE TRANSMISSION OF PROTO-CULTURAL BEHAVIORS

Monkeys and apes are known to be inclined to imitate certain behaviors, especially during their early years. The prolonged period of maturation provides them with a rich opportunity to learn by example from other conspecifics. But they do not imitate indiscriminately. Itani (10) reported that the acquisition of the potato-washing behavior, the effect of which was to remove sand prior to ingestion, spread from the originator, Imo--then 1/2 years old, to playmates and to the closer family units. Young chimpanzees who lose their mothers are unlikely to develop "termiting" behaviors, relatively common among the chimpanzees of the Gombe Stream (29).

The remarkable use of ladders by chimpanzees, reported by Menzel (14), spread directly from the prime innovator, to his two closest companions. Friendship and close family bonds appear to have prime influence upon both which animal and what will be imitated.

There is much to be learned about how primates, including the human child, learn via the process of imitation. Irwin Bernstein of the Yerkes Primate Center is currently investigating the possibilities of "taboo" transmissions across generations of monkeys, maintained in captive settings at the Field Station of that center. Hopefully he will succeed in shedding light on this topic of prime interest. As it appears to be a critical process for the transfer of highly innovative behaviors among nonhuman primates, it behooves us to determine its dynamics and functions in the life of the young child.

SUMMARY

This chapter has been a partial review of current trends and topics in the broad field of comparative primate learning in relation to a variety of behavioral phenomena. The interested reader is directed to two volume series, Schrier, Harlow, and Stollnitz (26) and Rosenblum (17), for reviews of additional studies of this order.

In the present chapter I have attempted to highlight reasons for the necessary development of the comparative behavioral primatology which we now have and so vitally need for coping with problems faced by modern man. In addition, I have underscored certain procedures that hold promise for the conduct of comparative-psychological studies in a manner that warrants the drawing of defensible, hence warranted, conclusions. In brief, the method advanced is one in which a performance criterion is required prior to the procurement of the measurements upon which prime emphasis is placed for comparative assessments. This method has proved of value in the development of the Transfer Index testing procedure, in the conduct of studies designed to assess the nature of the learning processes whereby criterional mastery is achieved, and in the assessment of what diverse primate species attend to in a formal test situation in which irrelevant cues are introduced. No experimental approach is beyond reproach, but better that we approximate reality than to refuse all inquiry as to its nature. Other methods, as proposed by Bitterman (2) through use of the method known as systemic variation and as inherent in the extensive comparative work by Warren (30) in which differential transfer effects between training and test conditions have revealed differences for cats as opposed to monkeys, bear refinement and should be used, in addition to those advanced in the present chapter, to the end of building a truly comparative psychology of primate behavior.

References

1. Beach, F.A. 1960. American Psychologist, 15, 1-18

2. Bitterman, M.E. 1964. American Psychologist, 19, 396-410

3. Breland, K. and Breland, M. 1961. American Psychologist, 16, 681-684

4. Bruner, J.S. 1972. American Psychologist, 27, 687-708

5. Davenport, R.K. and Rogers, C.M. 1970. Differential rearing of the chimpanzee. In G.H. Bourne (Ed.) The Chimpanzee. New York, S. Karger, Vol. 3, Pp. 337-360

6. Gardner, B.T. and Gardner, R.A. 1971. Two-way communication with an infant chimpanzee. In A.M. Schrier and F. Stollnitz (Eds.) Behavior of Nonhuman Primates. New York, Academic Press, Vol. 4, Pp. 117-184

7. Harlow, H.F. 1949. Psychological Review, 56, 51-65

8. Harlow, H.F., Harlow, M.K., Schiltz, K.A. and Mohr, D.J. 1971. The effect of early adverse and enriched environments on the learning ability of Rhesus monkeys. In L.E. Jarrard (Ed.) Cognitive Processes of Nonhuman Primates. New York, Academic Press, Pp. 121-148

9. Hodos, W. and Campbell, C.G.G. 1969. Psychological Review, 76, Pp. 337-350

10. Itani, J. 1958. Primates, 1, 84-98

11. Le Gros Clark, W.E. 1959. The Antecedents of Man. Edinburgh and London, Edinburgh University

12. Lockard, R.B. 1971. American Psychologist, 26, 168-179

13. Mason, W.A. 1971. Motivational factors in psychosocial development. In Wm. J. Arnold and Monte M. Page (Eds.) Nebraska symposium on motivation. Lincoln, University of Nebraska Press, Pp. 35-67

14. Menzel, E.W., Jr. 1972. Folia Primatologica, 17, 87-106

15. Mitchell, G. 1970. Abnormal behavior in primates. In L. Rosenblum (Ed.) Primate Behavior. New York, Academic Press, Vol. 1, Pp. 196-249

16. Premack, D. 1970. On the assessment of language competence in the chimpanzee. In A.M. Schrier and F. Stollnitz (Eds.) Behavior of Nonhuman Primates. New York, Academic Press, Vol. 4, Pp; 185-229

17. Rosenblum, L.A. (Ed.) 1970-71. Developments in Field and Laboratory Research, Vol. 1 and 2, Academic Press

18. Rosenblum, L.A. 1971. The ontogeny of mother-infant relations in

macaques. In H. Moltz (Ed.) The Ontogeny of Vertebrate Behavior. New York, Academic Press, Pp. 315-367

19. Rumbaugh, D.M. 1968. The learning and sensory skills of the squirrel monkey in phylogenetic perspective. In L.A. Rosenblum and R.C. Cooper (Eds.) The Squirrel Monkey. New York, Academic Press, Pp. 255-317

20. Rumbaugh, D.M. 1970. Learning skills of anthropoids. In L.A. Rosenblum (Ed.) Primate Behavior. New York, Academic Press, Pp. 1-70

21. Rumbaugh, D.M. 1971. Evidence of qualitative differences in learning among primates. Journal of Comparative and Physiological Psychology, 76, 250-255

22. Rumbaugh, D.M. and Gill, T.V. 1973. The learning skills of great apes. Journal of Human Evolution, 2, 171-179

23. Rumbaugh, D.M., Gill, T.V. and Wright, S.C. 1973. Readiness to attend to visual foreground cues. Journal of Human Evolution, 2, 181-188

24. Rumbaugh, D.M., Riesen, A.H. and Wright, S.C. 1972. Creative responsiveness to objects: A report of a pilot study with young apes. Folia primatologica, 17, 397-403

25. Sackett, G.P. 1970. Unlearned responses, differential rearing experiences, and development of social attachments by Rhesus monkeys. In L. Rosenblum (Ed.) Primate Behavior. New York, Academic Press, Vol. 1, Pp. 112-140

26. Schrier, A.M., Harlow, H.F. and Stollnitz, F. (Eds.). Behavior of Nonhuman Primates. Modern Research Trends, Vols. 1-4, Academic Press, 1965-1972

27. Seay, B.M. 1972. Peer social behavior. Paper presented at the Southeastern Psychological Association Convention, Atlanta, Georgia

28. Suomi, S.J. and Harlow, H.F. 1972. Developmental Psychology, 6, 487-496

29. van Lawick-Goodall, J. 1968. Animal Behavior Monographs, 1, 165-301

30. Warren, J.M. 1965. Primate learning in comparative perspective. In A.M. Schrier, H.F. Harlow and F. Stollnitz (Eds.) Behavior of Nonhuman Primates. New York, Academic Press, Vol. 1, Pp. 250-281

31. Young, F.A. and Farrer, D.N. 1971. Visual similarities of nonhuman and human primates. In E.I. Goldsmith and J. Moor-Jankowski (Eds.) Medical Primatology 1970. Basel, S. Karger, Pp. 316-328

32. Zimmerman, R.R. and Torrey, C.C. 1965. Ontogeny of learning. In A.M. Schrier, H.F. Harlow and F. Stollnitz (Eds.) Behavior of Nonhuman Primates. New York, Academic Press, Vol. 2, Pp. 405-447.

PRINCIPLES OF PRIMATE GROUP ORGANIZATION

I.S.Bernstein
Department of Psychology
University of Georgia
Athens, Georgia , U.S.A.

No single statement about social organization can be made which will apply to all of the taxa within the order Primates. Whereas many general statements apply to most primates, we must always consider the exceptions. Not only will no single formula describe the organization of all primate societies, but we must remember that many of the prosimians and perhaps some of the New World primates live essentially solitary lives. Even were we to restrict ourselves to Old World monkeys and apes, the wealth of variety in organizations precludes simplistic notions of unitary tendencies. A quick examination of the variety of organizations should discourage us from seeking one or two single universal social mechanisms to account for primate social behavior; and yet the literature is replete with arguments concerning the identification of the principle basis of primate sociality. Sexual attraction, dominance, predator protection and other mechanisms have all had their advocates and some have enjoyed temporary vogue as the dogma of the day. Is it not possible that all of these mechanisms are operable and that their differential contribution may vary from one taxon to another?

Admission of variety and denial of singular universal principles does not, however, require us to examine and explain each species order independently of all others. We may still search for principles of social organization common to many if not all of the animals in related taxa. Perhaps the variety of organizational outcomes only reflects alternative solutions to the same basic problems common to

( ) The author's references to ongoing personal research refer to projects supported by grant MH I3864 from the National Institute of Mental Health, USPHS, and in part by PHS grant RR 00I65 from NIH.

the members of the order Primates, or common to the organization of any social group.

In analyzing the available data, we must first admit the paucity of definitive information in the literature for the vast majority of primate taxa. Furthermore, studies such as Baldwin and Baldwin (3), DeVore and Hall (24), Durham (25), Gartlan and Brian (28), Jay (36), Neville (47), Rowell (53), Stoltz and Saayman (6I), Struhsaker (62, 63,64), Sugiyama (65, 66) and Yoshiba (7I), clearly indicate that variation in social organization may be found even among populations within the same taxon, and it has been suggested that some taxa possess mechanisms which allow the information of alternative social systems as conditions require (18, 26, 30, 34, 35, 40, 50, 51, 54, 66).

It has been suggested that specific ecological variables may determine the form of social organization (I9, I8, 22, 23, 26, 32, 4I, 64). Such theoretical formulations remain to be proved and simply reflect the greater sophistication we have achieved since the earlier models which focused on only one or a few determining variables. As our knowledge of the primates increases, both relevant to the number of taxa sampled and to the detail available for any particular taxon, we may expect continued refinement and amplification of explanatory models. At this juncture it is perhaps best to review the range of information presently available to us.

NUMERICAL DATA

Numerical descriptions usually begin with counts of the animals typically associating together. Even discounting ordinarily solitary taxa, the living primates may be found in social units as small as a single mated pair and their immature offspring, or as large as many hundreds of individuals. No consistent phylogenetic tendencies can be identified inasmuch as single mated pairs are well documented for such diverse forms as Callicebus and Hylobates (II, 27), and social organizations numbering several hundred individuals can be found for Saimiri (3) and Papio (40). Furthermore, absolute group size may vary significantly with local conditions as described for Saimiri, as suggested by the data on hamadryas (40) and gelada (I9) and as seen for Macaca mulatta in the diversity of habitats it occupies (58, 59).

A more useful numerical description of groups has been suggested to be the age sex ratios within the group. The socionomic sex ratios as proposed by Carpenter (I4), has been compiled for a variety of species and certain central tendencies have been reported. Some troops may be adequately described as monogamous, others as composed of one or more one male units and others as consisting of multiple adult males and females. More precise breakdowns are confusing, however, due once again to extreme within species variability. Altmann and Altmann (I) summarize data for various taxa of savannah baboons and find variability from ratios in which males somewhat exceed females to ratios of more than ten females per adult male. Rowell (54) presents some evidence for ecological influence on such ratios, but other variations are perhaps due to sampling errors but more likely indicate the flexibility possible within a taxon. Causal determinants for descrepancies from chance sex ratios have been proposed which include differential mortality and growth rates, and in some cases possible differential birth ratios. It is clear that despite the variability apparent, groups frequently tend to have more adult female than adult male members. Only part of this can be accounted for by differential tendencies for individuals to become solitary or to form unisexual bands. The presence of such bands and the implied difference in social organization has frequently been ignored in theories accounting for primate sociability. However, although one need not deny that the degree of cohesion and sociability in an all male band is different from that in a heterosexual troop, one cannot deny that such bands do exist and persist with some degree of cohesion in various species. Furthermore, solitary males are not the outcasts of societies but are frequently adult males in the prime of life. Lindburg (42), Neville (48), and Vandenbergh (70), among others, all report rhesus monkey solitaries as a function of breeding seasons and suggest that males may voluntarily leave troops in which they have high rank position only to join another troop shortly later. The mechanism for genetic exchange among semi-closed population units is quite clear, although understanding of the social mechanisms involved is more difficult to come by.

In considering age ratios within primate societies, one is immediately struck by the long developmental periods common to the primates and associated with somewhat longer life spans than seen in other mammals of comparable size. The ratio of mature to immature individuals is influenced both by breeding success and the life span of adult members, but in any case one can assume a relatively long association between an immature primate and adults prior to independence. The obvious speculation is that the long association period enforced by slow development insures ample opportunity for learning in an order characterized by the plasticity of individual behavior.

As implied in the discussion of numerical data, the basic social unit for some primates is a mated adult pair and their young. We do not, however, have sufficient evidence to suggest that this is the primitive primate condition, nor that it is the basic association pattern from which all others may be derived. Although it is easy to explain other forms of society as being generated when offspring fail to leave the parental pair at maturity, or as deriving from combinations of pairs or extensions of pairing by banding with multiple females, the evidence from just the primates alone indicates that these are unwarranted assumptions. One could just as easily argue to defend another social condition such as a multianimal bisexual aggregate as the basic material out of which various primate societies evolved.

Eisenberg (26) clearly depicts one logical schema for the evolution of primate social groups.

Just as monogamous pairs do not seem to be restricted to phyletically closely related taxa, one male units may be found as typical in such diverse groups as Presbytis (8, 49), and Cercopithecus (64). One male units, however, may occur in man-groups simply as an extreme condition of disproportionate male female sex ratios in a small unit. Small groups of Alouatta (17) may be one example. In other cases, one male units may be typical but not the only type of unit possible for the taxon (Presbytis examples). In other cases, one male units may be basic to higher level social organizations. That such one male units are not simply subgroups is indicated by the permanent association of animals within a one male unit whereas more fluid associations at other levels are possible and subgrouping within a social unit is ordinarily of a temporary nature.

Multiple male and multiple female troops seem characteristic of the well known macaques and savannah baboons as well as many other New World and Old World primate taxa. These troops are best characterized by the absence of any smaller consistent basic social units, although this is not to deny the existence of various patterns of subgrouping. Subgrouping may consist of genealogically related individuals, or other clusters of animals which are recognized by their high probability of association, or continuous association during a given temporal period. Such subgroups may consist of mother and offspring, consort pairs, juvenile play groups, sleeping associations or other such relationships within a larger permanent social unit. Among New World monkeys at least, it has been suggested that subgroups may form during the day (12, 38) and that some Old World monkeys at least, may subdivide during the night (16). In his analysis of subgroups, Cohen indicates that these combinations are not numerically random, nor are they composed of random individuals (1). It is possible that consistent subgrouping memberships may in fact represent another level of social organization, but insufficient data exists in most cases to specify individual identities in subgroups, and in some cases, subgroups are clearly not viable independent social units inasmuch as they have no breeding potential. Such considerations are applied to Saimiri, for example, where one might otherwise argue that females with young and adult males constitute separate social units which coalesce only temporarily during the breeding season (2).

Multi-level social organizations do quite clearly exist in such animals as gelada (19) and hamadryas (40) and possibly in Mandrillus (29). The chimpanzee community of perhaps fifty individuals is considered as the basic social unit because of: the fluidity of membership among the bands, their frequent lack of reproductive potential, and also the fact that the entire community does apparently operate as a single unit under certain conditions. Gorilla troops may or may not exist within a multi-level social structure. Mutual toleration of troops and temporary mixing is otherwise difficult to explain.

The hamadryas wase is especially intriguing because three levels of organization are suggested, the one male units, a band of such units and the sleeping troop, (40). The non-random formation of larger units by smaller units suggests that these larger units are organized and not simply aggregations of mutually tolerant units. Furthermore, Kummer (4I) indicates that special social relationships among males within the higher level social units prevents competition for females which is readily expressed when an unfamiliar one male unit is introduced to a sleeping troop unit. Similarly, Crook (I9) describes special troop travel patterns and organization for geladas which precludes a simple assemblage of totally independent one male units. The organization of bachelor males relative to the one male bands is a case in point.

One more level of social organization should also be considered. Among certain New World and Old World monkeys, there is frequent interspecific association. In fact, troops of different primate taxa travel together and coordinate their activities relative to one another in a non-random pattern suggesting the existence of social mechanisms bonding certain taxa to each other at the troop level (7, 68, 69). These associations appear to be more than simple feeding aggregations, but aside from data indicating that some individual animals may become members of a social troop consisting of extraspecific individuals, definitive data are lacking.

THE USE OF SPACE

Primate social organization is presumeably related to the use of space and, in particular, to spacing mechanisms relative to species density and the insurance of sufficient resources in an area for survival of the species inhabitants. No data exists for truly migratory or nomadic non-human primates and all groups studied to date appear to exist within a limited range during their entire life span. Slight changes in range tendencies may occur and individuals may occasionally exchange troop membership, but a troop is usually limited to an area of a few hectares, or at most a few square kilometers.

Among the primates, there is evidence for true territoriality for only a few taxa. Callicebus , gibbons (II, 27), siamang (I5), some vervets (62,63), Colobus (43, 57)and Presbytis cristatus (8), all defend specific geographic areas against extra troop conspecifics. Ritualized boundary defense at specific locations may become a prominent part of intertroop interactions. In such truly territorial taxa, the victor of an encounter is determined by the geographic location of the encounter, and reversals in chases may be seen as animals cross back and forth across a boundary. Such territorial defense is often a function of adult males, but both sexes may participate.

Most primate groups, however, live in geographic areas defined as home ranges. This constitutes a typical area within which a group's activities take place. Some seasonal shifts may occur related to availability of resources such as seasonal foods or water (I). The travel pattern of a group in any one day is usually limited to only a portion of the home range and it is only by plotting multiple successive day ranges that the boundaries of a home range are revealed. In most such taxa, adjacent troops show some home range overlap, the exact percentage are, however, apparently quite variable. The precision of the measure and definition obviously effects the determination of degree of overlap, but it is possible that overlap may include 100% of a troop's home range, not all of this necessarily with a single other troop. Alternatively, some troops may be found which do not show overlap with any conspecifics and yet the concept of home range rather than territory is applied because we have reason to believe such lack of overlap, in these particular cases, is due to some mechanism other than defense of geographical areas.

The use of space within a home range is not random and probably reflects clumping of resources within the home range. Certain heavily frequented areas are sometimes used exclusively by a single troop and have been designated as core areas within a home range. Once again, however, exclusive use in itself should not be taken as sufficient evidence for territorial defense.

A frequent source of confusion, in determination of territoriality in the strict sense, is generated by the phenomena of conspecific spatial intolerance. Whereas many animals have individual flight and fight distances, some troops of primates also apparently avoid the immediate proximity of other troops of conspecifics. This mutual avoidance of proximity is further confused by the development of specific social mechanisms which permit close association under particular circumstances. Thus, troops of baboons may congregate at water holes or near certain sleeping rocks (56), or tolerate certain particular troops but not others in their immediate vicinity (53). Furthermore, a troop may exercise dominance over another troop and consistently drive the other before it on successive encounters (60). Such intertroop dominance may be due to the presence of a particular individual (14) or individuals (44, 45). Intertroop dominance is, however, readily differentiated from territorial behavior only by repeated observation, and intertroop intolerance will be distinguished only by even more extensive data collected at a number of interaction sites.

The protection of the immediate space by a social unit or individual clearly has many aspects in common with territoriality. An intergradation between the two might be attempted, but due to the widespread occurrence of territoriality in the strict sense in such a wide variety of vertebrates, this concept appears worthy of distinction and should be preserved and conserved for use when defense of a specific geographic area can be demonstrated. Modifiers such as "moving territories" or combinatorial terms such as "territorial ranges" are thus considered confusing. Appropriate independent terminology for each phenomena should be sought along with identifying definitions. The existence of only some of the identifying criteria should not in themselves be sufficient evidence for the existence of a phenomenon. Questionable circumstances such as the howling exchanges in Alouatta should be carefully considered and decisions held in abeyance until definite evidence can be produced to show the existence of territorial defense, despite the tempting analog with gibbon morning calls. Whoop displays in langurs is another case in point.

It slould be made clear that many auditory displays as well as prominent visual signals, such as vigorous locomotion and conspicuous display, are certainly good cues for individual or troop locations, and whereas such signals may serve to locate a troop and aid in avoidance of conspecifics, they may or may not be associated with territoriality. All spacing mechanisms are not the same, even though they may all serve the same biologically relevant function of dispersing animals in space.

One main point needs to be considered and that is the specification of conspecific intolerance or territoriality. We may find that territoriality relates to all nongroup conspecifics, only to same sexed individuals, or only to certain conspecific social units. Thus, a one male unit leader may act only to exclude other males from the vicinity, and his response may be limited to adults either in the presence or absence of an accompanying heterosexual group.

We have repeatedly emphasized conspecific intolerance because of the frequent association of and tolerance among interspecific social groups. Even closely related forms may be sympatric and the usual explanation invoked is that in such cases there is minimal overlap of ecological niche and hence little competition. Such explanations are attractive but frequently unsubstantiated. Cogeneric primates (8) have been reported to coexist extensively despite considerable overlap in utilization of forest resources. Such competition does not seem to result in eventual elimination of one of the competitors and we cannot explain this failure as due to repeated in-migration of additional troops or individuals, in the absence of any such evidence. Furthermore, there is no basis for claiming the area of overlap as marginal for any of the sympatric species as all may appear at density levels typical of those throughout the ranges of the associated taxa. Some selective use of resources in areas of sympatry may exist but we should probably be safer if we assumed that the resources for which cogeneric sympatric forms compete are not those critical for the habitat. In other words, the limiting resources for the species involved are not identical and not among those for which there is extensive interspecific competition.

Furthermore, it must be remembered that not only are these cogeneric forms sympatric, but that they often orient their activities with respect to each other showing common or even synchronous travel patterns, feeding association and even predator responses. Rather than passively sharing the habitat, cogeneric forms may actively interact and there may even be cross species social facilitation of resource utilization.

SOCIAL MECHANISMS

No analysis of cospecific social organization would be complete without an analysis of dominance structure; not because it is necessarily so central to primate societies as because it is so central to many theories of primate social organization. The concept received wide attention in part due to historical accident. Macaques and baboons were among the first non-human primates to receive extensive study in laboratories and were available for study in some numbers due to their relatively large natural populations, as well as their hardy constitutions, which allowed them to survive the often abusive conditions of captivity. There is no question but that aggresive dominance assertion is one of the guiding principles in rhesus monkey social relationships, either with cospecifics or people. So dramatic is this behavior and so pervasive in its influence that one might readily assume it to be present in at least some form in all primates; and yet this is an unproven and perhaps dangerous assumption. The danger lies in misleading an investigator into designing measures and descriptions of a behavior he is certain he must be able to find.

One result has been the vast proliferation of definitions of dominance such that today hardly any two authors use the term in the same way and many use it with multiple meanings, even in the same article. As a result, various types of dominance have been described, but perhaps this only serves to obscure the original very real phenomenon. Bernstein (9) reviewed much of the literature on primate status hierarchies and tested several definitions for correlations. Indeed, it was demonstrated that multiple social mechanisms were functioning within each of the groups examined and that whereas these mechanisms relate to one another within a social context, none could be claimed to be determinant for the others.

The directionality of dyadic agonistic interactions was posed as closest to the original descriptions of "peck-orders" which underlie the concept of dominance relationships. Only if one individual may aggress against another with impunity can he be said to be dominant over that other individual. Furthermore, if dominance exists as an organizing principle, then a knowledge of prior interactions should allow one to predict the outcome of future interactions. Put another way, if dominance is cited as a principle of social organization, it must be consistent and predictive. To be sure, social organizations are dynamic rather than static structures, and changes will take place, but if social relationships are so fluid that no predictions are possible, then we either do not understand the aspects which are organized, or no organization exists.

The measurement of agonistic dominance or status relationships has included a multitude of factors, many of which correlate very highly with the directionality of dyadic agonistic encounters. Various incentives have been offered and successes in competition for such items has often been demonstrated to be a reasonably highly accurate reflection of dominance relationships. Since these successes lend themselves to ready objective measurement and since test sessions are directly controlled by the investigator, such measures are often the technique of choice. It would be a mistake, however, to confuse the measure with the phenomenon. This sometimes does occur when investigators assume the incentive value of an object, measure successes in obtaining the object and then wonder why other measures do not relate to their "dominance" structure.

One of the biggest pitfalls in such measurements is that they preclude the possibility of nonexistence of the mechanism. Someone will drink an arbitrary amount of water first, or eat the raisin, or copulate with a female, or escape the chamber first. If the "victor" is consistent this is taken as proof of the stability of the dominance relationship. If the "victor" is not consistent then an "unstable" dominance relationship is claimed. In any event the relationship is assumed and measured. It is possible, however, that in some species, agonistic dominance relationships are not only not central to the organization of social groups, but that they are even absent. One could, of course, claim that the differences among animals are so slight that we were simply unable to detect dominance relationships. A difference which makes no difference, is no difference.

Other unitary explanations of social organization have focused on sexual bonds within the group. Much discussion has ensued and the only clear result is that sex may be important in bonding some individual for unspecified periods , but sexual attraction alone is insufficient to account for all of primate social behavior or social organization. Sex should be viewed in proper perspective. It is neither all important, nor inconsequential. It is simply one of the many forces which shape primate societies.

ROLE ANALYSIS

The organization of primate societies is perhaps best understood not by searching for single explanatory principles, but by careful analysis to determine what functions are being served within the context of primate societies. Identification of individual social responses is perhaps a first beginning, but the study of social exchanges must follow and we must search for order in social relations. When we detect such order, we may assign higher order labels, such as roles, to identify consistent patterns of behavior elicited in identifiable individuals under specified circumstances. Bernstein and Sharpe (10), Bernstein (4, 5), and Reynolds (52), among others, have all applied elements of role theory to the analysis of primate units and have been able to clearly identify some consistent roles. The interrelationships among roles is more difficult to describe, except for reciprocal or directly interactive cases as in mother-infant, consort pairs, etc. Rowell et al. (55), Kummer (39) and others have clearly described certain interactive roles. A complete role description and analysis of a society is a more elusive goal, perhaps unattainable. Inasmuch as more than one individual may serve in a role capacity, and inasmuch as any one individual may serve in multiple role capacities as circumstances develop, it would take complete knowledge of all social events to ever attain the ideal. Nevertheless, the quest may in itself add sufficient insight into the workings of primate society to compensate for our inability to ever attain the unattainable ideal.

In the analysis of primate social organization, many investigators have chosen to "begin at the beginning". For some, this beginning in the identification of mating patterns. For most, the beginning starts with birth. In troops consisting of multiple males and females, only the biological mother is known for sure, and even then errors may be made if one simply assumes that the animal carrying a baby is its mother. Such errors are obvious and comical for marmosets where the father typically carries the infants, but such errors can occur in other taxa, for it is by no means a general rule that the biological mother will carry her baby exclusively, or even for the majority of the time. Among langurs, Jay (36) and others have clearly demonstrated that babies may be passed from one female to another even within hours of birth and that even young juvenile females may carry infants. Rowell et al.(55) and Kaufman and Rosenblum (37) demonstrated that infant transfers may even occur routinely in some macaques.

Infant care patterns are not limited to females and "peculiar" primates like male marmosets. Indeed, male care for infants runs a considerable range both among taxa and within a particular species. Deag and Crook (21) describe specific social mechanisms termed "agonistic buffering" which rely upon male carriage of immatures in *Macaca sylvana*. Kummer (40) describes the special relationship which may develop between juvenile females and adult male hamadryas, and my own work demonstrates clear role related behavior involving immature animal care by "bachelor" male geladas. In some taxa where social organization does not so rigidly specify male infant care roles, considerable variability may be observed. Itani (33) describes one such pattern in some troops of *Macaca fuscata* and I have seen considerable variation in male attention to infants within *Macaca nemestrina*, *M. fascicularis*, *M. arctoides*, *Cercocebus atys* and others.

Various labels have been applied to specific infant care roles displayed by other than the biological parents of primates. "Paternal" or male care roles, and "aunt" role behavior have all been described and various labels have been criticized. Criticism of the label in no way nullifies the data, and in some cases the labels may be fortuitously descriptive. "Aunt" patterns have received much notice and the label has been roundly criticized as implying an unwarranted genetic relationship. My own data suggests, however, that among pigtail macaques, "aunts" are indeed often the biological aunt of an infant although they may also be older sisters, grandmothers and even just "family friends". The strong genealogical bonds which persist for an indefinite period in this taxon often result in the new infant receiving far more attention from a closely related animal than from any other animal in the group. To be sure, even an older brother may be inordinately active in the care of a new infant, but inasmuch as adolescent females are most often involved in "aunt" behavior, they are most often "aunts" to the infants.

The discussion of infant care and the associated roles perhaps serves best to emphasize the rich interplay of roles in serving any of the vital functions of a group. As we examine other such important functions we should find similar variety of role patterns and interrelationships.

The maintenance of a group includes defense against extragroup disruption in the form of predators or hostile conspecifics. Such defense begins with vigilance. Hall (31) describes such role behavior in baboons and it is clear that whereas the function is served, the role is not exclusive to any one individual, but more likely to appear in some individuals rather than others. Bernstein (4, 5, 6) analyzed the "control animal role" in rhesus, pigtail and capuchin groups and stressed that whereas the alpha male in a dominance hierarchy frequently serves in this role capacity, the existence of the role was not dependent upon the presence of a dominance hierarchy. Recent work has also demonstrated that even in the presence of such a hierarchy, the control role may be most actively displayed by an individual other than the alpha male.

The control role function not only is called into play in defending the group against extragroup disturbance, but may also be a significant social mechanism to limit intragroup agonistic disruptions. Numerous authors have commented on the activities of certain individuals in response to intragroup aggressive episodes and the effectiveness of such individuals in controlling within group aggression and limiting its consequences. Such a role has clear implications for the maintenance of a group's social structure and the preservation of the existing social order.

GROUP FUNCTION

The logical question then is what purpose does a primate group serve? Put in such a fashion, most biologists immediately deny the validity of the question as reflecting a teological error. Asking about group function is only a slight improvement, but turning the question around to ask why social groups are positively adaptive, or why living as a member of a particular social organization has positive selective value for a primate, is a more acceptable question. Obviously the non-human primates do not plan their social organizations and organize to meet foreseen problems. Rather, the social organizations have evolved because of their value in aiding the survival of individual members, their progeny and the species. The answers to the different questions posed may in fact, vary very little, but the questions reflect the theoretical framework from which the query was derived. Many people thus ask the right questions for the wrong reasons - but in so doing often go on to ask irrelevant questions as well.

In considering evolutionary time scales we are clearly reduced to speculation, frequently called insights. Obviously, crucial experiments demonstrating the selective value of particular social mechanisms are easier to design than to conduct. Nonetheless, by careful observation of animals living under natural conditions, we may hope to understand the relevance of aspects of social organization to various factors which have operated within the context in which social orders have evolved.

One can immediately seize on such aspects as the utilization of natural resources and the controlled use of such resources, as having obvious survival implications. Social mechanisms relating to such factors as selection of sleeping sites, food and water sources and their use, infant care and survival, mating and possible seasonal control; are clearly important for species survival. Indirectly related features, such as spacing mechanisms, competition and predator recognition and defense are also easily recognized. More subtle features, such as the forms of communication and other elements of social behavior, may be more difficult to relate to survival, but within the context of the natural setting, and with the aid of field experiments which create theoretically relevant critical conditions, the importance of such social behavior may be revealed. It is sufficient at this point that we apply ourselves to the study and understanding of social behavior and group organization patterns. With knowledge may come insight and even recognition of the relationship between the study of one specie's solution to the problems of social living and the relevance of such solutions, and their variety to solving the various problems of our own most complex societies.

References

1. Altmann, S.A. and Altmann, J. 1970. Bibl. primat., 12, 1-220

2. Baldwin, J.D. 1971. Folia primat., 14, 23-50

3. Baldwin, J.D. and Baldwin, J.I. 1971. Primates, 12, 45-61

4. Bernstein, I.S. 1964. J. comp. physiol. psychol., 57, 404-406

5. Bernstein, I.S. 1966. Tulane Studies in Zool., 13, No. 2, 49-54

6. Bernstein, I.S. 1966. Primates, 7, No. 4, 471-480

7. Bernstein, I.S. 1967. Folia primat., 7, 198-207

8. Bernstein, I.S. 1968. Behaviour, 32, 1-16

9. Bernstein, I.S. 1970. Primate Status Hierarchies. In Primate Behavior. (Developments in Field and Laboratory Research). L.A. Rosenblum, Ed., Vol. 1, Academic Press, New York, Pp. 71-109

10. Bernstein, I.S. and Sharpe, L.G. 1966. Behaviour, XXVI, 91-104

11. Carpenter, C.R. 1940. Comp. Psychol. Monogr., 16, No. 5, 1-212

12. Carpenter, C.R. 1935. J. mammal., 16, (3), 171-180

13. Carpenter, C.R. 1942. Biol. Symposia., 8, 117-204

14. Carpenter, C.R. 1942. J. comp. Psychol., 33, 113-142

15. Chivers, D.J. 1971. Spatial relations within the siamang group. In Proceedings of III international congress of primatology, Zurich, 1970, Vol. 3, Basel, S. Karger, Pp. 14-21

16. Cohen, J.E. 1968. Grouping in a vervet monkey troop. Rand Corp. Rep. No. P-3896 Rand Corporation, Santa Monica, California (Dept. Commerce Clearinghouse AD-672 744), 8 pp.

17. Collias, N. and Southwick, C. 1952. Proc. Amer. Philosophical Soc., 96, 143-156

18. Crook, J.H. and Gartlan, J.S. 1966. Nature, 210, 1200-1203

19. Crook, J.H. 1966. Symp. zool. Soc. London, No. 18, 237-258

20. Crook, J.H. 1970. Anim. Behav., 18, 197-209

21. Deag, J.M. and Crook, J.H. 1971. Folia primat., 15, 183-200

22. Denham, W.W. 1971. Amer. Anthrop., 73, 77-104

23. DeVore, I. 1963. A comparison of the ecology and behavior of monkeys and apes. Classification and Human Evolution, S.L. Washburn, Ed., Aldin Publ. Co., Chicago, 301-319

24. DeVore, I. and Hall, K.R.L. 1965. Baboon ecology. In Primate Behavior. Field Studies of monkeys and apes. I. DeVore, Ed., Holt, Rinehart and Winston, Pp. 20-52

25. Durham, N.M. 1971. Effects of Altitude Differences on Group Organization of Wild Black Spider Monkeys (Ateles paniscus) Proc. 3rd int. Congr. Primat., Zurich 1970 (Karger-Basel 1971), vol. 3, Pp. 32-40

26. Eisenberg, J., Muekenhirn, N. and Rudran, R. 1972. Science, 176, 863-874

27. Ellefson, J.O. 1968. Territorial behavior in the common white-handed gibbon, Hylobates lar Linn., In Primates, P.C. Jay, Ed., Holt, Rinehart and Winston, Pp. 180-199

28. Gartlan, J.S. and Brian, C.K. 1968. Ecology and social variability in Cercopithecus aethiops and C. mitis. In Primates, P.C.Jay, Ed., Holt, Rinehart and Winston, Pp. 253-292

29. Gartlan, J.S. 1970. Preliminary notes on the ecology and behavior of the drill, Mandrillus leucophaeus, Ritgen, 1824. In Old World Monkeys: Evolution, Systematics, and Behavior. J.R. Napier and P.H. Napier, Eds., Academic Press, New York and London, Pp. 445-480

30. Goodall, J. 1965. Chimpanzees of the Gombe Stream Reserve. Ch. 12, In: Primate Behavior, Field Studies of monkeys and apes. I. DeVore, Ed., New York, Holt, Rinehart and Winston, Pp. 425-473

31. Hall, K.R.L. 1960. Behaviour, XVI, Part 3-4, 261-294

32. Hall, K.R.L. 1965. Symp. Zool. Soc. London, 14, 265-289

33. Itani, J. 1963. Paternal care in the wild Japanese monkey (Macaca fuscata fuscata). In Southwick's Primate Social Behavior, Ovan Nostrand Co., Princeton, Pp. 191

34. Itani, J. and Suzuki, A. 1967. Primates, 8, 335-381

35. Ozawa, K. 1970. Primates, 11, 1-46

36. Jay, P. 1965. The common langur of North India. Ch. 7, In Primate Behavior. Field Studies of monkeys and apes. I. DeVore, Ed., New York, Holt, Rinehart and Winston, Pp. 197-249

37. Kaufman, I.C. and Rosenblum, L.A. 1969. Ann. N.Y. Acad. Sci., 159, 681-695

38. Klein, L. and Klein, D. 1971. Int. Zoo. Yearb., 1971, 11, 175-181

39. Kummer, H. 1967. Tripartite relations in hamadryas baboons. In Social Communication among Primates. Stuart A. Altmann, Ed., Univ. of Chicago Press, Chicago and London, Pp. 63-71

40. Kummer, H. 1968. Social organization of hamadryas baboons. Univ. of Chicago Press, Chicago and London, 189 pp.

41. Kummer, H. 1971. Immediate causes of primate social structures. Pp. 1-11 in Proceedings of the third international Congress of

Primatology, Zurich, 1970, Vol. 3, H. Kummer, Ed., Basel, S. Karger

42. Lindburg, D.G. 1969. Science, 166, 1176-1178

43. Marler, P. 1969. Science, 163, 93-95

44. Marsden, H.M. 1968. Folia primat., 8, 240-246

45. Marsden, H.M. 1971. Intergroup relations in rhesus monkeys (Macaca mulatta). In Behavior and environment: The use of space by animals and men. A.H. Esser, Ed., Plenum Press, New York, Pp. 112-113

46. Marsden, W.A. 1966. Tulane Stud. Zool., 13, 23-28

47. Neville, M.K. 1968. Ecology, 49, 110-123

48. Neville, M.K. 1969. Primates, 9, 13-27

49. Poirier, F.E. 1968. Primates, 9, 351-364

50. Reynolds, V. 1963. Folia primat., 1, 95-102

51. Reynolds, V. and Reynolds, F. 1965. Chimpanzees of the Budongo Forest. Ch. 11, In Primate Behavior. Field Studies of monkeys and apes. I. DeVore, Ed., New York, Holt, Rinehart and Winston, Pp. 368-424

52. Reynolds, V. 1970. Man, Vol. 5, No. 3

53. Rowell, T.E. 1966. J. Zool., 149, 344-364

54. Rowell, T.E. 1969. Folia primat., 11, 241-254

55. Rowell, T.E., Hinde, R.A. and Spencer-Booth, Y. 1964. Anim. Behav., XII, 219-226

56. Saayman, G.S. 1971. Afr. wild Life, 25, 25-259

57. Schenkel, R. and Schenkel-Hulliger, L. 1967. On the sociology of free-ranging Colobus (Colobus guereza caudatus) Thomás 1885. In Progress in primatology, D. Starck, R. Schneider and H.-J. Kuhn, Eds., Stuttgart, Gustav Fischer Verlag., Pp. 185-194

58. Southwick, C.H., Beg, M.A.and Siddiqi, M.R. 1961. Ecology, 42, 538-547

59. Southwick, C.H., Beg, M.A. and Siddiqi, M.R. 1961. Ecology, 42, 698-710

60. Southwick, C.H., Beg, M.A. and Siddiqi, M.R. 1965. Rhesus monkeys in North India. Ch. 4, Primate Behavior. Field Studies of monkeys and apes. I. DeVore, Ed., Holt, Rinehart and Winston, New York, Pp. 111-159

61. Stoltz, L.P. and Saayman, G.S. 1970. Ann. Transv. Mus., 26, 99-143, pls. 10-17

62. Struhsaker, T.T. 1967. Ecology, 48, 891-904

63. Struhsaker, T.T. 1967. Behaviour, 29, 83-121

64. Struhsaker, T.T. 1969. Folia primat., 11, 80-118

65. Sugiyama, Y. 1964. Primates, 5, (3-4), 7-37

66. Sugiyama, Y. 1968. Primates, 9, 225-258

67. Sugiyama, Y., Yoshiba, K. and Pathasarathy, M.D. 1965. Primates, 6, 73-106

68. Thorington, R.W., Jr. 1967. Feeding and activity of Cebus and Saimiri in a Colombian forest. In Progress in primatology, D. Starck, R. Schneider and H.J. Kuhn, Eds., Stuttgart, Gustav Fischer Verlag., Pp. 180-184

69. Thorington, R.W., Jr. 1968. Observations of squirrel monkeys in a Colombian forest. Ch. 3, In The squirrel monkey, L.A. Rosenblum and R.W. Cooper, Eds., New York, Academic Press

70. Vandenbergh, J.G. 1967. Behaviour, 29, 179-194

71. Yoshiba, K. 1968. Local and intertroop variability in ecology and social behavior of common Indian langurs. In Primates, P.C. Jay, Ed., Holt, Rinehart and Winston, Pp. 217-242.

NONHUMAN PRIMATES: A VULNERABLE RESOURCE

Barbara Harrisson
Dept. of Anthropology
Cornell University

The exploitation of primate animals has become one of the most serious causes for their depletion. This is possible because the institutions which regulate acquisition, trade and usage of wild life do not take account of changes which have taken place in the recent past, and which are now part of an ungoing process.

Existing controls pertain to a situation before the advent of animal traffic by air, and of world-wide demands for captives to serve research, industry, and the public.

Resource nations have seen rapid developments in the vast areas rendered accessible to international communication since then. Now technologies fostered wide-scale destruction and alteration of habitats, especially of lowland tropical forests. Most influenced by these events were specialized primate animals with limited distributions. More generalised forms with wider distributions were also affected by uncontrolled killing as agricultural pests and extensive harvesting for exportation. Rare and endangered species the international commerce in spite of protection laws because of the high values placed on them outside resources countries. New species were listed as endangered as time went on. But the controls at the point of initial resource remained ineffective in this rapid development. The improvements which are now clearly necessary require investments and maneuvres resource nations cannot support or undertake single-handed. These nations insist that the problem is international; and that, if their interest in primate animals is valid, users must play an essential role in supporting change.

Researchers have clarified the validity of their interest. Nonhuman primates are unique because of their close relationship to man. The disappearance of any such speciesis a loss to man, particularly since nonhuman primates are growing increasingly valuable for biomedical research (2). But the legal, administrative and research framework under which this interest operates, is not designed to support and protect the resource on which it is based. New legislation cannot proceed because basic information is lacking. Existing controls cannot operate because they do not apply internationally. Veterinary and health controls cannot adequately eliminate hazards which are connected with the increased usage and movements of captive primate animals. Only one aspect received adequate regulation: international transportation. Airline carriers now provide facilities which ensure against excessive loss and cruelty to animals on route. But this leaves major areas of initial and subsequent wastage and abuse unaccounted for. The consequence is serious overexploitation and a general rapist attitude to the resource.

This paper identifies users and the trade in broad terms. It then describes the main activities which evolved in support of conservation, of resource management and of trade control. It finally advocates how these may be improved.

USERS AND TRADERS OF PRIMATE ANIMALS

The users of primate animals may be separated into three groups. Starting from a point in or near natural habitats, there is firstly a native interest in these animals as a source of protein or other physical properties with anthropomorphic cash or status values in local terms, and the relatively new one in captive live animals as a source of cash. This group of users connects with the second, that of traders. Linking supply and demand, traders are concerned mainly with the captive live animals. Their network operates at national and international levels. They make use of transporters, a subsidiary group. The third group combines ultimate users of captive live animals: researchers, biomedical industry, ecological parks, entertainent industry and private fanciers. All operate within frameworks of national interest: governments who derive economic benefits from the exportation of captives,from the turnover in trade, or the ultimate usage of captives.

A variety of causes allow for overexploitation of the resource --expanding human populations and advancing technology accounting for the most serious deteriorations. Other less acceptable reasons are lack of information and concern, and the impact a few individuals have who use the resource as a convenient stepping-stone for short--term economic gain to themselves in the absence of control. The large group of ultimate users makes this possible. Far removed from the resource, they inadvertently prompt irresponsible activities and developments contrary to the long-term interests of the resources countries and of themselves. This is possible because the majority of users are satisfied to exercise demand without enquiry into the status of the resource and into the methods by which traders produce and distribute supplies.

The trade in live primate animals is complex. Only few generalizations are possible. The majority of trappers are individual entrepreneurs. Many operate only incidentally or seasonally. In some countries their activities are licenced. A few are organised in commercial businesses with substantial investments. The majority of trapped animals percolate through many hands to a local dealer who has marketing capability to an exporter or wholesaler. Exporters and wholesalers who assemble their stock from many dealers connect with colleagues elsewhere and with prominent customers to whom they extend price and promotion. Some of the larger institutions and industries buy from foreign wholesalers. But most connect with dealers in their own countries, many of whom perform only brokerage functions.

An unknown portion of the trade is illicit. Protected species are captured and illegally transported over national borders and then exported from a neighbouring country (or state, in the case of federal nations where protection laws may differ from one state to another). They are also exported direct, with false declarations of identification. The commercial incentive for such operation is highest for the most vulnerable and valuable species; for apes, for rare species with limited distribution, for species that do not establish well in captivity. These operations succeed because law enforcement cannot effectively penetrate inland borders or extended coast-lines, and because the capacity to identify species and subspecies at customs points is very limited.

The interest in rare and vulnerable species is by tradition, a major feature of trade promotion. Wholesalers and brokers invite interest primarily by advertising their capability to produce them, since this has shown to foster demands. This then increase the pressure on species most in need of protection. Another type of advertising is addressed to researchers, laboratories and industries. Here the capability to deliver disease-free specimens of desirable genera and species with determined ages, sexes and weights is stressed. This produces the impression that advertisers exercise controls at the point of initial resources which is not the case in fact, generally. The specimens which cannot enter the market because they do not meet established criteria are wasted. Stress, disease and neglect take additional tolls during trapping, transportation and holding by chains of unskilled individuals prior to exportation, all losses remaining unaccounted for in trade statistics. Dealers buying stock on speculation take additional high losses if they cannot find immediate buyers. Strong competition among dealers intensifies and perpetuates abuses of which most customers are unaware.

Prices for imports increase as species become overexploited and wasted. The point when imports are as costly as captive-bred specimens may be that of no return for a number of species in their natural environment, under the present system.

ACTIVITIES SUPPORTING RESOURCE MENAGEMENT AND CONTROL

A variety of programs have evolved. The main issues are summarised below, listed in the order of importance accorded by user groups.

Captive Breeding

Captive breeding has been acclaimed as appropriate conservation in the absence of other -easures, and to prevent gene-stock of wildlife from being lost. This claim is valid so long as captive breeding is a long-term concern and supports genera and species which are most vulnerable in their natural habitats.

Importers primarily promoted captive breeding of primate animals which are most used in research and industry, and which establish reasonably well in captivity. But these genera and species are also under least pressure from a conservation point of view. For endangered and vulnerable genera and species, including for many useful in research, the future is insecure. A recent study which assesses the long-term breeding potential of endangered species in zoos (inclusive of 8 primate species (3)) indicates grave problems, mainly owing to the thin scatter of populations and individuals, lack for the accomodation of surplus, and difficulties regarding successful breeding of captive-born individuals. The situation in research laboratories is more favourable in the case of apes, where concentrated collections exist. Laboratory breeding of New World monkeys and forest-dwelling Old World monkeys has not reached the stage when adequate surplus can be expected to be available for experimentation. An animal models and genetic stocks program conducted by the Institute of Laboratory Animal Resources (ILAR) at the National Academy of Sciences provides reference data and description of colonies in the U.S. (4).

Research and breeding programs assisting the promotion of work in the advancement of sociobiology and of genetic and ecological theory, are disadvantaged by fragmentation and by the lack of data regarding the natural environments and behaviors of species used. All studies which identify physical and behavioral specialisation as conditions of adaptations to environmental phenomena must ultimately turn from the laboratory to the natural habitat. As the situation of endangered and vulnerable species deteriorates, the recovery of information is increasingly urgent.

Recovery of Information in Resource Countries

The information researchers require to support ongoing laboratory investigations and breeding programs of endangered and vulnerable species is that which habitat countries also need for the promotion of their protection. The passage of laws, and appropriate measures for their enforcement, which prevent wildlife from being seriously reduced or lost, by destruction or overexploitation, is consistently urged. But governments who deal with these problems need basic information regarding the distribution and coexistence of genera and species; their ranges, troup and group sizes, food habits, life statistics and adaptive properties -- to various types of habitat and to hunting and trapping pressures -- to arrive at conclusions which enable administrators to promote the formation of parks and preserves, and of ceilings of exploitation, with reasonable assurance at a political level. For apes and for forest-dwelling nonhuman primates, including for species used extensively in research, this information is spotty. The ecological studies which became available in the past (5) have identified some species in few limited ranges. They indicate that generalizations are not possible, that considerable differences may apply from region to region under different environmental conditions even within species, and that group and individual adaptations may introduce additional important variants.

The drastic upsets of populations of apes and monkeys through timber falling followed by tree poissoning, and through hunting, trapping and killing ar forest edges where populations concentrate in prime habitat, is known. Struhsaker, in a recent field investigation in Africa (6) states in this connection that the scientific interest in rain forests is at present grossly underrated, and that the support of scientists, institutions and universities who have a stake in this interest is essential to make the promotion of forest parks and preserves a better reality.

Resource nations have an economic interest in their rain forests. They manage and exploit them for timber and agricultural purposes in accordance with established traditions, policies and expertise. This is detrimental to the populations of nonhuman primates in them. To promote their survival requires the replacement of man's present economic stake in rain forests by an alternative. Struhsaker calls this an environmental problem of international proportions. "It seems apparent", he states, "that the precarious position of Africa's rain forests necessitates a carefully planned and ecologically founded multipurpose use, which, by definition, would include areas completely protected against human exploitation" (6).

Rain forests produce more opportunities for coexistence, competition and specialisation of fauna than do open, less varied ecosystems. Investigations of the progression of specialisations and adaptations, and of the inter-dependence of nonhuman primate populations across species in a flora-dominant environment have not proceeded well because they depend on comprehensive, inter-disciplinary research. Intensive studies which concentrate on a small area or problem and on a single species are easier to handle. But they cannot make significant impacts so long as the broader view remains clouded by uncertainties. If primatologists expand to combine their research interest with that of other disciplines and points of view, such fresh assessments of economic potentials as nations need to preserve forests in which primate animals can survive, may result. International broadly-oriented platforms and agencies where diverse research interests can be accomodated and promoted present opportunities. The facilities of ongoing UNESCO, IBP and FAO programs of the United Nations where they relate to tropical forestry (7) present opportunities for studies and surveys of nonhuman primates to contribute to habitat preservation and fresh concepts of resource management.

International, national and institutional biological field station where facilities and expertise can be shared by researchers have also produced good results. But some are now too removed from the areas where scientific presence is most needed. If the protection of habitat and controlled acquisition in strategic resource areas now under pressure of exploitation is intended, the initiation of ecological survey in them which provide basic information must become an urgent concern.

One vital area, the upper Amazon, developed such activities over the past two years with the cooperative support of user-interests. Aware of the possibility that consumer demands may drain the upper Amazon in significant areas rendered accessible through air and road communication over the past decade, U.S. researchers coordinated under ILAR of the National Academy of Sciences, which resulted in grants for biological and status survey. The Colombian government, through the auspices of the Instituto de Desarollo de los Recursos Naturales Renovables (INDERNA) is the first country to be placed in a position where it can recover the information it needs for the promotion of habitat preservation and resource management. Supported by the Smithsonian Institution, fieldwork has gone ahead. A biological station near Leticia, where the emphasis will be placed on the development of knowledge and techniques necessary for sound management of economically and biologically important renewable natural resources is planned. The urgency of many tasks, including that of employes training and public education which Colombia faces, require outside participation and support (8).

The regions of the Old World which qualify for the export trade in endangered and vulnerable species are widely scattered. Ecological studies proceed in forests of Southeast Asia and tropical Africa. But they are isolated and cannot fulfill the needs the rapidly deteriorating situation demands. Lack of funds is the main deterrent. A cooperative user-concern by importing nations outside the U.S., which accounted for approximately 41% of the total trade in nonhuman primates during 1971, could provide the basis for realistic promotion. Current user-responsibility is reliably documented in the U.S.(9). That of Eurasia has been computed from some government statistics, from airline carriers and from research opinions as below:

Imports of nonhuman primates during I97I

	New World sp.		Old World sp.		Total	
	numbers	%	numbers	%	numbers	%
U.S.A.	54.500	83	32.000	39	86.500	59
Eurasia	I0.500	I7	50.000	6I	60.500	4I
Total	65.000		82.000		I47.000	

Trade Control

The 3rd Congress of Primatology held in Switzerland in I970, recommended that Pan troglodytes be declared in the Red Data Book of the International Union for Conservation of Nature and Natura Resources (IUCN) (I0). The listing of Pan troglodytes as endangered could have introduced import restrictions in countries which model their controls on IUCN recommendations, irrespective of export restrictions applicable in resource countries. Decreased outflows from Africa and reduced demands could have resulted.

The editor of the Red Data Book requires for action data which prove that the continued survival of Pan troglodytes is unlikely without early implementation of protective measures, and that numbers are reduced to a critically low level. Such data is available for parts of its range. But because Pan troglodytes has a very wide distribution, and remains secure, or unidentified, in other parts of its range, the declared present status is vulnerable -- that is endangered if further deterioration takes place. If the data is evaluated for each sub-species of Pan troglodytes separately, it turns out that Pan troglodytes verus, the western race, is endangered. But the Red Data Book limits breakdown into subspecies, as a matter of policy, where this is possible.

One of the major functions of IUCN's Red Data Book is to assist governments in the protection of wildlife as it passes through the trade. The listing of sub-species extends and complicates data to a point where the information becomes counter-productive in practice. A young captive Chimpanzee cannot be readily identified through customs or veterinary inspection into any one sub-species of Pan troglodytes. He can be controlled from two points of view. The validity of documents which accompany him can be considered with reference to prevailing national or state regulations, including temporary or permanent import bans that may exist from a health point of view.

Then the documents must be linked to the animal itself. The capacity to confirm his identity or otherwise is important, False identification is a common feature in illicit trade operations.

A recent example may serve to illustrate another complicating aspect. Leontideus, listed as endangered in the Red Data Book, occurs only in Brazil where its protection receives intensive support through the Brasilian government and the Instituto de Desinvolvimento Florestal (II). The captive population in the world's zoos numbered 84 in I970, including 34 captive-born specimens -- total numbers declining because imports of new stock have been cut off from Brazil (3). Golden Marmosets continue to be offered on overseas markets with false claims that they are captive-bred. Recently 37 specimens were illegally exported to Paraguay from where they may enter international trade. They may be distributed to countries which do not require documentation of legal origin (i.e. a Brasilian export permit) or alternatively, appropriate verification of claims that an animal was bred outside Brazil (by an authority other than a zoo or trader). Parallel cases involving other genera and species may be complicated further if they occur in several countries where protection laws and if their propagation capacity is less well ascertained (I2).

IUCN's Draft Convention on the Export, Import and Trade of Certain Species of Wild Animals and Plants (I3) proposes a treaty which coordinates and streamlines national controls on an international platform. It suggests a central authority, empowered to the trade on the basis of current statistics and other information which it receives. The Convention distinguishes two categories of wildlife; species which are already threatened with world-wide extinction, and species which are approaching that condition. It excludes specimens which were bred in captivity.

The species are listed in two Appendicies to the main text. Appendix I defines species the acquisition and trade of which must be limited to exceptional circumstances. Such nonhuman primates as are listed endangered in the Red Data Book are here concerned. Appen dix II lists species not yet threatened by the export, import and transit of which must be controlled in order to avoid undue exploitation incompatible with survival. All species of primate animals, excluding those listed in Appendix I are in this category. They are to be marketed under the Convention subject to regulation primarily at the point of origin, through the institution of standard export permits. Permits are to be issued by the competent national authority only for specimens lawfully captured, or in the case of re-export, after proof of lawful import has been provided. Description and marks must enable to permit to be linked with the animal subject to trade. Copies of all permits issued are to be submitted to a central authority.

For the primate animals listed in Appendix I, two controls are executed. A prospective importer has to apply for a standard import permit first, from a national scientific authority. This authority may grant permits only for scientific purposes -- educational purposes included -- and only when adequately justified. The next step is lawful capture on the basis of the import permit, provided of course the authorities in the country of origin are agreed. The animal is then traded with import and export permits attached, copies being passed to the central authority.

It is obvious that the Convention can only succeed if a majority of exporters are convinced that this type of control can support the enforcement of their national laws, and if the main importing countries conscious of their responsibility as users for the protection of wildlife other than their own, support it. The Convention defines that contracting exporting states may only trade with other contracting states; but an assumption enables contracting states who are importers to acquire specimens from states that are not signatories -- for scientific purposes only, and provided the specimens were legally exported.

A large number of states are agreed with the fundamental principle of the Convention. Exporting countries favour the support they will derive through international controls. But in some major importing countries objections persist, mainly because difficult legal and administrative mechanisms are required the ultimate effectiveness of which they question.

Although the Convention does not affect the right of contracting states to adopt in their national legislation stricter measures regarding the trade and measures extending the Convention to national species not at present listed in the Appendicies; and although national provisions pertaining to customs, public health, veterinary or other fields are not affected, the requirements of the Convention must be lined up with the legal and administrative realities of every state and a minimum must to be met. Regulations to prohibit and penalize any export, import and transit in violation of the Convention must be passed. The possession, exhibit and offering for sale of illegally imported specimens must be prohibited. Machineries must be provided for the confiscation and care of specimens which may be obtained under the terms of the Convention. Any state must ensure that the clearance of specimens through customs, including verification of identification of species, is rapid and carried out by persons with appropriate qualifications. The designation of special ports of exit and entry may be a necessity. A national scientific authority, qualified to rule on requests for the import and export of specimens and on procedures regarding confiscated shipments must be appointed. The national scientific authority must be able to cooperate with the central authority where it may require representation. And since the Convention excludes specimens of species which are captive-bred, an authority must be available to deal with such claims. Some nations may have to overcome on all points at issue, legal and administrative obstacles at federal, state, county and municipal level.

International treaties are slow to come alive. The Convention, which involves complex issues and commands low priorities in human and economic terms, has been in circulation since 1964. Various national mechanisms have produced in the meantime, from conservation and public health points of view. Most comprehensive coverage exists to date in the U.S.A. The Endangered Species Act (14) uses Red Data Book guidance and combines this with the requirement of proof of legal acquisition from the country of origin in accordance with the Lacey Act (18 U.S. Code 43 and 44 as amended) which prohibits the importation of any wildlife that was collected, killed or exported illegally from its country of origin.

These national measures provide important relief, and assist exporting countries. But the fact remains that such shipments as cannot now enter the U.S. may be redirected to major European markets where less comprehensive restrictions apply, none from a regional point of view. This would be impossible with an international treaty.

There are two advantages to the Convention. One is standardised documentation. The other, a central authority which can monitor restrictions, in accordance with the current situation. Designation of species in the Appendices is flexible. If central statistics and other information make it opportune, a species may be deleted from Appendix I and placed into Appendix II or vice versa.

The scientific committees designed to cooperate with the central authority at national levels are conceived as interest groups of specialists. Since all nonhuman primates are covered by the Convention, national advisory bodies representative of scientific usage are required. This poses advantages. Coordinated, effective controls of the movements of nonhuman primates and other exotic animals over national borders is very desirable from a public health point of view (15). The support of the Convention by scientists able to project this particular viewpoint may play a significant part in its promotion.

Rationalisation of Usage

Proposals which assist conservation of primate animals through rationalisation are in two categories. Those which promote enlightened self-restraint within one group of users, and others where one group seeks to exercise restraint over another. Scientists recommend a code of ethics for themselves which requires a judicious choice of species and maximum usage of individual animals (2). There are more detailed formulations, some involving subtle argumentation among scientists with different orientations regarding the rationale or merit of research programs where primate animals feature as test subjects. Zoo associations recommend, and with some success exercise, pooling of resources (16), and restraint in the choice of traders. Private users recommend and promote, long-term responsibility for individual pets by their owners (17).

The merits of these promotions are obvious. They create awareness which produces pressure and finally new policies where glaring faults exist. Scientists who take only a limited interest in their test subjects or in the continued availability of the species they use, or in nonhuman primates as a whole, can be guided through policies which funding institutions, industries and associations may adopt. Policies which require research proposals to include detailed justifications for the choice of endangered and vulnerable species, and for the numbers of individual animals to be used, can be very beneficial, particularly if this is combined with a requirement to identify how the acquisition of test animals is compatible with the current status of the species in its country of origin. Policies which restrict acquisition of primate animals to traders who supply proof of legal origin and disclose the methods by which specimens are obtained, could have excellent effects. The screening of breeding projects by policies which require assurance of long-term maintenance and provision for the accommodation of surplus, as well as pooling of resources with existing projects would eliminate damaging fragmentation and short-term concerns.

A proposal which suggests the elimination of one large group of users is likely to be very effective in conservation terms. A significant number of primate animals are pets (18). Most pets are wild-caught and load short, single lives; a majority are vulnerable New World species. Scientists who themselves are involved in the usage of primate animals to a significant degree have difficulties in promoting a ban on pets because this inevitably leads into opinions which query the validity of scientific usage of primate animals, in particular in experimentation, and to request that equal restraints become operative in all areas if this is warranted from a conservation point of view.

The divergent views are unlikely to carry helpful publicity and constructive cooperation -- both necessary for the promotion of conservation issues. However, the issue of primates as pets is readily diffusable. Uncontrolled private possession, especially in urban and suburban areas, involve hazards to public health which remain unaccounted for, in the present system of trade and distribution of imports.

ECONOMICS AND POLITICS OF CONSERVATION IN PRACTICE

Conservation is practiced effectively in the field, where status surveys and baseline research recover information which leads to the preservation of habitat and controlled exploitation of animals for export; and in areas away from the field where trade controls and rational distribution ensure that economic benefits and research results derived from captives feed back into the resource.

The manpower and techniques to execute fieldwork is available. Scientists who are overburdened with other duties have students and staff able to proceed. But the opportunities to get them into the field are severely restricted. In resource countries, finance is unavailable. Prices for wildlife are generally based on overexploitation, levels rising in accordance with diminishing availability. The levy of royalties or dues which pay for the establishment or maintenance of effective resource-management -- that is, the fielding of researchers and staff -- would price wildlife out of international markets. An investment from outside sources is necessary to make a beginning.

The help of scientists and students from other countries can form part of this investment. They are unfortunately tied into systems which make their participation difficult. Research which enquires into the availability of primate animals takes very low priority because this is generally taken for granted. Most research is developed and funded for a wide range of specialised topics, primarily relevant to human health, where the necessary emphasis is on laboratory experimentation. Although the argument is accepted that laboratory studies are enriched by baseline data on the ecology and behavior of primate animals in natural habitats, the cost of recovering these data -- in time and money for the development and execution of field projects which support laboratory problems -- tend to frustrate attempts, particularly if alternative options to proceed with intensified or specialized laboratory experimentations remain.

The constitution of parks and preserves is dependent on manpower which assesses their potential. National interest and private capital which invest in developments within the boundaries of parks and preserves precludes their formation or acquisition, and can only be attracted if expectation of benefits can reasonably compete with other opportunities. Viable ecosystems of which populations of nonhuman primates form a part, are sizeable units. Pioneers able and willing to experiment with primate animals under these conditions, are very few (I9).

International agencies which qualify for the support of developments have their present orientation towards improvements of established systems of exploitation, such as the timber industry and agriculture. Alternatives which allow for the continued existence of primate animals can only be promoted with adequate data. Conservation foundations can qualify for their recovery only in exceptional circumstances. Their support falls short of many needs, particularly if other potential backers have a vested interest in the resource.

Research foundations and pharmaceutical industries where this interest applies cannot readily acknowledge that investments in field projects is in their own long-term interest. But re-orientation is necessary to ensure the continued availability of the resource: intensified, more liberal direct promotion of field investigations as necessary or relevant to laboratory research; or indirect promotion of such investigations and other conservation measures through the extension of earmarked funds.

Institutional and industrial researchers are to a large degree concerned with relatively common species. They are reluctant to accept responsibility for the whole spectrum of species. They rationalize usage of imports mainly in accordance with rising prices, through the developments of techniques which reduce the numbers of test animals required in experimentation, and through intensified laboratory breeding. This is valid conservation in a controlled trade where prices of imports reflect their availability on a sustained yield basis.

In uncontrolled trade, where market prices are based on over-exploitation, conservation can only be effective if it extends back into the trade itself and beyond. A concern for the whole spectrum of species is necessary. International and national measures which seek to control the traffic of all wildlife have been described. A more direct and effective approach can separate treatment of primate animals from a public health point of view. Public health is insufficiently guarded under present systems. The introduction of licences for all traders and users of primate animals is justified. Licences which are made dependent on the application of strict professional standards (including in trade, the ability to prove legal acquisition of specimens in countries of initial origin), on the existence of insurance covers against all public hazards (inclusive of unknown diseases) and on the payment of fees which cover costs of supervised administration and include a prevision earmarked for rehabilitation of the resources could price most pet owners and many marginal traders and exhibitors of primate animals out of the market. Such a system can eliminate all dangers and abuses of the current trade. It cannot inconvenience users who have a valid interest in the resource.

But such proposals must be formulated and promoted. Governments cannot be expected to act until they become motivated by lobbies, public appeals or emergencies. Professionals in developed countries are in key position for action. They present necessary data, as reflected in current research; they can formulate recommendations which fit the institutional realities of their countries; they can influence governments at a policy-making level.

APPEAL FOR CONSERVATION OF NONHUMAN PRIMATES

The scientists of many disciplines and different countries participating in the Advanced Study Institute on Comparative Biology of Primates, held in Montaldo (Torino), June 7-19, 1972:

- being aware of the unique value of nonhuman primates as man's closests relatives and as models serving the biological and medical sciences in the advancement of human health, welfare and knowledge;
- being aware of their responsibility to preserve the existence of the whole spectrum of contemporary primate species;
- being aware that expanding human populations and the growing usage of nonhuman primates have threatened some species with extinction while others have become drastically reduced.

URGENTLY SUBMIT to International and National Organizations the following appeal that:

1. Scientists be selective in the usage of nonhuman primates and to employ other animal models when they are appropriate, and require that endangered and rare species be limited to investigations in which other species of nonhuman primates are unsuitable.
2. Scientists contribute to the conservation needs of nonhuman primates by:
   - introducing and insisting upon, humanitarian and efficient procedures for their capture, translocation and maintenance prior to and during their use;
   - educating the public about the real health hazards presented by nonhuman primates which carry many transferable diseases (especially viral infections), and recommending that the use of Monkeys and Apes as pets be prohibited;
   - insisting upon methods of acquisition which ensure the enforcement of national laws covering capture and transportations and the international support of these laws;
   - promoting the development of knowledge on the distribution and status of nonhuman primate populations in the areas where they occur and of management and husbandry methods which ensure their survival in natural ecosystems and as economic and scientific resources;
   - promoting the development of permanent breeding programmes according to a long term requirement for different species;
   - encouraging the urgent cooperation and financial support and investments from research institutions, pharmaceutical and other industries to accomplish these aims for ensuring the continued existence and development of nonhuman primate populations in their natural ecosystems, or in especially designed environments, so that their availability for human use is ensured for posterity.

Notes and References

I. This paper which evaluates current activities supporting the long-term availability of nonhuman primates, is based on a comprehensive analysis of the "Conservation of nonhuman primates in I970", Primates in Medicine 5:I-99 (Karger, Basel/New York I97I) where an extensive list pf references is included. It is also based in part, on unpublished materials and discussion of issues, within the Survival Service Commission of the International Union for Conservation of Nature and Natural Resources(IUCN; Morges/ Switzerland) where the author is represented as secretary of the Primate Specialist Group.

2. Goldsmith E.I. and Moor-Jankowski J. (eds.) I969. Ann.N.Y.Acad. Sci. I62:I-704.

3. Perry,J., Bridgwater, D.D. and Horsemen, D.D. I972 in press. Zoologica.

4. Lab.Prim.Newsl. I972. II, I:20-2I.

5. Crook, J.H. I970. The Socioecology of Primates, Crook,J.H.(ed.): Social Behavior in Birds and Mammals, Acad. Press London.

6. Struhsaker, T.T. I972 in press. Rainforest Conservation in Africa. Primates.

7. The UNESCO and FAO HQS in Paris and Rome for general information . UNESCO's Field Science Office for Africa (P.O. Box 30592) and the Pacific Science Association (Bernice P.Bishop Museum P.O.Box 6037, Honolulu, Hawaii) which conects with regional Committees on Conservation and tropical forestry in Southeast Asia can also provide links and information.

8. Hernandex,J., Asesor de Investigation de Fauna Silvestrel, Institute de Desarrello de los Resources Naturales Renovables, Bogota; D.E.Correro I4:25A-66, Apdo.Aereo I3458.

9. U.S. Department of the Interior, Fish and Wildlife Service, Bureau of Sport Fisheries and Wildlife, Washington D.C. 20240. WL-498; February I972.

IO. International Union for Conservation of Nature and Natural Resources: 'Red Data Book', Mammalia, vol.I rev.(Arts graphiques Heliographica, Lausanne I966).

II. Dr.A.Magnanini, Jardin Botanica and Instituto Brasileira de Desinvolvimento Florestal, Rio de Janeiro. A Golden Marmoset Conference was held in Washington in February I972; its results are in preparation. For summary of background see Perry J. 'The Golden Lion Marmoset' ii, i (May I97I); and DuMond,F.V.'Comments on minimum requirementsof the husbandry of the Golden Marmose'; Lab. Prim.Newsl. IO,2 (April I97I).

I2. A recent case involved advertisement of 'Orang-utan twins born and bred in captivity', from Taiwan. Enquiries resulted in photographs showing animals of different sizes between I-3 years old, and a 'birth certificate' from a local zoo. The gradual accumulation of stud-books greatly facilitates tracing of information.

I3. International Union for Conservation of Nature and Natural Resources "Draft Convention on the Export, Import and Transit of Certain Species of Wild Animals and Plants", rev.Feb.I97I, distributed internationally to governments through embassies, consulates and commissions from its lig at IIIO Morges/Switzerland.

I4. United States Department of the Interior, Fish and Wildlife Service: Endangered Species Conservation; Notice of Proposed Rule making. Fed. Regist. 35(72): 6069-6075 (I970).

I5. A discussion of New World primates supply by Middleton C.C.; Moreland A.F. and Cooper R.W. outlines potentially severe hazards; the Center for Disease Control, Atlanta, Georgia, and also reports cases which come to the attention of authorities. For both see Lab.Prim.Newsl. II,2(Apr.I972).

I6. In the United States, the Wild Animal Propagation Trust (WAPT) Secretary: J.Perry, National Zoo, Washington D.C.) and its Committees ensure maximum coordination and efficiency in the placement and propagation of endangered species.

I7. The creation of sanctuaries for unwanted and hazardous pets, the placement of which is increasingly difficult, is a prominent feature in the U.S.

I8. The percentage of imports for pet markets varies greatly from country to country. It is largest in the U.S., between 40 and 50% (I).

I9. Mr.M.Tsalickis of Leticia, Colombia, purchased an I000-acre island in the upper Amazon and stocked it with 5000 Saimiri, over 3 years. He estimates the population to have increased to about 20,000, since introductions started in I967. Supported by Tarpon Zoo Inc., Tarpon Spring, Florida, U.S.A.

List of Contributors

Albright B.C. - Dept. of Anatomy, Medical College of Virginia, Richmond, Virginia, USA

Bernstein I.S. - Dept. of Psychology, University of Georgia, Athens, Georgia, USA

Chiarelli B. - Institute of Anthropology, University of Turin, Turin, Italy

Goode G.E. - Dept. of Anatomy, Medical College of Virginia, Richmond, Virginia, USA

Haines D.E. - Dept. of Anatomy, West Virginia University, School of Medicine, Morgantown, West Virginia, USA

Harrisson B. - Dept. of Anthropology, Cornell University, Ithaca, New York, USA

Kalter S.S. - Division of Microbiology and Infectious Diseases, Southwest Foundation for Research and Education, San Antonio, Texas, USA

Mukherjee A.B. - Dept. of Pediatrics, State University of New York, School of Medicine, Buffalo, New York, USA

Murray H.M. - Dept. of Anatomy, West Virginia University, School of Medicine, Morgantown, West Virginia, USA

Ruffie J. - Centre National de la Recherche Scientifique, Centre d'Hémotypologie, Toulouse, France

Rumbaugh D.M. - Dept. of Psychology, Emory University, Atlanta, Georgia, USA

Snodderly D.M. - Dept. of Retina Research, Retina Foundation, Boston, Massachusetts, USA

INDEX